Analytical Design of PID Controllers

Iván D. Díaz-Rodríguez ·
Sangjin Han · Shankar P. Bhattacharyya

Analytical Design of PID Controllers

Springer

Iván D. Díaz-Rodríguez
McAllen Higher Education Center
Texas A&M University
Mcallen, TX, USA

Sangjin Han
Electrical and Computer Engineering
Texas A&M University
College Station, TX, USA

Shankar P. Bhattacharyya
Electrical and Computer Engineering
Texas A&M University
College Station, TX, USA

ISBN 978-3-030-18227-4 ISBN 978-3-030-18228-1 (eBook)
https://doi.org/10.1007/978-3-030-18228-1

MATLAB® is a registered trademark of The MathWorks, Inc., 1 Apple Hill Drive, Natick, MA 01760-2098, USA, http://www.mathworks.com.

Mathematics Subject Classification (2010): 9, 93C05, 93C35, 93C55, 93C80, 93B36, 93B50, 93B51, 93B52, 93D09, 93D15, 93D25

This Springer imprint is published by the registered company Springer Nature Switzerland AG.
The registered company address is: Gewerbestrasse 11, 6330 Cham, Switzerland

Dedicated to Prof. J. B. Pearson (1930–2012).

Photo courtesy of Jane Pearson

Preface

The Proportional–Integral–Derivative (PID) controller dominates the control industry across the traditional fields of Aerospace, Electrical, Mechanical, and Chemical Engineering as well as emerging fields such as driverless cars, autonomous robots, and unmanned aerial vehicles. Indeed, they account for 99% of all the controllers in use in the world.

This universal presence of PID controllers in applications contrasts sharply with the relative lack of interest in them from the control theory community which, until recently, was mainly focused on designing high-order optimal controllers by state-space methods. This situation began to change in 1997 when it was demonstrated that high-order controllers rendered the closed-loop system dysfunctionally fragile with respect to controller parameters even if they were robust with respect to plant parameters. This helped to usher in a renewed interest in low-order controllers.

The PID controllers are the simplest of low-order controller structures providing servo and disturbance rejection capabilities provided closed-loop stability can be achieved. In the last 20 years, significant progress has been made in computing the complete set of stabilizing PID controllers for linear time-invariant continuous- and discrete-time plants of arbitrary order. These were reported in the monographs "Structure and Synthesis of PID Controllers" by A. Datta, Ming-Tzu Ho, and S. P. Bhattacharyya and "PID Controllers for Time-Delay Systems" by Guillermo Silva, A. Datta, and S. P. Bhattacharyya. The present monograph is the third in this sequence.

The main results presented here demonstrate how multi-objective designs can be carried out using PID controllers, by exploiting the availability of the stabilizing set. By superimposing gain margin, phase margin, H_∞, and time domain specifications calculated in terms of design parameters, on the stabilizing set, in a systematic and constructive manner one can effectively execute hitherto impossible multi-objective designs and determine the limits of achievable performance. The results are presented here for continuous-time and discrete-time systems in a unified and self-contained manner and are illustrated by examples. A recent extension of these results to multivariable systems is also given.

We expect the concepts and design methods presented here to be useful in engineering and other applications and to further drive research on PID controllers into Adaptive Control, Machine Learning, Computer Science, Biological Systems, and other areas.

The results given here would not have been possible without the support and collaboration of numerous colleagues. Especially, we would like to express our gratitude to L. H. Keel, A. Datta, Ming-Tzu Ho, Guillermo Silva, and Navid Mohsenizadeh for their contributions.

Iván D. Díaz Rodríguez would like to thank his beloved parents for their love, support, encouragement, and sacrifices. Sangjin Han would like to thank his parents and brother.

College Station, TX, USA
February 2019

Iván D. Díaz-Rodríguez
Sangjin Han
Shankar P. Bhattacharyya

Contents

Chapter 1
Introduction to Control

Abstract In this chapter, we describe control systems informally, emphasizing the key elements of tracking, disturbance rejection, stability, and robustness. Next, we show why integral control driven by tracking error provides the correct feedback architecture to try to achieve these goals. This leads naturally to the Proportional–Integral–Derivative (PID) controller structure, where the proportional, integral, and derivative gains k_p, k_i, and k_d now become the design parameters which need to be tuned to achieve robust stability and time domain response specifications. A brief description is given of some classical and existing tuning approaches. We conclude the chapter with an examination of why optimal control, and in particular quadratic optimization, is absent from PID design theory and show that the reason lies in the inherent fragility of the high-order controllers invariably produced by optimization. The contents of this chapter should serve as background, perspective, and motivation for the rest of the book.

1.1 Introduction

Control theory and control engineering deal with a variety of dynamic systems such as aircraft, spacecraft, ships, trains, automobiles, and robots. They also deal with industrial processes such as distillation columns and rolling mills, electrical systems such as motors, generators, and power systems. Nowadays, they are ubiquitous in biomedical applications, power electronics, driverless cars, and autonomous robots.

In each case, the *setting* of the control problem is represented by the following elements:

Sections 1.1, 1.2, 1.4 and 1.5.4 are reproduced from S. P. Bhattacharyya, A. Datta, L. H. Keel, *Linear System Theory: Structure, Robustness, and Optimization. Taylor & Francis LLC Books,* with permission © 2008 Taylor & Francis LLC Books.

I. D. Díaz-Rodríguez et al., *Analytical Design of PID Controllers*,
https://doi.org/10.1007/978-3-030-18228-1_1

1. There are dependent variables, called *outputs*, to be controlled, which must be made to behave in a prescribed way. For instance, it may be necessary to *assign* the temperature and pressure at various points in a process, or the position and velocity of a vehicle, or the voltage and frequency in a power system, to given desired fixed values, despite uncontrolled and unknown variations at other points in the system.
2. Specific independent variables, called *inputs*, such as a voltage applied to the motor terminals, or valve position, are available to regulate and to control the behavior of the system. Other dependent variables, such as position, velocity, or temperature, are accessible as dynamic *measurements* on the system.
3. There are unknown and unpredictable *disturbances* impacting the system. These disturbances could be, for example, the fluctuations of a load in a power system, disturbances such as wind gusts acting on a vehicle, external weather conditions acting on an air conditioning plant, or the fluctuating load torque on an elevator motor, as passengers enter and exit.
4. The equations describing the plant dynamics, and the parameters contained in these equations, are not known at all or are known imprecisely. This uncertainty can arise even when the physical laws and equations governing a process are known well, for instance, because these equations were obtained by linearizing a nonlinear system about an operating point. As the operating point changes so do the system parameters.

The previous considerations suggest the following general representation of the *plant* or system to be controlled. In Fig. 1.1, the inputs or outputs shown could be representing a vector of signals. In such cases, the plant is said to be a *multivariable plant* as opposed to the case where the signals are scalar, in which case the plant is said to be a *scalar or monovariable plant.*

Control is exercised by feedback, which means that a device, driven by the available measurements, generates the corrective control input to the plant. Thus, the *feedback* or *closed-loop system* in Fig. 1.2 represents the controlled system.

The control design problem is the problem of determining the characteristics of the controller so that the controlled outputs can be

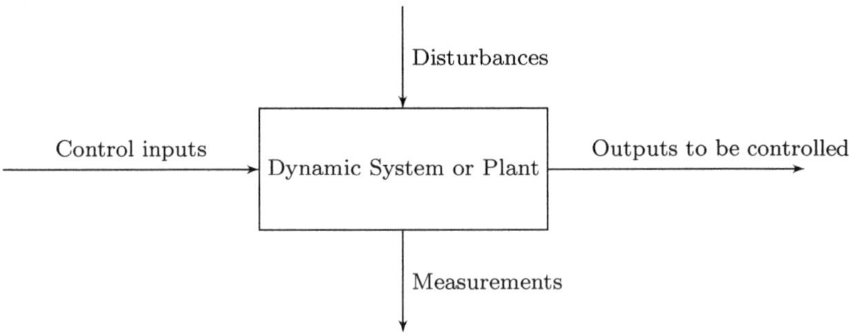

Fig. 1.1 A general plant. © Taylor & Francis LLC Books. Reproduced from [8] with permission

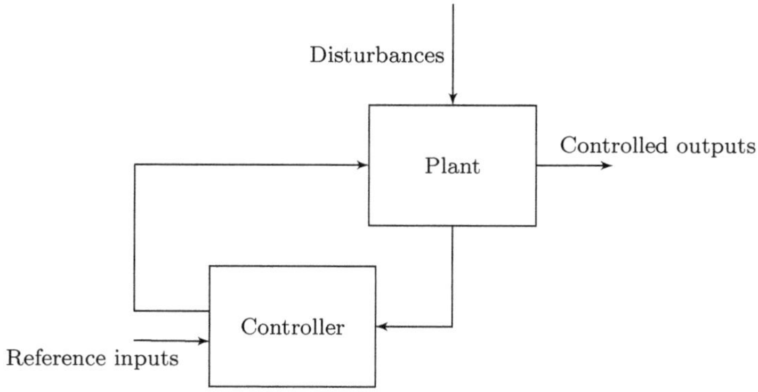

Fig. 1.2 A feedback control system. © Taylor & Francis LLC Books. Reproduced from [8] with permission

1. Set to prescribed values called *references*;
2. Maintained at the reference values despite the unknown disturbances;
3. Conditions (1) and (2) are met despite the inherent uncertainties and changes in the plant dynamic characteristics.

The first requirement above is called *tracking*. The second is called *disturbance rejection*. The third condition is called the *robustness* of the system. The simultaneous satisfaction of (1), (2), and (3) is called *robust tracking and disturbance rejection*, and the control systems designed to achieve this are called *servomechanisms*.

In the next section, we discuss how integral control is useful in the design of servomechanisms.

1.2 The Magic of Integral Control

Integral control is used almost universally in the control industry to design robust servomechanisms. Computer control most easily implements integral action. It turns out that hydraulic, pneumatic, electronic, and mechanical integrators are also commonly used elements in control systems. In this section, we explain how integral control works in general to achieve robust tracking and disturbance rejection.

Let us first consider an integrator as shown in Fig. 1.3.

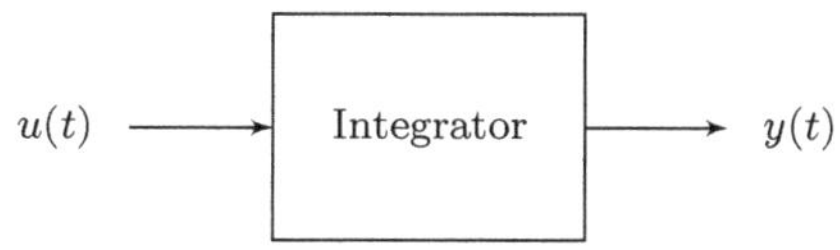

Fig. 1.3 An integrator. © Taylor & Francis LLC Books. Reproduced from [8] with permission

The input–output relationship is

$$y(t) = K \int_0^t u(\tau)d\tau + y(0) \tag{1.1}$$

or, in differential form,

$$\frac{dy(t)}{dt} = Ku(t), \tag{1.2}$$

where K is a nonzero real number called the integrator gain.

Now *suppose* that the output $y(t)$ is a *constant* for a segment of time $[t_1, t_2]$. It follows from (1.2) that

$$\frac{dy(t)}{dt} = 0 = Ku(t) \text{ for } t \in [t_1, t_2]. \tag{1.3}$$

Equation (1.3) proves the following essential facts about the operation of an integrator:

Fact 1 *If the output of an integrator is* constant *over a segment of time, then the input must be identically* zero *over that same segment.*

Fact 2 *The output of an integrator changes as long as the input is nonzero.*

The simple facts stated above suggest how an integrator can be used to solve the servomechanism problem. If a plant output $y(t)$ is to track a constant reference value r, despite the presence of unknown constant disturbances, it is enough to

A. attach an integrator to the plant and make the error

$$e(t) = r - y(t) \tag{1.4}$$

the input to the integrator;

B. ensure that the closed-loop system is asymptotically stable so that under constant reference and disturbance inputs, all signals, including the integrator output, reach constant steady-state values.

This structure is depicted in the block diagram shown in Fig. 1.4.

If the feedback system, shown in Fig. 1.4, is asymptotically stable, and the inputs r and d (disturbances) are constant, it follows that all signals in the closed loop will converge to constant values. In particular, the integrator output $v(t)$ tends to a constant value. Therefore, by the fundamental fact about the operation of an integrator established in Fact 1 above, it follows that the integrator input tends to *zero*. Since we have arranged that this input is the tracking error, it follows that $e(t) = r - y(t)$ goes to zero and hence $y(t)$ tracks r as $t \to \infty$.

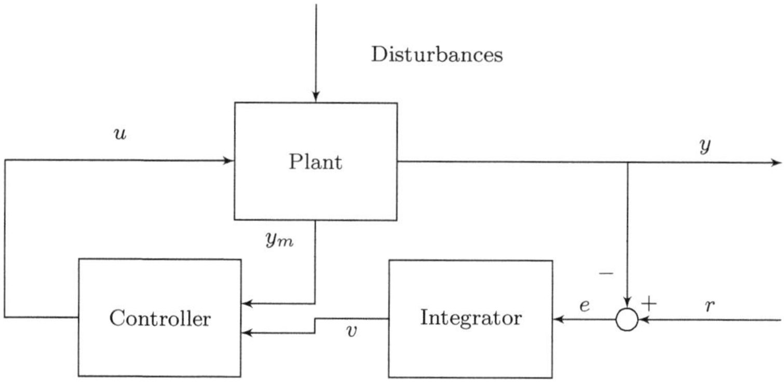

Fig. 1.4 Servomechanism. © Taylor & Francis LLC Books. Reproduced from [8] with permission

We emphasize that the steady-state tracking property established above is *very robust*. It holds as long as the closed loop is asymptotically stable and is (1) independent of the particular values of the constant disturbances or references, (2) independent of the initial conditions of the plant and controller, and (3) independent of whether the plant and controller are linear or nonlinear. Thus, the tracking problem is reduced to guaranteeing that stability is assured. In many practical systems, stability of the closed-loop system can even be ensured without detailed and exact knowledge of the plant characteristics and parameters; this is known as *robust stability*.

We next discuss how several plant's outputs $y_1, y_2, \ldots, y_m$ can track prescribed but arbitrary constant reference values $r_1, r_2, \ldots, r_m$ in the presence of unknown but constant disturbances $d_1, d_2, \ldots, d_q$. The previous argument can be extended to this multivariable case by attaching m integrators to the plant and driving each integrator with its corresponding error input $e_i = r_i - y_i$, $i = 1, \ldots, m$. This is shown in the configuration in Fig. 1.5. This requires the existence of $u_1, u_2, \ldots, u_r$ that makes $y_i = r_i, i = 1, \ldots, m$ for arbitrary $r_i, i = 1, \ldots, m$. Therefore, the plant's equations relating $y_i, i = 1, \ldots, m$ to $u_j, j = 1, \ldots, r$ must be invertible for constant inputs. In the case of Linear Time-Invariant (LTI) systems, this is equivalent to the requirement that the corresponding transfer matrix be right invertible or equivalently possess rank equal to m at $s = 0$. Sometimes, this is restated as two conditions: (1) $r \geq m$ or at least as many control inputs as outputs to be controlled and (2) $G(s)$ has no transmission zero at $s = 0$. The architecture of the block diagram of Fig. 1.5 is easily modified to handle servomechanism problems for more general classes of reference and disturbance signals such as ramps or sinusoids of a specified frequency. The only modification required is to replace the integrators by the corresponding signal generators of these external signals.

In general, the addition of an integrator to the plant tends to make the system less stable. This is because the integrator is inherently an unstable device; for instance, its response to a step input, a bounded signal, is a ramp, an unbounded signal. Therefore,

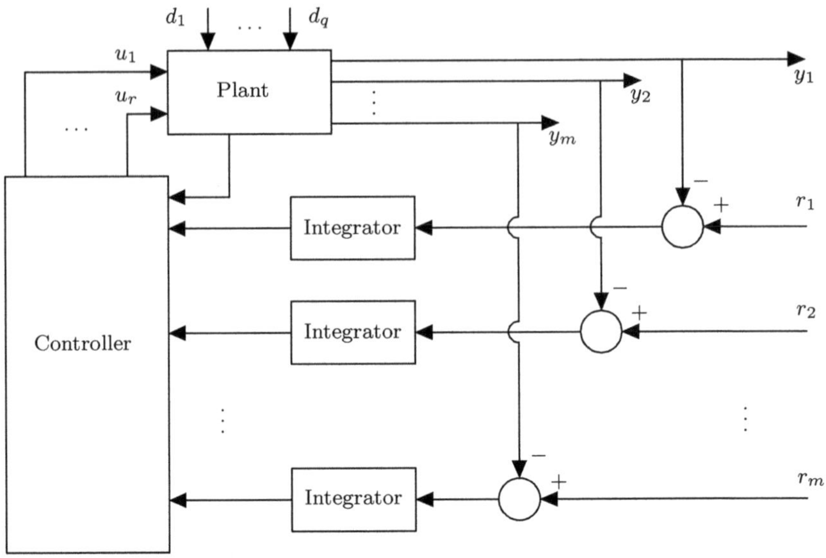

Fig. 1.5 Multivariable servomechanism. © Taylor & Francis LLC Books. Reproduced from [8] with permission

the problem of stabilizing the closed loop becomes a critical issue even when the stand-alone plant is stable.

Since the integral action and thus the attainment of zero steady-state error is *independent* of the particular value of the integrator gain K, we can see that this gain can be adjusted to try to stabilize the system. This single degree of freedom is sometimes insufficient for attaining stability and acceptable transient response, and additional gains are introduced as explained in later sections. The addition of gains naturally leads to the PID controller structure commonly used in industry.

1.3 Overview of PID Controller Design Approaches

PID controllers are the most widely used controllers in the control industry in motion control, process control, power electronics, hydraulics, pneumatics, and manufacturing. In fact, in process control, more than 95% of the control loops are of PID type with most loops using PI control. Their popularity is due to their simple structure, easy implementation, and straightforward maintenance. Also, they provide satisfactory performance with a cost/benefits ratio that is hard for other types of controllers to match. For this same reason, they are also popular in modern applications such as driverless cars, unmanned aerial vehicles, and autonomous robots.

1.3.1 PID Controller Structure

The PID controller is the name given to a controller which consists of the addition of three control actions (see Fig. 1.6). These actions are an action proportional to the error, an action proportional to the integral of the error, and an action proportional to the first derivative of the error in (1.4).

- **Proportional (P) controller**. The proportional action deals with the present values of the error signal; it is proportional to the size of the process error signal increasing the magnitude of the control variable when the error signal increases. When using only a P controller, we notice that increasing the proportional gain k_p may in general speed up the time response. However, it is possible that a steady-state error will occur. In general, under proportional control the steady-state error is zero if and only if k_p is very large.
- **Integral (I) controller**. The integral action is used to reduce the steady-state error to zero. When using an integral gain, increasing the value of k_i can give a broad range of response types in addition to the elimination of the offset error. The control signal is

$$u(t) = k_i \int e(t)dt. \tag{1.5}$$

  The integral of the error $e(t)$ is proportional to the area under the error curve. The control signal u will continuously change depending on whether the error signal is positive or negative. If the control signal $u(t)$ is constant, then the error signal must be identically zero, as expressed in Fact 1 in Sect. 1.2.
- **Derivative (D) controller**. The derivative action is often used to improve damping and closed-loop stability. It deals with the possible future values of the error signal based on its current rate of change, anticipating the trend of the error. The control signal here is

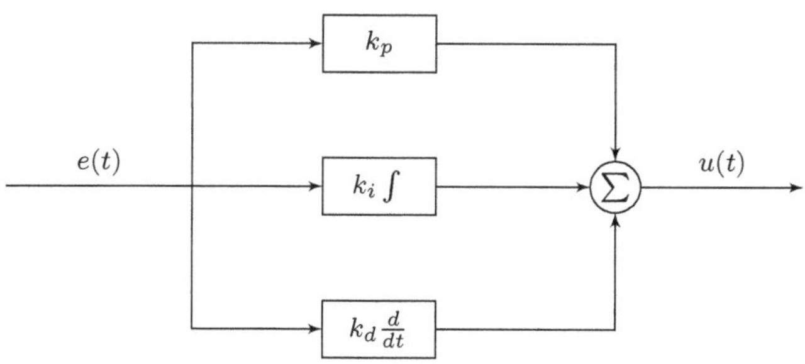

Fig. 1.6 PID controller block diagram

$$u(t) = k_d \frac{de(t)}{dt}. \tag{1.6}$$

The derivative part is proportional to the predicted error.

1.3.2 PID Controller Representations

Commonly used PID controller structures are of parallel and series types.

- **Parallel type**. This controller type has the following control law:

$$u(t) = K_c \left(e(t) + \frac{1}{T_i} \int e(t)dt + T_d \frac{de(t)}{dt} \right) \tag{1.7}$$

where $k_p = K_c$ is the proportional gain, T_i is the integral time of the controller with $k_i = \frac{K_c}{T_i}$, and T_d is the derivative time of the controller with $k_d = K_c T_d$. This representation is known as *ideal*.

- **Series type**. This controller type has the following control law:

$$\begin{aligned} e_1(t) &= e(t) + T_d \frac{de(t)}{dt}, \\ u(t) &= K_c \left(e_1(t) + \frac{1}{T_i} \int e(t)dt \right). \end{aligned} \tag{1.8}$$

In this case, all three portions of this PID structure are affected by the gain K_c. However, the proportional term is also affected by the values of the integral and derivative tuning parameters T_d and T_i. Therefore, adjusting T_i affects both the I and P actions, adjusting T_d affects both the D and P actions, and adjusting K_c affects all three actions.

1.3.3 Classical PID Controller Tuning

Due to the popularity of PID controllers in industry and their widespread use, there exist many approaches for their design and implementation developed over the years. The classical methods found in the literature can be classified as follows.

- **Trial and Error Method**. This method is applied when there is no systematic approach to follow when designing the controller. The method is based on experience about the effects of adjusting the individual k_p, k_i, k_d gains trying to get a better time response regarding speed and closed-loop stability. The typical effects of increasing each gain are represented in a table as in Table 1.1.

Table 1.1 Effects of adjusting individual PID gains on the system

Parameter	Steady-state error	Speed	Stability
k_p	Reduces	Increases	Decreases
k_i	Eliminates	Reduces	Increases
k_d	No effect	Increases	Increases

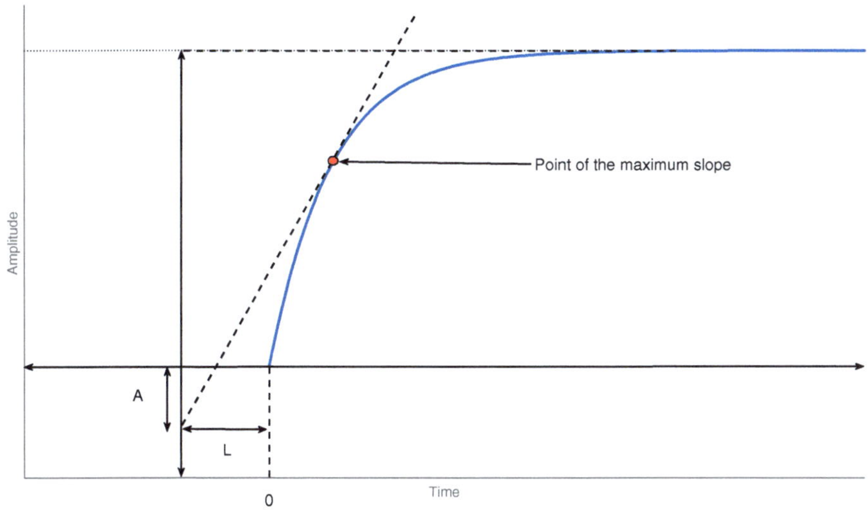

Fig. 1.7 Ziegler–Nichols step response method

The advantage of the trial and error method is that it does not require any mathematical model or mathematical derivation. However, it requires some experience to adequately adjust the controller gains to satisfy the desired performance regarding the speed of response and stability margins.

- **The Ziegler–Nichols step response method**. This PID tuning method was developed between 1941 and 1942 at the Taylor Instrument Company, USA. Since that time, this method has been extensively used in its original form and with some variations. The method is based on the measured step response of the open-loop stable system. For instance, see Fig. 1.7. The procedure is the following:

1. Calculate or determine experimentally the step response of the open-loop system.
2. Draw a tangent line with the maximum slope possible from the step response, see Fig. 1.7.
3. Calculate L, which is the distance from the intersection of the slope and vertical axis to the starting point of the step response.
4. Calculate A, which is the distance from the intersection of the slope and the vertical axis to the horizontal axis.

5. Compute the PID gains from the following formulas:

$$
\begin{aligned}
k_p &= \frac{1.2}{A}, \\
k_i &= \frac{0.6}{AL}, \\
k_d &= \frac{0.6L}{A}.
\end{aligned}
\tag{1.9}
$$

- **The Ziegler–Nichols frequency response method**. This PID tuning method considers a proportional controller attached to the system in a closed-loop configuration. The objective is to find the ultimate frequency where the phase of the process is $-180°$. That is the ultimate gain, where the system reaches the stability boundary. The tuning procedure is the following.

1. Connect a proportional controller to the system in a closed-loop configuration.
2. Slowly increase the proportional gain until the output starts oscillating. This gain is called ultimate gain K_u.
3. Measure the period of the oscillation in the output. This period is called the ultimate period T_u.
4. Compute the PID gains from the following formulas:

$$
\begin{aligned}
k_p &= 0.6K_u. \\
k_i &= \frac{1.2K_u}{T_u}. \\
k_d &= 0.075K_uT_u.
\end{aligned}
\tag{1.10}
$$

This tuning method is capable of finding the PID controller gains for the system. However, it requires some experience and skill because the system is taken to its limits of instability and it may become very close to getting damaged. It is emphasized that the Ziegler–Nichols design procedure assumes that the plant is of first-order cascaded with a delay.

- **Relay Tuning Method**. This PID tuning method was developed by K. Åström and T. Hägglund as an alternative to the Ziegler–Nichols frequency response PID tuning method. This method is very similar to the Ziegler–Nichols method, but instead of increasing a proportional gain until the system's output oscillates, a relay is used to generate an oscillation in the output, see Figs. 1.8 and 1.9. The relay connected to the system generates a square wave signal with specific amplitude and frequency. Then, a signal in the output approximated to a sinusoid is generated. The tuning procedure is the following:

1. The system should be working at the operating point.
2. Set the amplitude of the square signal in the relay.
3. Calculate the ultimate period T_u, see Fig. 1.9.

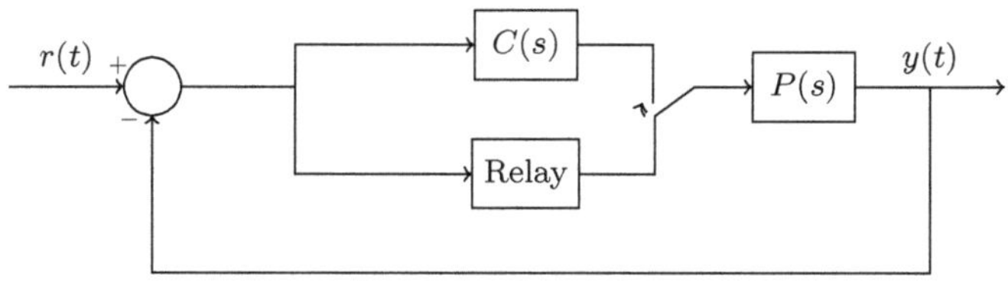

Fig. 1.8 Unity feedback block diagram with a relay

4. Calculate the controller parameters k_p, k_i, and k_d using the Ziegler–Nichols table using $K_u = K_e$, where $K_e = A_u/A_e$. Where $A_u = 4A/\pi$ and $A_e = E$ with E being the amplitude of the oscillations in the control error signal.

The advantage of this method is that it does not require one to force the system to be close to instability. Therefore, it keeps the system safer and reduces the possibility of damage. Also, this relay method can be automated since the output oscillation amplitude is proportional to the amplitude of the relay signal.

- **The Cohen-Coon Method**. This is an open-loop PID tuning method, which follows the same procedure as the Ziegler–Nichols step response method. In Fig. 1.10, we show the step response of the open-loop system for which the parameters k_p, L, and T can be determined. The gain k_p is determined by taking the ratio between the amplitude increment of the output and the increase in the control signal. That is

$$k_p = \frac{\Delta y}{\Delta u}. \tag{1.11}$$

The variables L and T can be found from Fig. 1.10, which represents the step response. Then, considering the PID controller of parallel type as in (1.7), the Cohen-Coon method recommends the following formulas to calculate the PID gains:

$$\begin{aligned} K_c &= \frac{1}{k_p}\left(0.25 + \frac{1.35T}{L}\right), \\ T_i &= \frac{2.5 + \frac{0.46L}{T}}{1 + \frac{0.61L}{T}} L, \\ T_d &= \frac{0.37}{1 + \frac{0.19L}{T}} L. \end{aligned} \tag{1.12}$$

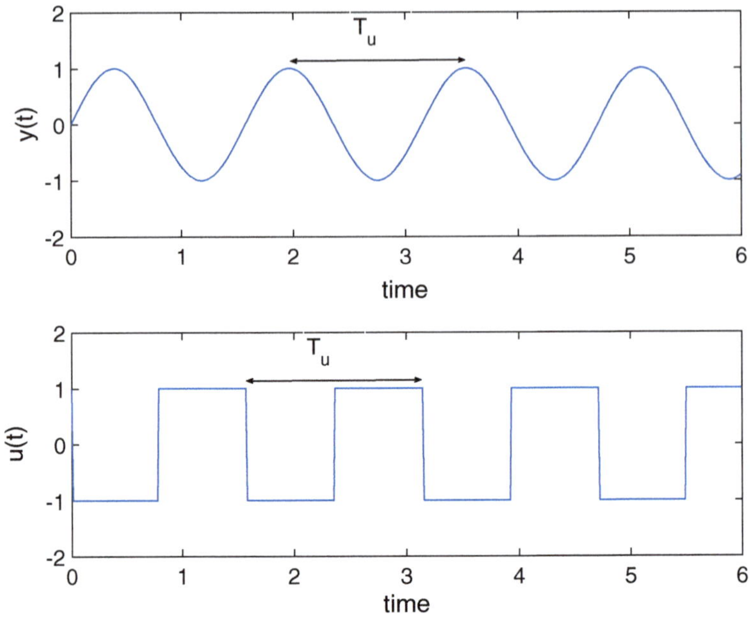

Fig. 1.9 Output oscillation and relay signals

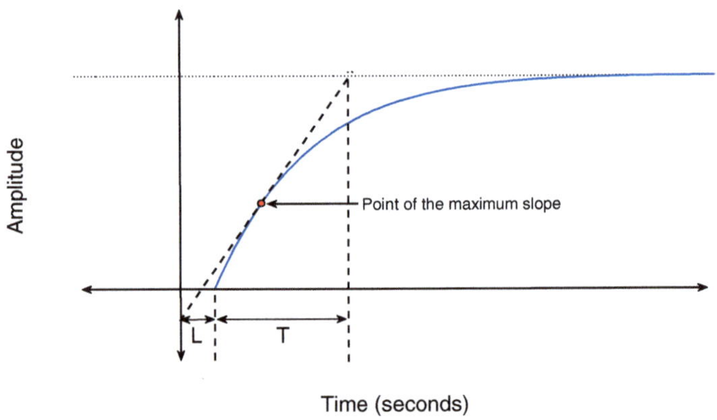

Fig. 1.10 Cohen-Coon method

1.3.4 PID Controller Tuning Methods

After the appearance of the classical PID controller tuning techniques, the complexity of the systems and performance demands from the control designer made necessary the development of new tuning design techniques. Over the years, many useful results were developed toward PID tuning methods for more performance- specific require-

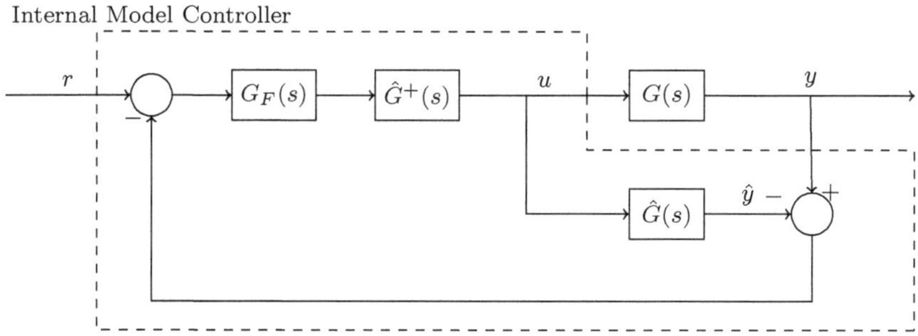

Fig. 1.11 Closed-loop system block diagram with internal model controller

ments and to deal with more complex systems. Some of these approaches are the following:

- **Internal Model Control design**
  This controller approach considers stable systems. Consider the closed-loop system block diagram presented in Fig. 1.11. Where $\hat{G}(s)$ is an approximation of the system $G(s)$, $G_F(s)$ is a low pass filter, and $\hat{G}^+(s)$ is the inverse of $\hat{G}(s)$. Then, the controller design objective is to cancel the poles and zeros of the original system $G(s)$ by connecting it in parallel with $\hat{G}(s)$. This approach is called internal model control because the controller contains a model of the system internally. The purpose of $G_F(s)$ is to make the system less sensitive to modeling errors. The controller $C(s)$ is given by

$$C(s) = \frac{G_F(s)\hat{G}^+(s)}{1 - G_F(s)\hat{G}^+(s)\hat{G}(s)}. \tag{1.13}$$

  Consider the case when this approach is applied to PI and PID controllers. For the case of plants which are first-order plus time-delay systems, we have that

$$P(s) = \frac{K}{1+sT}e^{-sL}. \tag{1.14}$$

$$\hat{G}^+(s) = \frac{1+sT}{K}. \tag{1.15}$$

$$G_F(s) = \frac{1}{1+sT_f}. \tag{1.16}$$

  Then, by a first-order Padé approximation for the time delay

$$e^{-sL} \approx \frac{1-sL/2}{1+sL/2}, \tag{1.17}$$

we have the controller of the PID form

$$C(s) = \frac{(1+sL/2)(1+sT)}{Ks(L+T_f+sT_fL/2)} \approx \frac{(1+sL/2)(1+sT)}{Ks(L+T_f)} = \frac{k_d s^2 + k_p s + k_i}{s}, \tag{1.18}$$

where

$$k_d = \frac{LT}{2K(L+T_f)}, \tag{1.19}$$

$$k_p = \frac{(L+2T)}{2K(L+T_f)}, \tag{1.20}$$

$$k_i = \frac{1}{K(L+T_f)}. \tag{1.21}$$

- **Pole Placement Design**
  Pole placement is a controller design method based on the knowledge of the system's transfer function, where the objective is to determine the closed-loop pole locations on the complex plane by setting the controller gains. It is known that the system's closed-loop pole locations influence the behavior of the system. Therefore, the designer can apply this method to place the locations of the poles for a desirable behavior of the closed-loop system.
  PI and PID controllers can be used with a pole placement design as long as the plant transfer function system is of the first or second order. For higher order systems, one way to use a PI or PID controller is to approximate the system's transfer function by a first- or second- order transfer function.
  For the first-order case, the system can be described by

$$P(s) = \frac{K}{1+Ts}, \tag{1.22}$$

where K is the system's gain and T is the time constant. Using a PI controller, we have

$$C(s) = K_c\left(1 + \frac{1}{T_i s}\right), \tag{1.23}$$

where K_c is the controller gain and T_i the integral time. The closed-loop transfer function is

$$G(s) = \frac{C(s)P(s)}{1+C(s)P(s)}. \tag{1.24}$$

The characteristic equation becomes of second order

$$\delta(s) = s^2 + \left(\frac{1 + KK_c}{T}\right)s + \left(\frac{KK_c}{TT_i}\right). \tag{1.25}$$

A second-order characteristic equation can be represented in terms of the relative damping ζ and the natural frequency ω_n as

$$\delta(s) = s^2 + 2\zeta\omega_n s + \omega_n^2, \tag{1.26}$$

where the parameters ζ and ω_n determine the time response of the second-order system.
Comparing (1.25) and (1.26) we must have

$$K_c = \frac{2\zeta\omega_n T - 1}{K}. \tag{1.27}$$

$$T_i = \frac{2\zeta\omega_n T - 1}{\omega_n^2 T}. \tag{1.28}$$

For a second-order plant without zeros, the plant can be described by

$$P(s) = \frac{K}{(1 + T_1 s)(1 + T_2 s)}. \tag{1.29}$$

Using the PID controller

$$C(s) = \frac{K_c\left(1 + T_i s + T_i T_d s^2\right)}{T_i s}, \tag{1.30}$$

the characteristic equation becomes of third order

$$\delta(s) = s^3 + \left(\frac{1}{T_i} + \frac{1}{T_2} + \frac{KK_cT_d}{T_1T_2}\right)s^2 + \left(\frac{1}{T_1T_2} + \frac{KK_c}{T_1T_2}\right)s + \frac{KK_c}{T_1T_2T_i}. \tag{1.31}$$

A third-order characteristic equation can also be represented in terms of the relative damping ζ and the natural frequency ω_n as

$$\delta(s) = (s + \alpha\omega_n)(s^2 + 2\zeta\omega_n s + \omega_n^2). \tag{1.32}$$

Combining (1.31) and (1.32) we have

Table 1.2 Cohen-Coon formulae for dominant pole placement controller design

Controller	K_c	T_i	T_d
PI	$\frac{0.9}{a}\left(1+\frac{0.92\tau}{1-\tau}\right)$	$\frac{3.3-3.0\tau}{1+1.2\tau}L$	
PID	$\frac{1.35}{a}\left(1+\frac{0.18\tau}{1-\tau}\right)$	$\frac{2.5-2.0\tau}{1-0.39\tau}L$	$\frac{0.37-0.37\tau}{1-0.81\tau}L$

$$K_c = \frac{T_1 T_2 \omega_n^2 (1 + 2\alpha\zeta) - 1}{K}. \quad (1.33)$$

$$T_i = \frac{T_1 T_2 \omega_n^2 (1 + 2\alpha\zeta) - 1}{T_1 T_2 \alpha \omega_n^3}. \quad (1.34)$$

$$T_d = \frac{T_1 T_2 \omega_n (\alpha + 2\zeta) - T_1 - T_2}{T_1 T_2 \omega_n^2 (1 + 2\alpha\zeta) - 1}. \quad (1.35)$$

- **Dominant Pole Placement Design**
  This controller design approach follows the same idea of the previous pole placement design. However, this method is focused on higher order systems. The objective is to select a pair of dominant poles, which have more influence on the behavior of the system time response than the rest of the closed-loop poles.
  For PI and PID controllers design using a dominant pole placement method, there is an approach developed by Cohen-Coon for first-order plus time-delay systems such as the one shown in (1.14). The central design criterion is the rejection of load disturbances by placing the dominant poles that give a quarter amplitude decay ratio in the time response. For PID controllers, two complex dominant poles and one real pole are placed to satisfy the quarter amplitude decay ratio in the time response. The following table presents some formulae to calculate the PI and PID controller gains (Table 1.2),
  where

$$a = \frac{KL}{T}. \quad (1.36)$$

$$\tau = \frac{L}{(L+T)}. \quad (1.37)$$

- **Time Domain Optimization Methods**
  In time domain optimization methods, the controller gains are calculated based on numerical optimization methods where an objective function is specified. For PID controllers, an objective function is defined by one of the forms

$$J(\theta) = \int_0^\infty t|e(\theta, t)|dt, \quad (1.38)$$

$$J(\theta) = \int_0^\infty |e(\theta, t)|dt, \quad (1.39)$$

$$J(\theta) = \int_0^\infty e(\theta, t)^2 dt, \tag{1.40}$$

where θ represents a vector with the PID gains and $e(\theta, t)$ is the error signal of the control system. The objective function in (1.38) is called integral time-weighted absolute error (ITAE); this function integrates the absolute error multiplied by time as a weight. The objective function (1.39) is called integral absolute error (IAE). This function integrates the absolute error without weights. The objective function (1.40) is called integral square error (ISE), which only integrates the square of the error.

The parameters of the controller are obtained after minimizing a selected objective function to obtain a better performance of the closed-loop system .

- **Gain and Phase Margin Based Design**

  Gain and phase margins can indicate how stable the system is. These margins are calculated from the open-loop system to determine how robust the closed-loop system is. The gain margin is the amount of gain necessary to make the system unstable and the phase margin is the amount of phase reduction necessary to make the system unstable. These margins are considered in classical control designs associated with the frequency response of the system. The gain and phase margins can be obtained from the Nyquist plot of $P(j\omega)C(j\omega)$ $\omega \in [0, \infty)$ as in Fig. 1.12. In Fig. 1.12, GM represents the gain margin, PM is the phase margin, ω_p is the phase crossover frequency, and ω_g the gain crossover frequency. Over time, there has been a research interest in developing new controller design approaches to achieve specific gain and phase margins for the closed-loop system. There is a significant number of research papers with different approaches for PI and PID controllers to achieve specific gain and phase margins. These different approaches for PI/PID controller design generally consider first-order or second-order plants cascaded with a time delay. In this monograph, we have presented a general approach for simultaneously achieving prescribed gain and phase margins for an arbitrary order plant.

- **Adaptive Control Design**

  In the adaptive controller design, the controller gains are to be readjusted in response to the changes in the system or due to the presence of perturbations. There are two types of adaptive control called direct and indirect methods. In the direct approach, the adaptive controller design approach known as *model reference adaptive control* is considered. In this a reference model, representing desired performance, is specified in terms of the characteristics of a dynamic system. Then, the difference between the output of the plant and the reference model, is used by an adaptation algorithm to *directly* adjust the parameters of the controller in real time to force the plant model error to zero. In the *indirect* approach, a model of the plant is estimated from the available input–output measurements. Then, the adaptive control scheme is called *indirect* because the readjustment of the controller parameters is made by first performing the estimation of the plant and then the

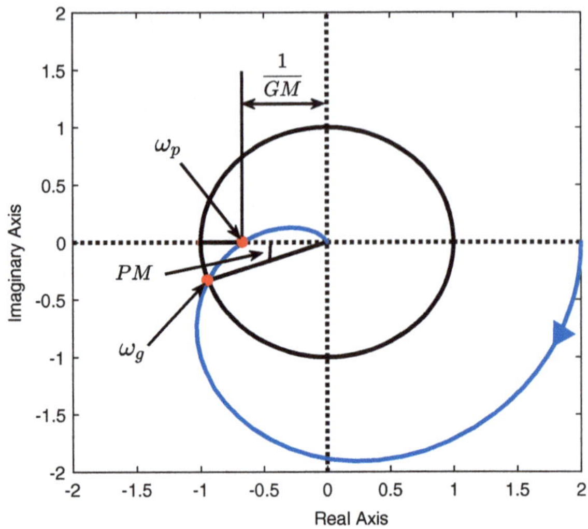

Fig. 1.12 Gain and phase margins from a typical Nyquist plot

computation of the controller parameters is based on the current estimated plant model. Recursive parameter estimation is used to update the process model. These types of adaptive controller techniques are widely used for PID controllers.

1.4 Integrator Windup

An essential element of the controller is the actuator, which applies the control signal u to the plant. However, all actuators have limitations that make them nonlinear elements. For instance, a valve cannot be more than fully opened or less than fully closed. During the regular operation of a control system, it can very well happen that the control variable reaches the actuator limits. When this situation arises, the feedback loop is broken, and the system runs as an open loop because the actuator will remain at its limit independently of the process output. If the controller is of the PID type, the error will continue to be integrated. This condition, in turn, means that the integral term may become very large, which is commonly referred to as *windup*. The error signal needs to have an opposite sign for an extended period to return to a normal state. As a consequence of all this, a system with a PID controller may give large transients when the actuator saturates.

The phenomenon of windup has been known for a long time. It may occur in connection with large setpoint changes or large disturbances or equipment malfunction may cause it. Several techniques are available to avoid windup when the integrator is in the controller. We describe some of these techniques in this section.

1.4.1 Setpoint Limitation

The easiest way to avoid integrator windup is to introduce limiters on the setpoint variations so that the controller output will never reach the actuator bounds. However, this approach has several disadvantages: (a) it leads to conservative bounds; (b) it imposes limitations on the controller performance; (c) it does not prevent windup caused by disturbances.

1.4.2 Back-Calculation and Tracking

This technique is illustrated in Fig. 1.13. We notice that the controller has an extra feedback path. This path is generated by measuring the actual actuator output $u(t)$ and forming the error signal $e_s(t)$ as the difference between the output of the controller $v(t)$ and the signal $u(t)$. This signal $e_s(t)$ is fed to the input of the integrator through a gain $1/T_t$.

When the actuator is within its operating range, the signal $e_s(t)$ is zero. Thus, it will not have any effect on the normal operation of the controller. When the actuator saturates, the signal $e_s(t)$ is different from zero. The regular feedback path around the process is broken because the process input remains constant. However, there is a new feedback path around the integrator due to $e_s(t) \neq 0$, and this prevents the integrator from winding up. The feedback gain $1/T_t$ governs the rate at which the controller output is reset. The parameter T_t can thus be interpreted as the time constant that determines how quickly the integral action is reset. In general, the smaller the value of T_t, the faster the integrator is reset. However, if the parameter T_t is chosen too small, spurious errors can cause saturation of the output, which accidentally resets the integrator.

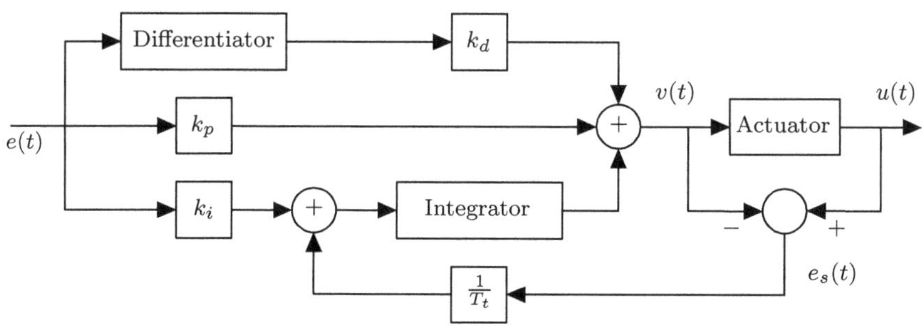

Fig. 1.13 Controller with antiwindup. © Taylor & Francis LLC Books. Reproduced from [8] with permission

1.4.3 Conditional Integration

Conditional integration is an alternative to the back-calculation technique. It merely consists of switching off the integral action when the control is far from the steady state. This procedure means that the integral action is only used when certain conditions are fulfilled; otherwise, the integral term is kept constant. We now consider two of these switching conditions.

A straightforward approach is to switch off the integral action when the control error $e(t)$ is big. Another one is to switch off the integral action when the actuator saturates. However, both approaches have a disadvantage: the controller may get stuck at a nonzero error $e(t)$ if the integral term has significant value at the time of switch off.

Because of the above disadvantage, a better approach is to switch off integration when the controller is saturated, and the integrator update is such that it causes the control signal to become more saturated. For example, consider that the controller becomes saturated at its upper bound. Integration is then switched off if the control error is positive, but not if it is negative.

1.5 The Rise and Fall of Optimal Control

This introduction to control would not be complete without discussing optimal control, its impact on control theory, its relation to robustness, and its conspicuous absence in the literature on PID control. Optimal control was born in the late 1950s with the contributions of Bellman (Dynamic Programming), Pontryagin (Maximum Principle), and Kalman (Linear-Quadratic Regulator (LQR)).

The evolution of control theory in the latter half of the twentieth century presents a fascinating story of seduction, deception, and betrayal. The story begins in 1960 when Kalman published his paper on the optimal linear-quadratic regulator which established that the optimal control consisted of state feedback and was stabilizing under mild conditions. The control community was immediately seduced by the mathematical elegance of this result and would choose to embrace this approach to control, obsessively, for the next 40 years. Some detractors and critics of quadratic optimal control pointed out that optimality was essentially subjective but robustness was not and the issue of nonmeasurability of all the states had to be solved.

Kalman answered the first question most elegantly in 1964 by showing that the optimal state-feedback control provided a guaranteed universal gain margin of $[1/2, \infty)$ and a phase margin of $60°$ in each channel regardless of the particular quadratic index chosen. The issue of nonmeasurability of states was also solved in 1966 when Luenberger proposed his dynamic observer, which could produce approximations of the state vector from input and output measurements. These "estimates" could substitute the actual state in the optimal control law without significant loss of optimality. It was shown that in such a system the closed-loop eigenvalues would consist of the eigenvalues of the optimal system along with those of the observer which could in fact be chosen arbitrarily.

More than a decade would pass before Doyle and Stein showed with a simple counterexample that the excellent aforementioned stability margins of the optimal state-feedback system could be drastically reduced when observers were used to implement state feedback (see Example 1.1). This result in fact was instrumental in reviving the interest of the control community in the issue of robustness of control systems. One outcome of this renewed interest was the formulation of the H_∞ optimal control problem, which took into account tolerance of plant uncertainty as a design objective from the outset. An elegant solution of the H_∞ optimal control problem in the state space was published by Doyle, Glover, Khargonekar, and Francis in 1989 and came to be known as the DGKF paper.

Almost another 10 years would pass before Keel and Bhattacharyya showed in 1997 that the high-order controllers typically produced by various optimization schemes were extremely fragile under coefficient perturbations of the controller. This was true for H_∞ optimal controllers, which were robust with respect to plant uncertainty as well. It was pointed out in the 1997 paper that the inverse relation between plant robustness and controller robustness was in fact expected from the theory of Blacks amplifier. This result on fragility was declared by Prof. J. B. Pearson to be "the last nail in the coffin of optimal control" (1997 Private Communication). In any case, this result finally convinced the control community that controller order could not be indiscriminately inflated to assign poles and zeroes like "salt and pepper" to feedback control systems. The price of high controller order was fragility. This ushered in renewed interest in low-order controllers and in particular the PID controller.

1.5.1 Quadratic Optimization and Robustness

In 1960, Kalman introduced the state-variable approach and quadratic optimal control in the time domain as new design approaches. This phase in the theory of automatic control systems arose out of the crucial new technical problems that were encountered at that time: the launching, maneuvering, guidance, and tracking of space vehicles. Much effort into and rapid developments in both theory and practice took place. Optimal control theory was developed under the influence of many great researchers such as Pontryagin, Bellman, Kalman, and Bucy. In the 1960s, Kalman introduced many key state-variable concepts. Among these were controllability, observability, optimal linear-quadratic regulator (LQR), state-feedback and optimal state estimation (Kalman filtering). The optimal state-feedback control produced by the LQR problem was guaranteed to be stabilizing for any quadratic performance index subject to mild conditions.

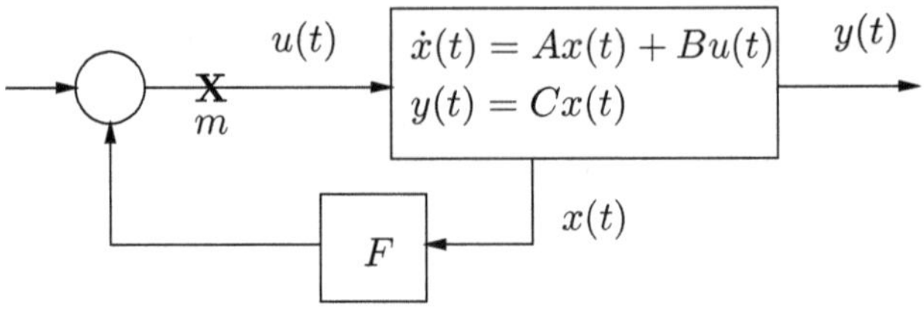

Fig. 1.14 State-feedback configuration

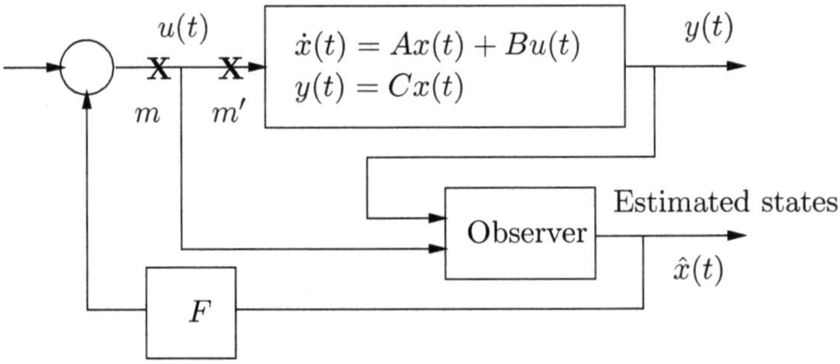

Fig. 1.15 Observed state-feedback configuration

In a 1964 paper by Kalman which demonstrated that the optimal LQR state-feedback control laws had some powerful guaranteed robustness properties, namely an infinite upper gain margin and a 60-degree phase margin in each channel, which also were independent of the particular quadratic index chosen. This concept is illustrated in Fig. 1.14, where the state-feedback system designed via LQR optimal control has the above-guaranteed stability margins at the loop breaking point "m".

For some time, control scientists were generally led to believe that the extraordinary robustness properties of the LQR state-feedback design were preserved when the control was implemented as an output feedback system through an observer. We depict this in Fig. 1.15, where the stability margin at the point m continues to equal that obtained in the state-feedback system. However, it was shown by Doyle that the margin at the point m', which is much more meaningful, could be drastically less. This observation ushered in a period of renewed interest in the robustness of closed-loop designs and led to the field of H_∞ optimal control.

Example 1.1 (*Loss of LQR Margins–Doyle and Stein's example*) Consider the plant:

$$\dot{x} = \begin{bmatrix} 0 & 1 \\ -3 & -4 \end{bmatrix} x + \begin{bmatrix} 0 \\ 1 \end{bmatrix} u,$$

$$y = \begin{bmatrix} 2 & 1 \end{bmatrix} x,$$

and controller:

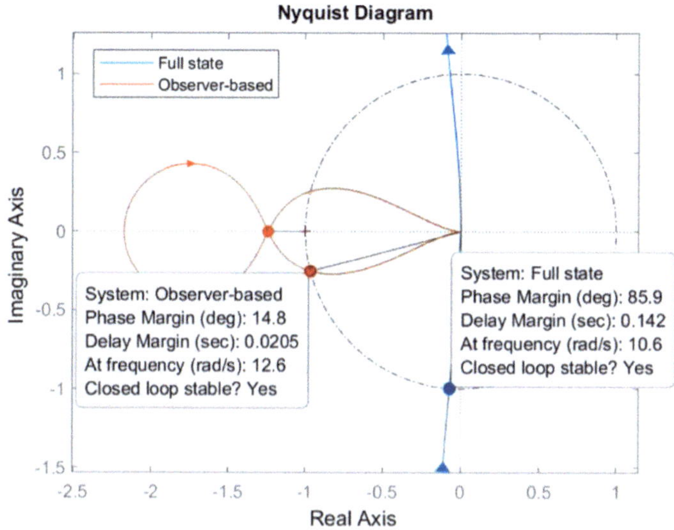

Fig. 1.16 Nyquist plot for Doyle and Stein's example

$$u = \begin{bmatrix} -50 & -10 \end{bmatrix} \hat{x} + \begin{bmatrix} 50 \end{bmatrix} r,$$

where $\hat{x}$ is the estimated variable of x. The plant is a stable system with transfer function

$$\frac{y(s)}{u(s)} = \frac{s+2}{(s+1)(s+3)}.$$

The controller is linear quadratic optimal, corresponding to the performance index

$$J = \int_0^\infty \left(x^T H^T H x + u^2\right) dt$$

with

$$H = 4\sqrt{5}\begin{bmatrix} \sqrt{35} & 1 \end{bmatrix}.$$

It places the closed-loop poles at

$$-7.0 \pm j\,2.0.$$

A Nyquist diagram is given in Fig. 1.16 to compare phase margins. It is clear that the phase margin of 85.9° achieved with full state-feedback collapses to 14.8° under observer-based implementation.

1.5.2 H_∞ Optimal Control

In this approach to feedback control, one considers a system subject to controls u and disturbances v as inputs , and with outputs required to be controlled and available for measurement denoted z and y, respectively:

$$\begin{aligned}
\dot{x} &= Ax + B_1 v + B_2 u,\\
z &= C_1 x + D_{11} v + D_{12} u,\\
y &= C_2 x + D_{21} v + D_{22} u.
\end{aligned}$$

The objective of feedback control is to minimize the effect of the disturbance on the controlled output and this performance specification is quantified by the H_∞ norm of the closed-loop disturbance transfer function. Let the size of a vector signal $x(t)$ be measured by the L_2-norm:

$$\|x(t)\|_2^2 = \int_0^\infty |x(t)|^2 \, dt, \tag{1.41}$$

where

$$|x(t)|^2 = \sum_{i=1}^{n} x_i^2(t) \tag{1.42}$$

corresponds to the standard Euclidean norm of the vector $x(t)$ at time t. The definition of the H_∞ norm of a transfer function $H(s)$ with input v and output z is

$$\|H(s)\|_\infty = \sup \left\{ \frac{\|z(t)\|_2}{\|v(t)\|_2} : v(t) \neq 0 \right\}. \tag{1.43}$$

The objective here is to find a dynamic controller so that the H_∞ norm of the disturbance v to output z transfer function is below a prescribed level $\gamma > 0$. Under some mild technical restrictions on the system, the solution can always be found for a large enough γ and proceeds as follows:

1. Let $X \geq 0$ be a solution to the algebraic Riccati equation

$$A^T X + XA + C_1^T C_1 + X(\gamma^{-2} B_1 B_1^T - B_2 B_2^T)X = 0 \tag{1.44}$$

with the eigenvalues of

$$A + (\gamma^{-2} B_1 B_1^T - B_2 B_2^T)X \tag{1.45}$$

in the left half-plane.

2. Let $Y \geq 0$ be a solution to the algebraic Riccati equation

$$AY + YA^T + B_1 B_1^T + Y(\gamma^{-2} C_1^T C_1 - C_2^T C_2)Y = 0 \tag{1.46}$$

with the eigenvalues of

$$A + Y(\gamma^{-2} C_1^T C_1 - C_2^T C_2) \tag{1.47}$$

in the left half-plane.
3. Let

$$\rho(XY) < \gamma^2, \tag{1.48}$$

where $\rho(A)$ denotes the spectral radius (magnitude of the eigenvalue with largest magnitude) of the square matrix A.
4. A controller rendering the H_∞ norm of the transfer function between v and z less than γ is given by

$$\begin{aligned}
\dot{\hat{x}} &= A\hat{x} + B_1\hat{v} + B_2 u + L(C_2\hat{x} - y), \\
u &= F\hat{x}, \\
F &= -B_2^T X, \\
L &= -(I - \gamma^{-2} YX)^{-1} YC_2^T, \\
\hat{v} &= \gamma^{-2} B_1^T X\hat{x}.
\end{aligned}$$

This was essentially a continuation of the quadratic optimization approach of Kalman with the additional feature of optimization under a worst-case disturbance input. The controllers obtained in the important work of Doyle et. al. (DGKF) in 1988 and solving this problem were of output feedback type and were robust to plant perturbations. However, the dynamic order of the controllers were invariably high, often several times the order of the plant. This feature would render them fragile as shown in the 1997 paper of Keel and Bhattacharyya.

1.5.3 Blacks Amplifier: Robustness with High Gain Feedback

In this subsection, we illustrate how robust systems can be built from highly unreliable components by using high gain and the feedback structure. These ideas were introduced by H. S. Black in 1926 when he invented the feedback amplifier. Specifically, we consider the problem of obtaining a precisely controlled value of gain, from a system containing large parameter uncertainty. Consider the system shown in Fig. 1.17.

Suppose that the system gain G is required to be 100. Due to the poor reliability of the components, this can vary by 50%. Then, the actual gain can range from 50 to

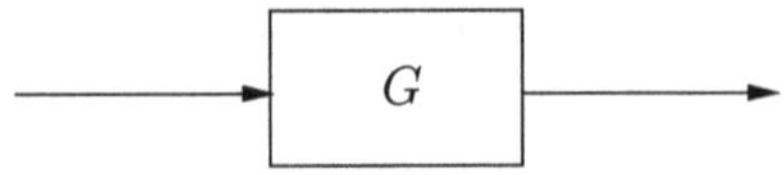

Fig. 1.17 An open-loop system

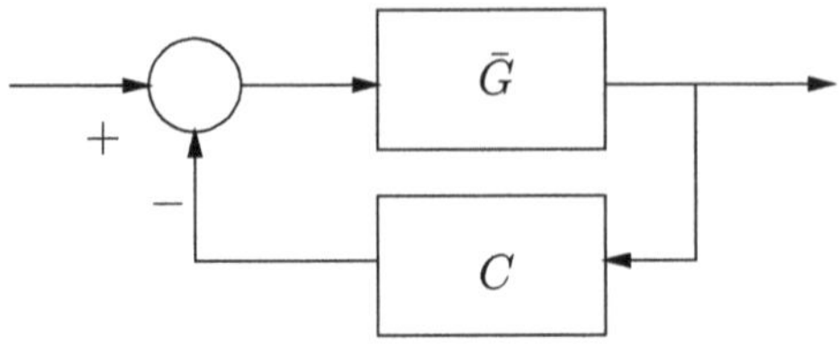

Fig. 1.18 A feedback system

150. To remedy this situation, Black proposed the feedback structure shown in Fig. 1.18.

The gain $\bar{G}$ is again made with the same unreliable components but with a nominal value much higher than 100, say 10,000, and we set $C = 0.01$. The overall gain of the feedback system is given by the expression

$$\frac{\bar{G}}{1 + C\bar{G}}. \tag{1.49}$$

With 50% variation in $\bar{G}$ (that is, $\bar{G}$ varies from 5,000 to 15,000) the gain of the feedback system varies from 98.039 to 99.338. This remarkable increase in robustness, corresponding to a reduction of uncertainty from 50 to 1%, is one of the main reasons for the widespread use of feedback in the control and electronics industry. As a final observation, it is not difficult to see that robustness to variations in G is obtained at the cost of *fragility* with respect to C.

1.5.4 Fragility of High-Order Controllers

In 1997, Keel and Bhattacharyya demonstrated that high-order controllers irrespective of the specific design approach (l^1 optimal μ or H_∞) were fragile with respect to controller parameters and could destabilize the closed loop under minuscule perturbations. Their 1997 paper also pointed out that this controller fragility of plant-robust closed-loop systems was consistent with the theory of Black's feedback amplifier developed in 1926, described in the previous subsection. The fragility of high-order controllers and the inability of quadratic optimization to produce fixed-order controllers, such as PIDs, severely limited the applicability of quadratic optimization to real-world problems and ultimately led to its demise.

In the sequel, we present a sample example of a high-order optimal controller and its fragility.

Example 1.2 (*H_∞ based optimum gain margin controller*) This example uses the YJBK parametrization and the machinery of the H_∞ Model Matching Problem to optimize the upper gain margin. The plant to be controlled is

$$P(s) = \frac{s-1}{s^2-s-2}$$

and the controller, designed to give an upper gain margin of 3.5, (the closed loop is stable for the gain interval [1, 3.5]) is obtained by optimizing the H_∞ norm of a complementary sensitivity function. The controller found is

$$C(s) = \frac{q_6^0 s^6 + q_5^0 s^5 + q_4^0 s^4 + q_3^0 s^3 + q_2^0 s^2 + q_1^0 s + q_0^0}{p_6^0 s^6 + p_5^0 s^5 + p_4^0 s^4 + p_3^0 s^3 + p_2^0 s^2 + p_1^0 s + p_0^0},$$

where

$$\begin{array}{ll} q_6^0 = 379 & p_6^0 = 3 \\ q_5^0 = 39383 & p_5^0 = -328 \\ q_4^0 = 192306 & p_4^0 = -38048 \\ q_3^0 = 382993 & p_3^0 = -179760 \\ q_2^0 = 383284 & p_2^0 = -314330 \\ q_1^0 = 192175 & p_1^0 = -239911 \\ q_0^0 = 38582 & p_0^0 = -67626. \end{array}$$

The poles of this nominal controller are

$$174.70, \quad -65.99, \quad -1.86, \quad -1.04, \quad -0.98 \pm j0.03$$

and the poles of the closed-loop system are

$$\begin{aligned} &-0.4666 \pm j14.2299, \quad -5.5334 \pm j11.3290, \quad -1.0002, \\ &-1.0000 \pm j0.0002, \quad -0.9998 \end{aligned}$$

and this verifies that the controller is indeed stabilizing.

The Nyquist plot of $P(s)C(s)$ is shown in Fig. 1.19 and verifies that the desired upper gain margin is achieved. On the other hand, we see from Fig. 1.19 that the lower gain margin and phase margin are

$$\begin{aligned} &\text{Gain Margin} = [1, 0.9992], \\ &\text{Phase Margin} = [0, 0.1681] \text{ degree.} \end{aligned}$$

The lower gain margin means roughly that a reduction in gain of one part in one thousand will destabilize the closed-loop system! Likewise, a vanishingly small phase perturbation is destabilizing.

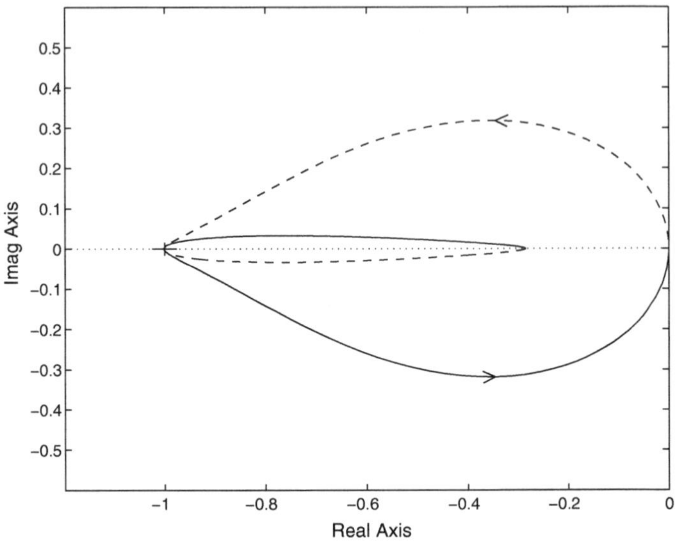

Fig. 1.19 Nyquist plot of $P(s)C(s)$. © Taylor & Francis LLC Books. Reproduced from [8] with permission

To continue with our analysis, let us consider the transfer function coefficients of the controller to be a parameter vector $\mathbf{p}$ with its nominal value being

$$\mathbf{p}^0 = \left[q_6^0 \ \cdots \ q_0^0 \ p_6^0 \ \cdots \ p_0^0\right]$$

and let $\Delta\mathbf{p}$ be the vector representing perturbations in $\mathbf{p}$. We compute the l^2 parametric stability margin around the nominal point. This margin comes out to be

$$\rho = 0.15813903109631.$$

The normalized ratio of change in controller coefficients required to destabilize the closed loop is

$$\frac{\rho}{\|\mathbf{p}^0\|_2} = 2.103407115900516 \times 10^{-7}.$$

This result shows that a change in the controller coefficients of less than 1 part in a million destabilizes the closed loop. This controller is anything but robust; in fact, we are certainly justified in labeling it as a *fragile* controller.

To verify this somewhat surprising result, we construct the destabilizing controller whose parameters are obtained by setting $\mathbf{p} = \mathbf{p}^0 + \Delta\mathbf{p}$ and are

$$\begin{aligned}
q_6 &= 379.000285811, & p_6 &= 3.158134748\\
q_5 &= 39382.999231141, & p_5 &= -327.999718909\\
q_4 &= 192305.999998597, & p_4 &= -38048.000776386\\
q_3 &= 382993.000003775, & p_3 &= -179760.000001380\\
q_2 &= 383284.000000007, & p_2 &= -314329.999996188\\
q_1 &= 192174.999999982, & p_1 &= -239910.999999993\\
q_0 &= 38582.000000000, & p_0 &= -67626.000000018.
\end{aligned}$$

The closed-loop poles of the system with this controller are

$$\begin{gathered}
0.000 \pm j14.2717, \quad -5.5745 \pm j10.9187,\\
-1.0067 \pm j\,0.0158, \quad -1.0044, \quad -0.9820,
\end{gathered}$$

which shows that the roots crossover to the right-half plane at $\omega = 14.27$, and the perturbed controller is indeed destabilizing.

A number of examples of the fragility of high-order optimal controllers designed by l^1, μ, H_∞, and H_2 methods were demonstrated by Keel and Bhattacharyya in 1997.

1.6 Modern PID Control

The subject of this monograph is the development of a new design theory for PID controllers. As opposed to most previous approaches, our methods are analytical and moreover suitable for computer implementation. They are based on the effective computation of the stabilizing set in the space of PID gains developed rigorously over the past 20 years. With this set in hand, it is possible to impose several performance requirements and find the subsets achieving them over this set. Finally, it is possible to conduct multiple performance attaining designs and determine whether a given vector of specifications can or cannot be attained over the stabilizing set. The performance specifications that can be effectively dealt with are gain margin, phase margin, and H_∞ norm bounds which are related to robustness. It is also possible to deal with some time domain specifications. These results are described for both continuous-time systems and discrete-time systems. Recent results on multivariable controller design are also presented. We hope that these results will find application to real-world control systems and also stimulate fundamental research on controller design to mitigate the fragility of high-order controller designs.

1.7 Notes and References

There is an extensive body of literature covering all aspects of PID control. We have not attempted to be complete in citing this literature. Instead, we have tried to cite all relevant publications related to the new results given later in this book.

The servomechanism problem was solved in Bhattacharyya and Pearson [9, 10] in 1970 and 1972. In [9], the single output problem was solved and the internal model was introduced. In [10], the multivariable servomechanism problem was solved by formulating it as a problem of zeroing the output of a nonstate-stabilizable system and using the results of Bhattacharyya, Pearson, and Wonham [11]. Necessary and sufficient conditions for output feedback tracking and disturbance rejection with internal stability were obtained in [10] under the assumption that signal modes are disjoint from plant zeros. These results were subsequently refined in Wonham and Pearson [73], where the reference and disturbances were allowed to have poles at the system zeros. In [33], Francis, Sebakhy, and Wonham introduced the Internal Model Principle wherein the necessity of error feedback and hence of internal models, was established for the existence of a solution that is insensitive to plant and stabilizing controller parameters. A controller incorporating the Internal Model had been presented for the multivariable case by Davison in [19, 20], where the term "robust servomechanism" was used to refer to the above-mentioned insensitivity. In Howze and Bhattacharyya [43], it was shown that error feedback is not necessary if insensitivity to *only* plant parameters is required. The definition of blocking zeros and the fact that Internal Models assign "blocking zeros" to the error transfer function were established in Ferreira [31] and Ferreira and Bhattacharyya [32]. A clear and self-contained treatment of the servomechanism problem is given in Desoer and Wang [21], where it is stated "we leave it to the science historian to describe fairly the history of the subject."

Applications of PID controllers in Sect. 1.3 can be found in the following references [30, 53, 59, 65]. For applications to process control see [4]. For more information about the PID controller actions in Sect. 1.3.1 see [4, 8, 58, 67, 68]. More details about PID controller representations in Sect. 1.3.2 are available in [67]. The classical control design approaches presented in Sect. 1.3.3 are discussed in [1, 4, 5, 61, 67, 75].

The Ziegler–Nichols methods were first presented in [75]. The alternative method using relay feedback is described in [3]. The relay feedback technique and its applications to automatic PID tuning can be found in [3, 4]. For a better understanding of describing functions, the book [51] is recommended. The Cohen-Coon method can be found in [14, 67]. A comprehensive survey on tuning methods for PID controllers and a description of antiwindup techniques can be found in [4]. For a more detailed explanation of the IMC structure and its applications in process control, the reader is referred to [18, 35, 41, 62]. For more details about dominant pole placement design see [4, 18, 44, 70, 74]. For optimization-based controller design methods see [18].

For gain and phase margin-based design approaches, the reader can consult [34, 67]. For recent results concerning PI/PID controller design approaches based on gain and phase margin design, the reader can refer to [22–28]. Related to achieving specific gain and phase margins, different approaches for PI/PID controller design consider first-order or second-order plus time-delay processes. Examples of these approaches are presented in [36, 39, 42, 56, 60, 66]. In [40, 54], the reader can refer to survey papers dealing with PID controller design considering gain and phase margin specifications. In [41, 49, 62], different methods are presented using IMC. In [39, 55, 56], optimization approaches are presented. In [42, 60, 72], unstable processes are considered for the design of PI/PID controllers. In [1, 16, 48, 71], controller design methods applying system identification are presented. In [17, 38, 64, 72], some graphical methods are applied to find controller gains. In [15, 36, 69], several different methods are presented to calculate the controller gains.

For adaptive PID controller design, the reader can find more information in [2–4, 6, 7, 13, 45, 63].

For more details about the integrator windup described in Sect. 1.4, the reader can see [3]. In Sect. 1.5.1, the reader can see more details of the Kalman quadratic optimization approach in [29, 46, 47]. The 1997 fragility results of Keel and Bhattacharyya are presented in [50]. For more information about the feedback amplifier invented in 1926 by H. S. Black see [52]. For more details about the example in Sect. 1.5.4 and other examples, the reader should refer to [50, p. 200].

Recent work on biological control systems has identified the existence and necessity of integral feedback mechanisms for robust regulation at the biomolecular level [12, 57].

In concluding these notes, we point out that, in addition to the approaches discussed above, there are many other approaches for tuning PID controllers [4]. Despite this, for plants having order higher than two, there were no approaches that could be used to determine the set of all stabilizing PID gain values. The principal contribution of this book to the PID literature is the development of a methodology that provides a complete answer to this long-standing open problem for both delay-free LTI plants as well as for plants with time delay. For the former class of plants, the results were first reported in [37]. In this book, we give results for determining, in a computationally efficient way, the complete set of PID controllers that stabilize a given linear time-invariant continuous- or discrete-time plant and achieve prescribed levels of performance.

References

1. Ang, K.H., Chong, G., Yun, L.: PID control system analysis, design, and technology. IEEE Trans. Control. Syst. Technol. **13**, 559–576 (2005)
2. Åström, K.J.: Theory and applications of adaptive control: A survey. Automatica **19**(5), 471–486 (1983)
3. Åström, K.J., Hägglund, T.: Automatic tuning of simple regulators with specifications on phase and amplitude margins. Automatica **20**(5), 645–651 (1984)
4. Astrom, K.J., Hagglund, T.: PID Controllers: Theory, Design, and Tuning, 2nd edn. International Society of Automation, North Carolina (1995)

5. Astrom, K.J., Hagglund, T.: The future of PID control. Control. Eng. Pract. **9**(11), 1163–1175 (2001)
6. Åström, K.J., Hägglund, T., Hang, C.C., Ho, W.K.: Automatic tuning and adaptation for PID controllers-a survey. Control. Eng. Pract. **1**(4), 699–714 (1993)
7. Åström, K.J., Wittenmark, B.: Adaptive Control. Courier Corporation, United States (2013)
8. Bhattacharyya, S.P., Datta, A., Keel, L.H.: Linear Control Theory Structure, Robustness, and Optimization. CRC Press Taylor and Francis Group, Boca Raton (2009)
9. Bhattacharyya, S.P., Pearson, J.B.: On the linear servomechanism problem. Int. J. Control **12**(5), 795–806 (1970)
10. Bhattacharyya, S.P., Pearson, J.B.: On error systems and the servomechanism problem. Int. J. Control **15**(6), 1041–1062 (1972)
11. Bhattacharyya, S.P., Pearson, J.B., Wonham, W.M.: On zeroing the output of a linear system. Inf. Control **2**, 135–142 (1972)
12. Briat, C., Gupta, A., Khammash, M.: Antithetic integral feedback ensures robust perfect adaptation in noisy biomolecular networks. Cell Syst. **2**(1), 15–26 (2016)
13. Chang, W.D., Yan, J.J.: Adaptive robust PID controller design based on a sliding mode for uncertain chaotic systems. Chaos, Solitons Fractals **26**(1), 167–175 (2005)
14. Cohen, G.H., Coon, G.A.: Theoretical consideration of retarded control. Trans. Am. Soc. Mech. Eng. **76**, 827–834 (1953)
15. Crowe, J., Johnson, M.A.: Automated PI control tuning to meet classical performance specifications using a phase locked loop identifier. In: American Control Conference, pp. 2186–2191 (2001)
16. Crowe, J., Johnson, M.A.: Towards autonomous pi control satisfying classical robustness specifications. IEE Proc.-Control Theory Appl. **149**(1), 26–31 (2002)
17. Darwish, N.M.: Design of robust PID controllers for first-order plus time delay systems based on frequency domain specifications. J. Eng. Sci. **43**(4), 472–489 (2015)
18. Datta, A., Ho, M.T., Bhattacharyya, S.P.: Structure and Synthesis of PID Controllers. Springer Science and Business Media, Berlin (2013)
19. Davison, E.J.: The output control of linear time-invariant multivariable systems with unmeasurable arbitrary disturbances. IEEE Trans. Autom. Control AC **17**(5), 621–630 (1972)
20. Davison, E.J.: The robust control of a servomechanism problem for linear time-invariant systems. IEEE Trans. Autom. Control AC **21**(1), 25–34 (1976)
21. Desoer, C.A., Wang, Y.T.: Linear time-invariant robust servomechanism problem: a self-contained exposition. Control. Dyn. Syst. **16**, 81–129 (1980)
22. Diaz-Rodriguez, I.D.: Modern design of classical controllers: continuous-time first order controllers. In: Proceedings of the 41st Annual Conference of the IEEE Industrial Electronics Society, Student Forum. IECON, pp. 000070–000075 (2015)
23. Diaz-Rodriguez, I.D., Bhattacharyya, S.P.: A one-shot approach to classical controller design: continuous-time PI controllers. In: Proceedings of International Conference on Advances in Engineering and Technology (AET) (2015)
24. Diaz-Rodriguez, I.D., Bhattacharyya, S.P.: Modern design of classical controllers: digital PI controllers. In: IEEE International Conference on Industrial Technology (ICIT), pp. 2112–2119 (2015)
25. Diaz-Rodriguez, I.D., Bhattacharyya, S.P.: PI controller design in the achievable gain-phase margin plane. In: IEEE 55th Conference on Decision and Control (CDC), pp. 4919–4924 (2016)
26. Diaz-Rodriguez, I.D., Han, S., Bhattacharyya, S.P.: Advanced tuning for Ziegler-Nichols plants. In: 20th World Congress of the International Federation of Automatic Control (IFAC 2017), pp. 1805–1810 (2017)
27. Diaz-Rodriguez, I.D., Han, S., Bhattacharyya, S.P.: Stability margin based design of multivariable controllers. In: IEEE Conference on Control Technology and Applications (CCTA), pp. 1661–1666 (2017)
28. Diaz-Rodriguez, I.D., Oliveira, V., Bhattacharyya, S.P.: Modern design of classical controllers: digital PID controllers. In: Proceedings of the 24th IEEE International Symposium on Industrial Electronics, pp. 1010–1015 (2015)

29. Doyle, J.C., Stein, G.: Robustness with observers. Technical report, DTIC Document (1979)
30. Dubonjić, L., Nedić, N., Filipović, V., Pršić, D.: Design of PI controllers for hydraulic control systems. Math. Probl. Eng. **2013**, 1–10 (2013)
31. Ferreira, P.M.G.: The servomechanism problem and the method of the state space in the frequency domain. Int. J. Control **23**(2), 245–255 (1976)
32. Ferreira, P.M.G., Bhattacharyya, S.P.: On blocking zeros. IEEE Trans. Autom. Control AC **22**(2), 258–259 (1977)
33. Francis, B.A., Sebakhy, O.A., Wonham, W.M.: Synthesis of multivariable regulators: the internal model principle. Applide Math. Optim. **1**, 64–86 (1974)
34. Franklin, G.F., Powell, J.D., Emami-Naeini, A.: Feedback Control of Dynamic Systems, 6th edn. Pearson Prentice Hall, New Jersey (2009)
35. Garcia, C.E., Morari, M.: Internal model control. A unifying review and some new results. Ind. Eng. Chem. Process Des. Dev. **21**(2), 308–323 (1982)
36. Hamamci, S.E., Tan, N.: Design of PI controllers for achieving time and frequency domain specifications simultaneously. ISA Trans. **45**(4), 529–543 (2006)
37. Ho, M.T., Datta, A., Bhattacharyya, S.P.: A linear programming characterization of all stabilizing PID controllers. In: Proceedings of American Control Conference, pp. 3922–3928 (1997)
38. Ho, M.T., Wang, H.S.: PID controller design with guaranteed gain and phase margins. Asian J. Control **5**(3), 374–381 (2003)
39. Ho, W., Lim, K., Xu, W.: Optimal gain and phase margin tuning for PID controllers. Automatica **34**(8), 1009–1014 (1998)
40. Ho, W.K., Gan, O.P., Tay, E.B., Ang, E.I.: Performance and gain and phase margins of well-known PID tuning formulas. IEEE Trans. Control Syst. Technol. **4**(4), 473–477 (1996)
41. Ho, W.K., Lee, T.H., Han, H.P., Hong, Y.: Self-tuning IMC-PID control with interval gain and phase margins assignment. IEEE Trans. Control Syst. Technol. **9**(3), 535–541 (2001)
42. Ho, W.K., Xu, W.: PID tuning for unstable processes based on gain and phase-margin specifications. IEE Proc.-Control Theory Appl. **145**(5), 392–396 (1998)
43. Howze, J.W., Bhattacharyya, S.P.: Robust tracking, error feedback and two degrees of freedom controllers. IEEE Trans. Autom. Control **42**(7), 980–984 (1997)
44. Hwang, S.H., Shiu, S.J.: A new autotuning method with specifications on dominant pole placement. Int. J. Control **60**(2), 265–282 (1994)
45. Ioannou, P.A., Fidan, B.: Adaptive Control Tutorial. Society for Industrial and Applied Mathematics, Philadelphia, Philadelphia, PA (2006)
46. Kalman, R.: Contributions to the theory of optimal control. Bol. Soc. Mat. Mexicana **5**(2), 102–119 (1960)
47. Kalman, R.: When is a linear control system optimal? J. Basic Eng. **86**(1), 51–60 (1964)
48. Kaya, I.: Tuning PI controllers for stable processes with specifications on gain and phase margins. ISA Trans. **43**(2), 297–304 (2004)
49. Kaya, I.: Two-degree-of-freedom IMC structure and controller design for integrating processes based on gain and phase-margin specifications. IEE Proc.-Control Theory Appl. **151**(4), 481–487 (2004)
50. Keel, L.H., Bhattacharyya, S.P.: Robust, fragile, or optimal? IEEE Trans. Autom. Control **42**(8), 1098–1105 (1997)
51. Khalil, H.K.: Nonlinear Systems. MacMillan, London (1992)
52. Kline, R.: Harold Black and the negative-feedback amplifier. IEEE Control Syst. **13**(4), 82–85 (1993)
53. Krohling, R.A., Jaschek, H., Rey, J.P.: Designing PI/PID controllers for a motion control system based on genetic algorithms. In: Proceedings of the IEEE International Symposium on Intelligent Control, pp. 125–130 (1997)
54. Lee, C.H.: A survey of PID controller design based on gain and phase margins. Int. J. Comput. Cogn. **2**(3), 63–100 (2004)
55. Lennartson, B., Kristiansson, B.: Robust and optimal tuning of PI and PID controllers. IEE Proc.-Control Theory Appl. **149**(1), 17–25 (2002)

56. Li, K.: PID tuning for optimal closed-loop performance with specified gain and phase margins. IEEE Trans. Control Syst. Technol. **21**(3), 1024–1030 (2013)
57. Lillacci, G., Aoki, S., Gupta, A., Baumschlager, A., Schweingruber, D., Khammash, M.: A universal rationally-designed biomolecular integral feedback controller for robust perfect adaptation. Nat. Biotechnol. (To appear)
58. Michael, A.J., Mohammad, H.M.: PID Control New Identification and Design Methods. Springer, London (2005)
59. Natarajan, K.: Robust PID controller design for hydroturbines. IEEE Trans. Energy Convers. **20**(3), 661–667 (2005)
60. Paraskevopoulos, P., Pasgianos, G., Arvanitis, K.: PID-type controller tuning for unstable first order plus dead time processes based on gain and phase margin specifications. IEEE Trans. Control Syst. Technol. **14**(5), 926–936 (2006)
61. Patel, H.B., Chaphekar, S.N.: Developments in PID controllers: literature survey. Int. J. Eng. Innov. Res. **1**(5), 425–430 (2012)
62. Rivera, D.E., Morari, M., Skogestad, S.: Internal model control: PID controller design. Ind. Eng. Chem. Process Des. Dev. **25**(1), 252–265 (1986)
63. Seborg, D.E., Edgar, T.F., Shah, S.L.: Adaptive control strategies for process control: a survey. AIChE J. **32**(6), 881–913 (1986)
64. Senthilkumar, M., Lincon, S.A.: Multiloop PI controller for achieving simultaneous time and frequency domain specifications. J. Eng. Sci. Technol. **10**(8), 1103–1115 (2015)
65. Singh, B., Payasi, R.P., Verma, K.S., Kumar, V., Gangwar, S.: Design of controllers PD, PI & PID for speed control of DC motor using IGBT based chopper. Ger. J. Renew. Sustain. Energy Res. (GJRSER) **1**(1), 29–49 (2013)
66. Srivastava, S., Pandit, V.S.: A PI/PID controller for time delay systems with desired closed loop time response and guaranteed gain and phase margins. J. Process Control **37**, 70–77 (2016)
67. Tan, K.K., Wang, Q.G., Hang, C.C.: Advances in PID Control. Springer Science and Business Media, Berlin (2012)
68. Visioli, A., Zhong, Q.: Control of Integral Processes with Dead Time. Springer Science and Business Media, Berlin (2010)
69. Wang, Q.G., Fung, H.W., Zhang, Y.: PID tuning with exact gain and phase margins. ISA Trans. **38**(3), 243–249 (1999)
70. Wang, Q.G., Zhang, Z., Astrom, K.J., Chek, L.S.: Guaranteed dominant pole placement with PID controllers. J. Process Control **19**(2), 349–352 (2009)
71. Wang, Y.G., Shao, H.H.: PID autotuner based on gain- and phase-margin specifications. Ind. Eng. Chem. Res. **38**(8), 3007–3012 (1999)
72. Wang, Y.J.: Determination of all feasible robust PID controllers for open-loop unstable plus time delay processes with gain margin and phase margin specifications. ISA Trans. **53**(2), 628–646 (2014)
73. Wonham, W.M., Pearson, J.B.: Regulation and internal stabilization in linear multivariable systems. SIAM J. Control **12**, 5–8 (1974)
74. Zhang, Y., Wang, Q.G., Astrom, K.J.: Dominant pole placement for multi-loop control systems. Automatica **38**(7), 1213–1220 (2002)
75. Ziegler, J.G., Nichols, N.B.: Optimum settings for automatic controllers. J. Trans. ASME, 759–768

Part I
Computation of PID Stabilizing Sets

Chapter 2
Stabilizing Sets for Linear Time-Invariant Continuous-Time Plants

Abstract In this chapter, we develop algorithms and procedures for the computation of the complete stabilizing set of PID controllers for continuous-time systems based on signature methods for root distribution determination. First, we present some basic results for the computation of stabilizing sets. Second, we provide justification and background for the computation of stabilizing sets. Then, we describe the procedure to compute the stabilizing set for LTI systems with P, PI, and PID controllers, and first-order controllers without delay. Finally, we present the computation of the PID stabilizing set which assigns closed-loop poles with real parts less than $-\sigma$, for prescribed σ.

2.1 Introduction

Consider the general feedback system that is shown in Fig. 2.1 where $r(t)$ is the command signal, $e(t)$ is the error signal, $u(t)$ is the control input signal, and $y(t)$ is the output. $P(s)$ is transfer function of the plant to be controlled and $C(s)$ is the controller transfer function to be designed. The controller $C(s)$ is assumed to be of PID type:

$$C(s) = k_p + \frac{k_i}{s} + k_d s, \tag{2.1}$$

where k_p, k_i, and k_d are the proportional, integral, and derivative gains, respectively. The pure derivative term $k_d s$ is not allowed in practice, mainly if the measured error $e(t)$ is a noisy signal. In such cases, we may consider the PID controller with a transfer function

---

I. D. Díaz-Rodríguez et al., *Analytical Design of PID Controllers*,
https://doi.org/10.1007/978-3-030-18228-1_2

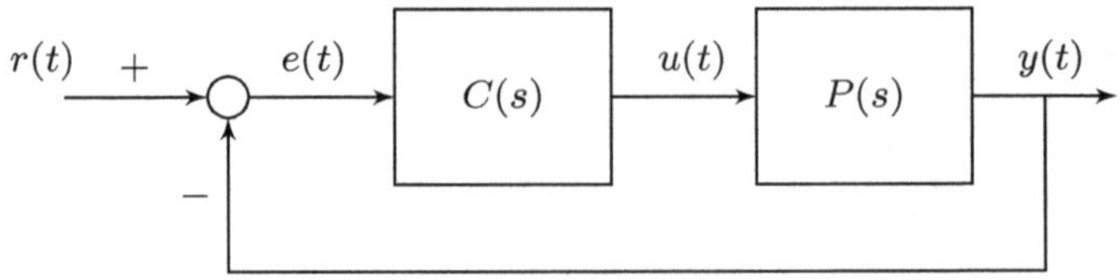

Fig. 2.1 Feedback control system

$$C(s) = \frac{sk_p + k_i + k_d s^2}{s(1+sT)}, \quad T > 0, \tag{2.2}$$

where T is usually fixed a priori at a small positive value.

Throughout this chapter, unless mentioned otherwise, the plant transfer function $P(s)$ is assumed to be of the rational form:

$$P(s) = \frac{N(s)}{D(s)}, \tag{2.3}$$

where $N(s)$ and $D(s)$ are polynomials in the Laplace variable s with real coefficients. Thus, the closed-loop characteristic polynomial with $C(s)$ in (2.1) becomes

$$\delta(s, k_p, k_i, k_d) = sD(s) + (k_i + k_p s + k_d s^2)N(s), \tag{2.4}$$

or with $C(s)$ in (2.2),

$$\delta(s, k_p, k_i, k_d) = s(1+sT)D(s) + (k_i + k_p s + k_d s^2)N(s). \tag{2.5}$$

The problem of stabilization using a PID controller is to determine the values of k_p, k_i, and k_d for which the closed-loop characteristic polynomial is Hurwitz or, in other words, the characteristic polynomial has all its roots in the open left half-plane. Since PID controllers cannot stabilize plants with a zero at the origin (N(0) = 0), we exclude such plants from the outset. Finding the complete set of stabilizing parameters is an essential first step in searching for subsets attaining various design objectives.

2.2 The Stabilizing Set

We define

$$\mathbf{k} := \left[k_p, k_i, k_d\right], \tag{2.6}$$

and let

$$\mathcal{S}^o := \{\mathbf{k} : \delta(s, \mathbf{k}) \text{is Hurwitz}\} \tag{2.7}$$

denote the set of PID controllers that stabilize the closed loop for the given plant characterized by the transfer function $P(s)$ as defined in (2.3). Due to the presence of integral action on the error, any controller in $\mathcal{S}^o$ automatically provides asymptotic tracking and disturbance rejection for step inputs. In general, additional design specifications on stability margins and transient response are also required, and subsets representing these must necessarily be sought within $\mathcal{S}^o$.

The three-dimensional set $\mathcal{S}^o$ is merely defined by (2.7), but not necessarily simple to calculate. For example, a naive application of the Routh–Hurwitz criterion to $\delta(s, \mathbf{k})$ will result in a description of $\mathcal{S}^o$ in terms of highly nonlinear and intractable inequalities, as the following example shows.

Example 2.1 Consider the problem of choosing stabilizing PID gains of $C(s)$ in (2.1) for the plant $P(s)$ in Fig. 2.1, where

$$\begin{aligned} D(s) &= s^5 + 8s^4 + 32s^3 + 46s^2 + 46s + 17 \\ N(s) &= s^3 - 4s^2 + s + 2. \end{aligned}$$

The closed-loop characteristic polynomial is

$$\begin{aligned} \delta(s, k_p, k_i, k_d) =& sD(s) + (k_i + k_p s + k_d s^2)N(s) \\ =& s^6 + (k_d + 8)s^5 + (k_p - 4k_d + 32)s^4 \\ &+ (k_i - 4k_p + k_d + 46)s^3 \\ &+ (-4k_i + k_p + 2k_d + 46)s^2 \\ &+ (k_i + 2k_p + 17)s + 2k_i. \end{aligned}$$

Using the Routh–Hurwitz criterion to determine the stabilizing values for k_p, k_i, and k_d, we see that the following inequalities must hold:

$$\begin{aligned} & k_d + 8 > 0 \\ & k_p k_d - 4k_d^2 - k_i + 12k_p - k_d + 210 > 0 \\ & k_i k_p k_d - 4k_p^2 k_d + 16k_p k_d^2 - 6k_d^3 - k_i^2 + 16k_i k_p \\ & \; + 63k_i k_d - 48k_p^2 + 48k_p k_d \\ & \; - 263k_d^2 + 428k_i - 336k_p - 683k_d + 6852 > 0 \\ & \; - 4k_i^2 k_p k_d + 16k_i k_p^2 k_d - 52k_i k_p k_d^2 \\ & \; - 6k_p^3 k_d + 24k_p^2 k_d^2 - 6k_p k_d^3 - 12k_d^4 + 4k_i^3 \\ & \; - 64k_i^2 k_p - 264k_i^2 k_d + 198k_i k_p^2 - 9k_i k_p k_d + 1238k_i k_d^2 \\ & \; - 72k_p^3 - 213k_p^2 k_d + 957k_p k_d^2 - 1074k_d^3 - 1775k_i^2 \\ & \; + 2127k_i k_p + 7688k_i k_d - 3924k_p^2 + 3027k_p k_d - 11322k_d^2 \\ & \; - 10746k_i - 31338k_p - 1836k_d + 206919 > 0 \end{aligned}$$

$$
\begin{aligned}
&- 6k_i^3k_pk_d + 24k_i^2k_p^2k_d - 84k_i^2k_pk_d^2 - 6k_ik_p^3k_d + 60k_ik_p^2k_d^2 \\
&- 102k_ik_pk_d^3 - 12k_p^4k_d + 48k_p^3k_d^2 - 12k_p^2k_d^3 - 24k_pk_d^4 \\
&+ 6k_i^4 - 96k_i^3k_p - 390k_i^3k_d + 294k_i^2k_p^2 - 285k_i^2k_pk_d \\
&+ 1476k_i^2k_d^2 - 60k_ik_p^3 + 969k_ik_p^2k_d - 1221k_ik_pk_d^2 \\
&- 132k_ik_d^3 - 144k_p^4 - 528k_p^3k_d + 2322k_p^2k_d^2 - 2250k_pk_d^3 \\
&- 204k_d^4 - 2487k_i^3 + 273k_i^2k_p - 2484k_i^2k_d + 5808k_ik_p^2 \\
&+ 10530k_ik_pk_d + 34164k_ik_d^2 - 9072k_p^3 + 2433k_p^2k_d \\
&- 6375k_pk_d^2 - 18258k_d^3 - 92961k_i^2 + 79041k_ik_p \\
&+ 184860k_ik_d - 129384k_p^2 + 47787k_pk_d - 192474k_d^2 \\
&- 549027k_i - 118908k_p - 31212k_d + 3517623 > 0 \\
&2k_i > 0.
\end{aligned}
$$

Clearly, the above inequalities are highly nonlinear, and there is no straightforward method for obtaining a complete solution set.

In the following sections, we present results on the effective computation of $\mathcal{S}^o$. These results are based on some root counting or signature formulas.

2.3 Signature Formulae

Let $\delta(s)$ denote a polynomial of degree n with real coefficients without zeros on the $j\omega$-axis. Write

$$\delta(s) := \underbrace{\delta_0 + \delta_2 s^2 + \cdots}_{\delta_{\text{even}}(s^2)} + s\,\underbrace{\left(\delta_1 + \delta_3 s^2 + \cdots\right)}_{\delta_{\text{odd}}(s^2)}, \tag{2.8}$$

so that

$$\delta(j\omega) = \delta_r(\omega) + j\delta_i(\omega), \tag{2.9}$$

where $\delta_r(\omega)$ and $\delta_i(\omega)$ are the polynomials in ω with real coefficients with

$$\delta_r(\omega) = \delta_{\text{even}}(-\omega^2), \tag{2.10}$$

$$\delta_i(\omega) = \omega\delta_{\text{odd}}(-\omega^2). \tag{2.11}$$

Definition 2.1 The standard signum function sgn : $\mathbb{R} \to \{-1, 0, 1\}$ is defined by

$$\text{sgn}[x] = \begin{cases} -1 & \text{if } x < 0 \\ 0 & \text{if } x = 0 \\ 1 & \text{if } x > 0, \end{cases}$$

where $\mathbb{R}$ represents the set of real numbers.

Definition 2.2 Let $\mathbb{C}$ denote the set of numbers in the complex plane. Let $\mathbb{C}^-$ denote the open left half-plane (LHP), $\mathbb{C}^+$ the open right half-plane (RHP), and l and r the numbers of roots of $\delta(s)$ in $\mathbb{C}^-$ and $\mathbb{C}^+$, respectively. Let $\angle\delta(j\omega)$ denote the phase or angle of $\delta(j\omega)$ and $\Delta_{\omega_1}^{\omega_2}\angle p(j\omega)$ the net change, in radians, in the phase of $\delta(j\omega)$ as ω runs from ω_1 to ω_2 $(\omega_2 \geq \omega_1)$.

Lemma 2.1

$$\Delta_0^{\infty}\angle\delta(j\omega) = \frac{\pi}{2}(l - r). \tag{2.12}$$

Proof Each LHP root contributes π and each RHP root contributes $-\pi$ to the net change in phase of $\delta(j\omega)$ as ω runs from $-\infty$ to ∞, and (2.12) follows from the symmetry about the real axis of the roots since $\delta(s)$ has real coefficients. □

We call $l - r$, the Hurwitz *signature* of $\delta(s)$, and denote it as

$$\sigma(\delta) := l - r. \tag{2.13}$$

2.3.1 Computation of $\sigma(\delta)$

By Lemma 2.1, the computation of $\sigma(p)$ amounts to a determination of the total phase change of $p(j\omega)$. To see how the total phase change may be calculated, consider typical plots of $p(j\omega)$ where ω runs from 0 to $+\infty$ as in Fig. 2.2. We note that the frequencies $0, \omega_1, \omega_2, \omega_3, \omega_4$ are the points where the plot cuts or touches the real axis.

In Fig. 2.2a, ω_3 is a point where the plot *touches* but does not cut the real axis.

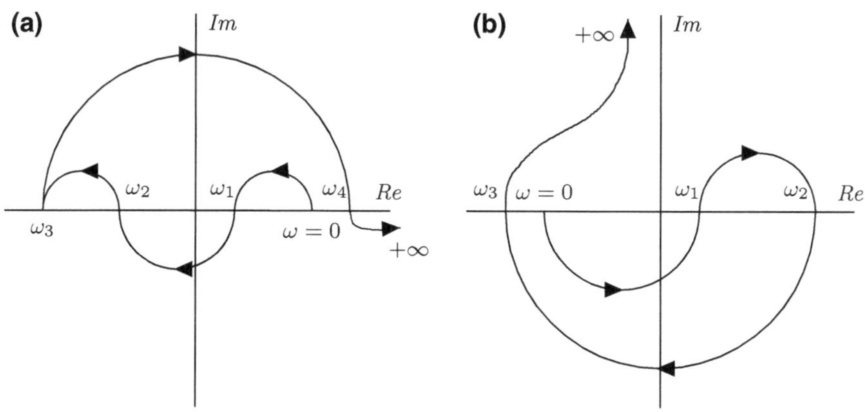

Fig. 2.2 **a** Plot of $p(j\omega)$ for $p(s)$ of even degree. **b** Plot of $p(j\omega)$ for $p(s)$ of odd degree. © Taylor & Francis LLC Books. Reproduced from [3] with permission

In Fig. 2.2a, we have

$$\Delta_0^\infty \angle\delta(j\omega) = \underbrace{\Delta_0^{\omega_1}\angle\delta(j\omega)}_{0} + \underbrace{\Delta_{\omega_1}^{\omega_2}\angle\delta(j\omega)}_{-\pi} + \underbrace{\Delta_{\omega_2}^{\omega_3}\angle\delta(j\omega)}_{0} + \underbrace{\Delta_{\omega_3}^{\omega_4}\angle\delta(j\omega)}_{-\pi} + \underbrace{\Delta_{\omega_4}^{\infty}\angle\delta(j\omega)}_{0}. \tag{2.14}$$

Observe that

$$\begin{aligned}
\Delta_0^{\omega_1}\angle\delta(j\omega) &= \operatorname{sgn}[\delta_i(0^+)]\big(\operatorname{sgn}[\delta_r(0) - \operatorname{sgn}[\delta_r(\omega_1)]\big)\frac{\pi}{2}\\
\Delta_{\omega_1}^{\omega_2}\angle\delta(j\omega) &= \operatorname{sgn}[\delta_i(\omega_1^+)]\big(\operatorname{sgn}[\delta_r(\omega_1) - \operatorname{sgn}[\delta_r(\omega_2)]\big)\frac{\pi}{2}\\
\Delta_{\omega_2}^{\omega_3}\angle\delta(j\omega) &= \operatorname{sgn}[\delta_i(\omega_2^+)]\big(\operatorname{sgn}[\delta_r(\omega_2) - \operatorname{sgn}[\delta_r(\omega_3)]\big)\frac{\pi}{2}\\
\Delta_{\omega_3}^{\omega_4}\angle\delta(j\omega) &= \operatorname{sgn}[\delta_i(\omega_3^+)]\big(\operatorname{sgn}[\delta_r(\omega_3) - \operatorname{sgn}[\delta_r(\omega_4)]\big)\frac{\pi}{2}\\
\Delta_{\omega_4}^{+\infty}\angle\delta(j\omega) &= \operatorname{sgn}[\delta_i(\omega_4^+)]\big(\operatorname{sgn}[\delta_r(\omega_4) - \operatorname{sgn}[\delta_r(\infty)]\big)\frac{\pi}{2}
\end{aligned} \tag{2.15}$$

and

$$\begin{aligned}
\operatorname{sgn}[\delta_i(\omega_1^+)] &= -\operatorname{sgn}[\delta_i(0^+)]\\
\operatorname{sgn}[\delta_i(\omega_2^+)] &= -\operatorname{sgn}[\delta_i(\omega_1^+)] = +\operatorname{sgn}[\delta_i(0^+)]\\
\operatorname{sgn}[\delta_i(\omega_3^+)] &= +\operatorname{sgn}[\delta_i(\omega_2^+)] = +\operatorname{sgn}[\delta_i(0^+)]\\
\operatorname{sgn}[\delta_i(\omega_4^+)] &= -\operatorname{sgn}[\delta_i(\omega_3^+)] = -\operatorname{sgn}[\delta_i(0^+)],
\end{aligned} \tag{2.16}$$

and note also that 0, ω_1, ω_2, and ω_4 are the real zeros of $\delta_i(\omega)$ of *odd* multiplicities, whereas ω_3 is a real zero of *even* multiplicity. From these relations, it is evident that (2.14) may be rewritten, skipping the terms involving ω_3 the root of even multiplicity so that

$$\begin{aligned}
\Delta_0^\infty\angle\delta(j\omega) &= \Delta_0^{\omega_1}\angle\delta(j\omega) + \Delta_{\omega_1}^{\omega_2}\angle\delta(j\omega) + \Delta_{\omega_2}^{\omega_4}\angle\delta(j\omega) + \Delta_{\omega_4}^{\infty}\angle\delta(j\omega)\\
&= \frac{\pi}{2}\Big(\operatorname{sgn}[\delta_i(0^+)]\big(\operatorname{sgn}[\delta_r(0)] - \operatorname{sgn}[\delta_r(\omega_1)]\big)\\
&\quad -\operatorname{sgn}[\delta_i(0^+)]\big(\operatorname{sgn}[\delta_r(\omega_1)] - \operatorname{sgn}[\delta_r(\omega_2)]\big)\\
&\quad +\operatorname{sgn}[\delta_i(0^+)]\big(\operatorname{sgn}[\delta_r(\omega_2)] - \operatorname{sgn}[\delta_r(\omega_4)]\big)\\
&\quad -\operatorname{sgn}[\delta_i(0^+)]\big(\operatorname{sgn}[\delta_r(\omega_4)] - \operatorname{sgn}[\delta_r(\infty)]\big)\Big).
\end{aligned} \tag{2.17}$$

Equation (2.17) can be rewritten as

$$\Delta_0^{\infty}\angle\delta(j\omega) = \frac{\pi}{2}\text{sgn}[\delta_i(0^+)]\big(\text{sgn}[\delta_r(0)] - 2\text{sgn}[\delta_r(\omega_1)] + 2\text{sgn}[\delta_r(\omega_2)] \\ -2\text{sgn}[\delta_r(\omega_4)] + \text{sgn}[\delta_r(\infty)]\big). \tag{2.18}$$

In the case of Fig. 2.2b, that is, when $\delta(s)$ is of odd degree, we have

$$\Delta_0^{\infty}\angle\delta(j\omega) = \underbrace{\Delta_0^{\omega_1}\angle\delta(j\omega)}_{+\pi} + \underbrace{\Delta_{\omega_1}^{\omega_2}\angle\delta(j\omega)}_{0} + \underbrace{\Delta_{\omega_2}^{\omega_3}\angle\delta(j\omega)}_{-\pi} + \underbrace{\Delta_{\omega_3}^{+\infty}\angle\delta(j\omega)}_{-\frac{\pi}{2}} \tag{2.19}$$

and $\Delta_0^{\omega_1}\angle\delta(j\omega)$, $\Delta_{\omega_1}^{\omega_2}\angle\delta(j\omega)$, $\Delta_{\omega_2}^{\omega_3}\angle\delta(j\omega)$ are as in (2.15), whereas

$$\Delta_{\omega_3}^{\infty}\angle\delta(j\omega) = \frac{\pi}{2}\text{sgn}[\delta_i(\omega_3^+)]\text{sgn}[\delta_r(\omega_3)]. \tag{2.20}$$

Similar to (2.16), we have

$$\text{sgn}[\delta_i(\omega_j^+)] = (-1)^j\text{sgn}[\delta_i(0^+)], \quad j = 1, 2, 3. \tag{2.21}$$

Combining (2.19)–(2.21) for Fig. 2.2b,

$$\Delta_0^{\infty}\angle\delta(j\omega) = \frac{\pi}{2}\text{sgn}[\delta_i(0^+)]\big(\text{sgn}[\delta_r(0)] - 2\text{sgn}[\delta_r(\omega_1)] + 2\text{sgn}[\delta_r(\omega_2)] \\ -2\text{sgn}[\delta_r(\omega_3)]\big). \tag{2.22}$$

We can now easily generalize the above formulae for the signature, based on Lemma 2.1.

Theorem 2.1 *Let $\delta(s)$ be a polynomial of degree n with real coefficients, without zeros on the imaginary axis. Write*

$$\delta(j\omega) = \delta_r(\omega) + j\delta_i(\omega)$$

and let $\omega_0, \omega_1, \omega_3, \ldots, \omega_{l-1}$ denote the real nonnegative zeros of $\delta_i(\omega)$ with odd multiplicities with $\omega_0 = 0$. If n is even,

$$\sigma(\delta) = \text{sgn}[\delta_i(0^+)]\left(\text{sgn}[\delta_r(0)] + 2\sum_{j=1}^{l-1}(-1)^j\text{sgn}[\delta_r(\omega_j)] + (-1)^l\text{sgn}[\delta_r(\infty)]\right).$$

If n is odd,

$$\sigma(\delta) = \text{sgn}[\delta_i(0^+)]\left(\text{sgn}[\delta_r(0)] + 2\sum_{j=1}^{l-1}(-1)^j\text{sgn}[\delta_r(\omega_j)]\right).$$

2.4 Computation of the Stabilizing Set for Delay-Free Systems with Proportional Controllers

In this section, we utilize the signature formulas to give a solution to the problem of feedback stabilization of a given linear time-invariant plant by a constant gain controller. Even though this problem can be solved using classical approaches such as the Nyquist stability criterion and the Routh–Hurwitz criterion, it is not clear how to extend these methods to the more complicated cases where PI or PID controllers are involved. By using the signature formulas, an elegant procedure is developed that can be extended to the cases mentioned above.

Consider the feedback system shown in Fig. 2.1.

$$P(s) = \frac{N(s)}{D(s)} \tag{2.23}$$

is the plant to be controlled, $N(s)$ and $D(s)$ are the coprime polynomials, and $C(s)$ is the controller to be designed. In the case of proportional controllers,

$$C(s) = k, \tag{2.24}$$

so that the closed-loop characteristic polynomial $\delta(s, k)$ is given by

$$\delta(s, k) = D(s) + kN(s). \tag{2.25}$$

Our objective is to determine those values of k, if any, for which the closed-loop system is stable, that is, $\delta(s, k)$ is Hurwitz.

If we consider the even and odd decomposition of $N(s)$ and $D(s)$

$$N(s) = N_{\text{even}}(s^2) + sN_{\text{odd}}(s^2), \tag{2.26}$$

$$D(s) = D_{\text{even}}(s^2) + sD_{\text{odd}}(s^2), \tag{2.27}$$

then (2.25) can be rewritten as

$$\delta(s, k) = kN_{\text{even}}(s^2) + D_{\text{even}}(s^2) + s(kN_{\text{odd}}(s^2) + D_{\text{odd}}(s^2)). \tag{2.28}$$

It is clear from this expression that both the even and odd parts of $\delta(s, k)$ depend on k. We will now construct a new polynomial for which only the even part depends on k.

Suppose that the degree of $D(s)$ is n while the degree of $N(s)$ is m and $m < n$. Multiplying $\delta(s, k)$ by

$$N(-s) = N_{\text{even}}(s^2) - sN_{\text{odd}}(s^2), \tag{2.29}$$

we obtain the following result.

Lemma 2.2 *$\delta(s,k)$ is Hurwitz if and only if*

$$\sigma(\delta(s,k)N(-s)) = n - (l(N(s)) - r(N(s))), \tag{2.30}$$

where $l(N(s))$ and $r(N(s))$ are the numbers of $\mathbb{C}^-$ and $\mathbb{C}^+$ roots of $N(s)$ including multiplicities, respectively.

Proof Since we know that for polynomials $a(s)$ and $b(s)$,

$$l(a(s) \cdot b(s)) = l(a(s)) + l(b(s)), \tag{2.31}$$
$$r(a(s) \cdot b(s)) = r(a(s)) + r(b(s)), \tag{2.32}$$

we have

$$\begin{aligned} &l(\delta(s,k)N(-s)) - r(\delta(s,k)N(-s)) \\ &= l(\delta(s,k) - r(\delta(s,k)) + l(N(-s)) - r(N(-s)) \quad (2.33)\\ &= l(\delta(s,k) - r(\delta(s,k)) - (l(N(s)) - r(N(s))). \quad (2.34)\end{aligned}$$

A polynomial $\delta(s,k)$ of degree n is Hurwitz if and only if $l(\delta(s,k)) = n$ and $r(\delta(s,k)) = 0$. Furthermore from (2.13),

$$\sigma(\delta(s,k)N(-s)) = l(\delta(s,k)N(-s)) - r(\delta(s,k)N(-s)). \tag{2.35}$$

Thus,

$$\sigma(\delta(s,k)N(-s)) = n - (l(N(s)) - r(N(s))). \tag{2.36}$$

This completes the proof. □

In order to solve our problem, we need to determine those values of k, if any, for which (2.30) holds. Notice that in this expression the values of n and $l(N(s)) - r(N(s))$ are known and fixed.

Using the even and odd decomposition of $N(s)$ and $D(s)$, we have

$$\delta(s,k)N(-s) = h_1(s^2) + kh_2(s^2) + sg(s^2), \tag{2.37}$$

where

$$h_1(s^2) = D_{\text{even}}(s^2)N_{\text{even}}(s^2) - s^2 D_{\text{odd}}(s^2)N_{\text{odd}}(s^2), \tag{2.38}$$
$$h_2(s^2) = N_{\text{even}}(s^2)N_{\text{even}}(s^2) - s^2 N_{\text{odd}}(s^2)N_{\text{odd}}(s^2), \tag{2.39}$$
$$g(s^2) = D_{\text{odd}}(s^2)N_{\text{even}}(s^2) - D_{\text{even}}(s^2)N_{\text{odd}}(s^2). \tag{2.40}$$

Substituting $s = j\omega$, we obtain

$$\delta(j\omega,k)N(-j\omega) = p(\omega,k) + jq(\omega), \tag{2.41}$$

where

$$p(\omega, k) = p_1(\omega) + kp_2(\omega) \tag{2.42}$$
$$p_1(\omega) = D_{\text{even}}(-\omega^2)N_{\text{even}}(-\omega^2) + \omega^2 D_{\text{odd}}(-\omega^2)N_{\text{odd}}(-\omega^2), \tag{2.43}$$
$$p_2(\omega) = N_{\text{even}}(-\omega^2)N_{\text{even}}(-\omega^2) + \omega^2 N_{\text{odd}}(-\omega^2)N_{\text{odd}}(-\omega^2), \tag{2.44}$$
$$q(\omega) = \omega\left(D_{\text{odd}}(-\omega^2)N_{\text{even}}(-\omega^2) - D_{\text{even}}(-\omega^2)N_{\text{odd}}(-\omega^2)\right). \tag{2.45}$$

Note that the zeros of the imaginary part $q(\omega)$ are independent of k. For clarity of the presentation, we first introduce some definitions before formally stating the main result of this section.

Definition 2.3 Let the integers m, n and the function $q(\omega)$ be as already defined. Let $\omega_1 < \omega_2 < \cdots < \omega_{l-1}$ be the real, nonnegative, distinct, and finite zeros of $q(\omega)$ with odd multiplicities and let $\omega_0 = 0, \omega_l = \infty$. Define a sequence of numbers $i_0, i_1, i_2, \ldots, i_l$ as follows:

(i)

$$i_0 = \begin{cases} \text{sgn}[p_1^{(k_n)}(0)] & \text{if } N(-s) \text{ has a zero of} \\ & \text{multiplicity } k_n \text{ at the origin} \\ \alpha & \text{otherwise,} \end{cases} \tag{2.46}$$

where $\alpha \in \{+1, -1\}$.

(ii) For $t = 1, 2, \ldots, l-1$:

$$i_t = \begin{cases} 0 & \text{if } N(-j\omega_t) = 0 \\ \alpha & \text{otherwise.} \end{cases} \tag{2.47}$$

(iii)

$$i_t = \begin{cases} \alpha & \text{if } n + m \text{ is even} \\ 0 & \text{if } n + m \text{ is odd.} \end{cases} \tag{2.48}$$

With $i_0, i_1, \ldots$ defined in this way, we define the string $\mathcal{I}$ as the following sequence of numbers:

$$\mathcal{I} := \{i_0, i_1, \ldots, i_l\}. \tag{2.49}$$

Define A to be the set of all possible strings $\mathcal{I}$ that can be generated to satisfy the preceding requirements.

Definition 2.4 Let the integers m, n and the function $q(\omega)$ be as already defined. Let $\omega_1 < \omega_2 < \cdots < \omega_{l-1}$ be the real, nonnegative, distinct, and finite zeros of $q(\omega)$ with odd multiplicities and let $\omega_0 = 0, \omega_l = \infty$. For each string $\mathcal{I} = \{i_0, i_1, \ldots\}$ in A, let $\gamma(\mathcal{I})$ denote the "signature" associated with the string $\mathcal{I}$ defined by

$$\gamma(\mathcal{I}) := (-1)^{l-1}\text{sgn}[q(\infty)] \cdot [i_0 - 2i_1 + 2i_2 + \cdots + (-1)^{l-1}2i_{l-1} + (-1)^l i_l]. \tag{2.50}$$

Definition 2.5 The set F of feasible strings for the constant gain stabilization problem is defined as

$$F = \{\mathcal{I} \in A | \gamma(\mathcal{I}) = n - (l(N(s)) - r(N(s)))\}. \tag{2.51}$$

We are now ready to state the main result of this section.

Theorem 2.2 (Constant Gain Stabilization) *The stabilizing set of delay-free systems with P controllers for a given plant $P(s)$ is nonempty if and only if the following conditions hold:*

(i) F is not empty, that is, at least one feasible string exists,
(ii) there exists a string $\mathcal{I} := \{i_0, i_1, \ldots\} \in F$ such that

$$\max_{\{t:i_t>0\}} L_t < \min_{\{t:i_t<0\}} U_t, \tag{2.52}$$

where

$$L_t := -\frac{p_1(\omega_t)}{p_2(\omega_t)} \quad \text{for } i_t \in \mathcal{I}, i_t > 0. \tag{2.53}$$

$$U_t := -\frac{p_1(\omega_t)}{p_2(\omega_t)} \quad \text{for } i_t \in \mathcal{I}, i_t < 0. \tag{2.54}$$

$p_1(\omega_t)$, $p_2(\omega_t)$ are given by (2.43) *and* (2.44)*, respectively, and $\omega_0, \omega_1, \ldots$ are as already defined.*

Furthermore, if the above conditions are satisfied by the feasible strings $\mathcal{I}_1, \mathcal{I}_2, \ldots \mathcal{I}_s \in F$, then the set of all stabilizing controllers is given by

$$K = \bigcup_{r=1}^{s} K_r, \tag{2.55}$$

where

$$K_r = \left(\max_{\{t:i_t>0, i_t \in \mathcal{I}\}} L_t, \min_{\{t:i_t<0, i_t \in \mathcal{I}\}} U_t \right). \tag{2.56}$$

Proof From Lemma 2.2, we know that $\delta(s, k)$ is Hurwitz if and only if

$$\sigma(\delta(s, k)N(-s)) = n - (l(N(s)) - r(N(s))). \tag{2.57}$$

Thus, $\delta(s, k)$ is Hurwitz if and only if $\mathcal{I} \in F$ where

$$F = \{\mathcal{I} \in A | \gamma(\mathcal{I}) = n - (l(N(s)) - r(N(s)))\} \tag{2.58}$$

and

$$\mathcal{I} = i_0, i_1, \ldots \tag{2.59}$$
$$i_0 = \text{sgn}[p^{(k_n)}(\omega_0, k)] \tag{2.60}$$
$$i_t = \text{sgn}[p(\omega_t, k)], \text{ for } t = 1, 2, \ldots, l-1 \tag{2.61}$$
$$i_l = \begin{cases} \text{sgn}[p(\omega_l, k)] & \text{if } n+m \text{ is even,} \\ 0 & \text{if } n+m \text{ is odd.} \end{cases} \tag{2.62}$$

Let us now consider two different cases.

Case 1: $N(-s)$ does not have any zeros on the imaginary axis. In this case, for all stabilizing values of the gain k, $\delta(s,k)N(-s)$ will also not have any zeros on the $j\omega$-axis so that $i_t \in \{-1, 1\}$ for $t = 0, 1, 2, \ldots, l-1$, and $i_l \in \{-1, 0, 1\}$. Next we consider two possibilities:

(a) If $i_t > 0$, then the stability requirement is

$$p_1(\omega_t) + kp_2(\omega_t) > 0. \tag{2.63}$$

From (2.44), we know that

$$p_2(\omega) = |N(j\omega)|^2 . \tag{2.64}$$

Since $N(-s)$ does not have any zeros on the $j\omega$-axis, it follows that $p_2(\omega_t) > 0$. Hence

$$k > -\frac{p_1(\omega)}{p_2(\omega)}. \tag{2.65}$$

(b) If $i_t < 0$, then the stability requirement is

$$p_1(\omega_t) + kp_2(\omega_t) < 0. \tag{2.66}$$

Similarly, since $p_2(\omega_t) > 0$, it follows that

$$k < -\frac{p_1(\omega)}{p_2(\omega)}. \tag{2.67}$$

Case 2: $N(-s)$ has one or more zeros on the $j\omega$-axis including a zero of multiplicity k_n at the origin. In this case, for all stabilizing values of the gain k, $\delta(s,k)N(-s)$ will also have the same set of $j\omega$-axis zeros. Furthermore, it is clear that these zero locations will be a subset of $\{\omega_0, \omega_1, \ldots, \omega_{l-1}\}$. Since the location of these zeros depends on $N(-s)$ and is independent of the gain k, it is reasonable to expect that such a zero, say at ω_m, will not impose any additional constraint on k. Instead, it will only mandate that $i_m \in \mathcal{I}$ be constrained to a particular value. We next proceed to establish rigorously these facts. We consider two possibilities:

(a) $m \neq 0$. Here, $N(-s)$ has a zero at $j\omega_m$, where $\omega_m \neq 0$. This implies that

$$N_{\text{even}}(-\omega_m^2) = N_{\text{odd}}(-\omega_m^2) = 0, \tag{2.68}$$

so that from (2.43) and (2.44) we obtain

$$p_1(\omega_m) = 0 \text{ and } p_2(\omega_m) = 0. \tag{2.69}$$

Thus, $i_m = 0$ independent of k and this constraint on $\mathcal{I}$ was already incorporated into the definition of A.

(b) $m = 0$. Here, $N(-s)$ has a zero at the origin of multiplicity k_n. Since

$$N(-s)\big|_{s=j\omega} = N_{\text{even}}(-\omega^2) - j\omega N_{\text{odd}}(-\omega^2), \tag{2.70}$$

it follows that $N_{\text{even}}(-\omega^2)$ and $\omega N_{\text{odd}}(-\omega^2)$ must have zeros at the origin of multiplicity at least k_n. Thus from (2.44), we see that $p_2(\omega)$ will have a zero at the origin of multiplicity $2k_n$ so that for $k_n \geq 1$,

$$p_2^{(k_n)}(0) = 0. \tag{2.71}$$

Since

$$p^{(k_n)}(0, k) = p_1^{(k_n)}(0) + kp_2^{(k_n)}(0) \tag{2.72}$$

it follows that for $k_n = 1$

$$p^{(k_n)}(0, k) = p_1^{(k_n)}(0) \tag{2.73}$$

independent of k. Thus, although no constraints on k appear, we must have

$$i_0 = \text{sgn}[p_1^{(k_n)}(0)]. \tag{2.74}$$

Once again, we note that this condition has been explicitly incorporated into the definition of the set A.

Of the two cases discussed above, only Case 1 imposes constraints on k as given by (2.65) and (2.67). This leads us to the conclusion that each $i_t > 0$ in the string $\mathcal{I} \in F$ contributes a lower bound on k while each $i_t < 0$ contributes an upper bound on k. Thus, if the string $\mathcal{I} \in F$ is to correspond to a stabilizing k, then we must have

$$\max_{i_t \in \mathcal{I}, i_t > 0} \left[-\frac{p_1(\omega_t)}{p_2(\omega_t)} \right] < \min_{i_t \in \mathcal{I}, i_t < 0} \left[-\frac{p_1(\omega_t)}{p_2(\omega_t)} \right], \tag{2.75}$$

which is condition (ii) in the theorem statement. This completes the proof of the necessary and sufficient conditions for the existence of a stabilizing k. The set of all stabilizing k's is now determined by taking the union of all k that are obtained from all the feasible strings that satisfy (ii). □

Remark 2.1 Since

$$P(s) = \frac{N(s)}{D(s)} = \frac{N_{\text{even}}(s^2) + sN_{\text{odd}}(s^2)}{N_{\text{even}}(s^2) + sD_{\text{odd}}(s^2)}, \tag{2.76}$$

we have

$$\frac{1}{P(j\omega)} = \frac{N_{\text{even}}(-\omega^2) + j\omega D_{\text{odd}}(-\omega^2)}{N_{\text{even}}(-\omega^2) + j\omega N_{\text{odd}}(-\omega^2)} \tag{2.77}$$

$$= \frac{\left[N_{\text{even}}(-\omega^2) + j\omega D_{\text{odd}}(-\omega^2)\right]\left[N_{\text{even}}(-\omega^2) - j\omega N_{\text{odd}}(-\omega^2)\right]}{\left[N_{\text{even}}(-\omega^2) + j\omega N_{\text{odd}}(-\omega^2)\right]\left[N_{\text{even}}(-\omega^2) - j\omega N_{\text{odd}}(-\omega^2)\right]} \tag{2.78}$$

$$= \frac{\left[N_{\text{even}}(-\omega^2)N_{\text{even}}(-\omega^2) + \omega^2 D_{\text{odd}}(-\omega^2)N_{\text{odd}}(-\omega^2)\right]}{\left[N_{\text{even}}(-\omega^2)N_{\text{even}}(-\omega^2) + \omega^2 N_{\text{odd}}(-\omega^2)N_{\text{odd}}(-\omega^2)\right]}$$
$$+ j\frac{\omega\left[N_{\text{even}}(-\omega^2)D_{\text{odd}}(-\omega^2) - N_{\text{even}}(-\omega^2)N_{\text{odd}}(-\omega^2)\right]}{\left[N_{\text{even}}(-\omega^2)N_{\text{even}}(-\omega^2) + \omega^2 N_{\text{odd}}(-\omega^2)N_{\text{odd}}(-\omega^2)\right]} \tag{2.79}$$

$$= \frac{p_1(\omega) + jq(\omega)}{p_2(\omega)}. \tag{2.80}$$

Since $q(\omega_t) = 0$ for finite ω_t, it follows that for all such frequencies,

$$-\frac{p_1(\omega_t)}{p_2(\omega_t)} = -\frac{1}{P(j\omega_t)}. \tag{2.81}$$

Remark 2.2 It is appropriate to point out that Theorem 2.2 parts (i) and (ii) do provide a characterization of all plants that are stabilizable by a constant gain. Also note that a necessary condition for F to be nonempty is that for $m+n$ even,

$$l \geq \frac{|n - (l(N(s)) - r(N(s)))|}{2} \tag{2.82}$$

and for $n+m$ odd,

$$l \geq \frac{|n - (l(N(s)) - r(N(s)))| + 1}{2}. \tag{2.83}$$

The following examples illustrate the usefulness of Theorem 2.2 when solving the constant gain stabilization problem.

Example 2.2 Consider a system described by

$$D(s) = s^4 + 5s^3 + 10s^2 + 4s + 6 \tag{2.84}$$
$$N(s) = s^3 + 3s^2 + 2s - 2. \tag{2.85}$$

The closed-loop characteristic polynomial is

$$\delta(s,k) = D(s) + kN(s). \tag{2.86}$$

Here $N_{\text{even}}(s^2) = 3s^2 - 2$ and $N_{\text{odd}}(s^2) = s^2 + 2$, so that

$$N(-s) = N_{\text{even}}(s^2) - sN_{\text{odd}}(s^2). \tag{2.87}$$

Therefore,

$$\begin{aligned}\delta(s,k)N(-s) = {} & (-2s^6 + 14s^4 - 10s^2 - 12) \\ & + k(-s^6 + 5s^4 - 16s^2 + 4) \\ & + s(-s^6 + 3s^2 - 24s^2 - 20),\end{aligned} \tag{2.88}$$

so that

$$\delta(j\omega)N(-j\omega) = p_1(\omega) + kp_2(\omega) + jq(\omega) \tag{2.89}$$

with

$$p_1(\omega) := 2\omega^6 + 14\omega^4 + 10\omega^2 - 12 \tag{2.90}$$

$$p_2(\omega) := \omega^6 + 5\omega^4 + 16\omega^2 + 4 \tag{2.91}$$

$$q(\omega) := \omega(\omega^6 + 3\omega^4 + 24\omega^2 - 20). \tag{2.92}$$

The real, nonnegative, distinct, and finite zeros of $q(\omega)$ with odd multiplicities are

$$\omega_0 = 0 \text{ and } \omega_1 = 0.8639. \tag{2.93}$$

Since $n + m = 7$, which is odd, and $N(-s)$ has no roots on the $j\omega$-axis, from Definition 2.3, the set A becomes

$$A = \begin{Bmatrix} \{-1,-1,0\} & \{1,-1,0\} \\ \{-1,1,0\} & \{1,1,0\} \end{Bmatrix}. \tag{2.94}$$

Since $l(N(s)) - r(N(s)) = 1$ and $(-1)^{l-1}\text{sgn}[q(\infty)] = -1$, it follows using Definition 2.5 that every string $\mathcal{I} = \{i_0, i_1, i_2\} \in F$ must satisfy

$$-(i_0 - 2i_1 + i_2) = 3 \tag{2.95}$$

for stability.

Hence, $F = \{\mathcal{I}_1\}$ where $\mathcal{I}_1 = \{-1, 1, 0\}$. Furthermore,

$$U_0 = -\frac{p_1(\omega_0)}{p_2(\omega_0)} = 3, \tag{2.96}$$

$$L_1 = -\frac{p_1(\omega_1)}{p_2(\omega_1)} = -0.2139. \tag{2.97}$$

Hence, from Theorem 2.2, we have

$$K_1 = (-0.2139, 3) \text{ for } \mathcal{I}_1. \tag{2.98}$$

Therefore, $\delta(s, k)$ is Hurwitz only for $k \in (-0.2139,\ 3)$.

Example 2.3 Consider the constant gain stabilization problem with

$$D(s) = s^5 + 11s^4 + 22s^3 + 60s^2 + 47s + 25 \tag{2.99}$$

$$N(s) = s^4 + 6s^3 + 12s^2 + 54s + 16. \tag{2.100}$$

The closed-loop characteristic polynomial is

$$\delta(s, k) = D(s) + kN(s). \tag{2.101}$$

Here, $N_{\text{even}}(s^2) = s^4 + 12s^2 + 16$ and $N_{\text{odd}}(s^2) = 6s^2 + 54$ so that

$$N(-s) = N_{\text{even}}(s^2) - sN_{\text{odd}}(s^2). \tag{2.102}$$

Therefore,

$$\begin{aligned}\delta(s, k)N(-s) = &(5s^8 + 6s^6 - 549s^4 - 1278s^2 + 400)\\ &+ k(s^8 - 12s^6 - 472s^4 - 2532s^2 + 256)\\ &+ s(s^8 - 32s^6 - 627s^4 - 2474s^2 - 598),\end{aligned} \tag{2.103}$$

so that

$$\delta(j\omega, k)N(-j\omega) = p_1(\omega) + kp_2(\omega) + jq(\omega), \tag{2.104}$$

with

$$p_1(\omega) := 5\omega^8 - 6\omega^6 - 549\omega^4 + 1278\omega^2 + 400 \tag{2.105}$$

$$p_2(\omega) := \omega^8 + 12\omega^6 - 472\omega^4 + 2532\omega^2 + 256 \tag{2.106}$$

$$q(\omega) := \omega(\omega^8 + 32\omega^6 - 627\omega^4 + 2474\omega^2 - 598). \tag{2.107}$$

The real, nonnegative, distinct, and finite zeros of $q(\omega)$ with odd multiplicities are

$$\omega_0 = 0,\ \omega_1 = 0.50834,\ \omega_2 = 2.41735,\ \omega_3 = 2.91515. \tag{2.108}$$

Since $n + m = 9$, which is odd, and $N(-s)$ has no roots on the $j\omega$-axis, from Definition 2.3, the set A is

$$A = \left\{ \begin{array}{cc} \{-1,-1,-1,-1,0\} & \{1,-1,-1,-1,0\} \\ \{-1,-1,-1,1,0\} & \{1,-1,-1,1,0\} \\ \{-1,-1,1,-1,0\} & \{1,-1,1,-1,0\} \\ \{-1,-1,1,1,0\} & \{1,-1,1,1,0\} \\ \{-1,1,-1,-1,0\} & \{1,1,-1,-1,0\} \\ \{-1,1,-1,1,0\} & \{1,1,-1,1,0\} \\ \{-1,1,1,-1,0\} & \{1,1,1,-1,0\} \\ \{-1,1,1,1,0\} & \{1,1,1,1,0\} \end{array} \right\}. \quad (2.109)$$

Since $l(N(s)) - r(N(s)) = 4$ and $(-1)^{l-1}\text{sgn}[q(\infty)] = -1$, it follows using Definition 2.5 that every string $\mathcal{I} = \{i_0, i_1, i_2, i_3, i_4\} \in F$ must satisfy

$$-(i_0 - 2i_1 + 2i_2 - 2i_3 + i_4) = 1 \quad (2.110)$$

for stability.

Hence, $F = \{\mathcal{I}_1, \mathcal{I}_2, \mathcal{I}_3\}$ are the feasible strings where

$$\mathcal{I}_1 = \{1, -1, -1, 1, 0\} \quad (2.111)$$
$$\mathcal{I}_2 = \{1, 1, 1, 1, 0\} \quad (2.112)$$
$$\mathcal{I}_3 = \{1, 1, -1, -1, 0\}. \quad (2.113)$$

Furthermore,

$$-\frac{p_1(\omega_0)}{p_2(\omega_0)} = -1.56250 \quad (2.114)$$
$$-\frac{p_1(\omega_1)}{p_2(\omega_1)} = -0.78898 \quad (2.115)$$
$$-\frac{p_1(\omega_2)}{p_2(\omega_2)} = 2.50345 \quad (2.116)$$
$$-\frac{p_1(\omega_3)}{p_2(\omega_3)} = 22.49390. \quad (2.117)$$

Hence, substituting (2.114), (2.115)–(2.117) into (2.56) in Theorem 2.2, we have

$$\begin{cases} K_1 = \varnothing & \text{for } \mathcal{I}_1 \\ K_2 = (22.49390, \infty) & \text{for } \mathcal{I}_2 \\ K_3 = (-0.78898, 2.50345) & \text{for } \mathcal{I}_3. \end{cases} \quad (2.118)$$

Therefore, $\delta(s, k)$ is Hurwitz for $k \in (-0.78898, 2.50345) \cup (22.49390, \infty)$.

2.5 Computation of the Stabilizing Set for Delay-Free Systems with PI Controllers

Consider the system configuration in Fig. 2.1 with an LTI delay-free system $P(s)$ and a PI controller $C(s)$ of the form

$$C(s) = k_p + \frac{k_i}{s}. \tag{2.119}$$

The procedure to compute the complete stabilizing set for delay-free systems with PI controllers is the following:

1. Calculate the characteristic equation derived from Fig. 2.1

$$\delta(s) = sD(s) + (k_p s + k_i)N(s). \tag{2.120}$$

2. Form the new polynomial

$$\nu(s) := \delta(s)N(-s). \tag{2.121}$$

3. The even and odd decomposition of the polynomial $\nu(s)$ in (2.121) is of the form

$$\nu(s) = \nu_{\text{even}}(s^2, k_i) + s\nu_{\text{odd}}(s^2, k_p). \tag{2.122}$$

4. Fix $k_p = k_p^*$ and let $0 < \omega_1 < \omega_2 < \cdots < \omega_{l-1}$ be the finite frequencies that are real and positive roots of

$$\nu_{\text{odd}}(-\omega^2, k_p^*) = 0 \tag{2.123}$$

   of odd multiplicities. Let $\omega_0 := 0$ and $\omega_l := \infty$.
5. Let

$$j = \text{sgn}[\nu_{\text{odd}}(0^+, k_p)], \tag{2.124}$$

   $\deg[D(s)] = n$, $\deg[N(s)] = m \leq n$, and let z^+ and z^- be the number of zeros of $N(s)$ in $\mathbb{C}^+$ and $\mathbb{C}^-$, respectively. Let $i_0, i_1, \ldots$ be the sequence of numbers with $i_t \in \{+1, -1\}$, $\forall t \in \{0, \ldots, l\}$. If $n + m$ is odd, the signature requirement for stability is

$$j(i_0 - 2i_1 + 2i_2 + \cdots + (-1)^{l-1}2i_{l-1} + (-1)^l i_l) = n - m + 1 + 2z^+. \tag{2.125}$$

   If $n + m$ is even, the signature requirement for stability is

$$j(i_0 - 2i_1 + 2i_2 + \cdots + (-1)^{l-1}2i_{l-1}) = n - m + 1 + 2z^+. \tag{2.126}$$

6. Let $I_1, I_2, I_3, \ldots$ be the distinct strings of $i_0, i_1, \ldots$ that satisfy the signature condition in (2.125) or (2.126). For a fixed $k_p = k_p^*$, the stabilizing sets in the space of k_i can be computed by solving the set of linear inequalities:

$$\nu_{\text{even}}(-\omega_t^2, k_i)i_t > 0 \tag{2.127}$$

for all $t \in \{0, \ldots, l\}$ if $n+m$ is odd or for all $t \in \{0, \ldots, l-1\}$ if $n+m$ is even.
7. Each string I_j in the previous step creates a stability region $\mathcal{S}_j^o(k_p^*)$ and the region is a convex set since it is the intersection of linear (or affine) inequalities. Thus, the stabilizing region for the fixed $k_p = k_p^*$ is a union of convex sets:

$$\mathcal{S}^o(k_p^*) = \cup_j \mathcal{S}_j^o(k_p^*). \tag{2.128}$$

8. The stabilizing set in the space of (k_p, k_i) can be computed by sweeping k_p over the real axis and by following the steps above.

Example 2.4 Let us consider the continuous-time system represented in Fig. 2.1 with the plant

$$P(s) = \frac{s-5}{s^2+1.6s+0.2} \tag{2.129}$$

and the controller

$$C(s) = k_p + \frac{k_i}{s}. \tag{2.130}$$

The closed-loop characteristic polynomial is

$$\delta(s, k_p, k_i) = s^3 + (k_p+1.6)s^2 + (k_i - 5k_p + 0.2)s - 5k_i. \tag{2.131}$$

Here, $n = 2$, $m = 1$, and $N(-s) = -s - 5$. Therefore, following the procedure in step 2 with Eq. (2.121), we obtain

$$\begin{aligned}\nu(s) &= \delta(s, k_p, k_i)N(-s)\\ &= -s^4 - (6.6+k_p)s^3 - (8.2+k_i)s^2 + (25k_p - 1)s + 25k_i,\end{aligned} \tag{2.132}$$

so that the even and odd decomposition shown in (2.122) are given by

$$\begin{aligned}\nu(j\omega, k_p, k_i) = &\underbrace{(-\omega^4 + (k_i+8.2)\omega^2 + 25k_i)}_{\nu_{\text{even}}(-\omega^2, k_i)}\\ &+ j\omega[\underbrace{(k_p+6.6)\omega^3 + (25k_p - 1)}_{\nu_{\text{odd}}(-\omega^2, k_p)}].\end{aligned} \tag{2.133}$$

Since $z^+ = 1$, the required Hurwitz signature $\sigma(\nu)$ for stability is

$$n - m + 1 + 2z^+ = 4. \tag{2.134}$$

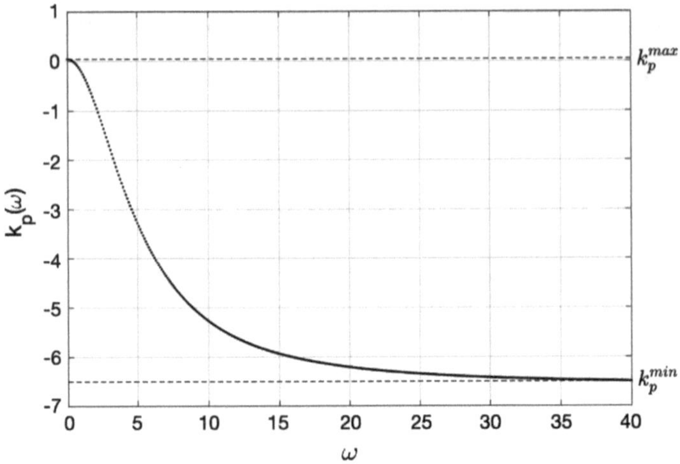

Fig. 2.3 Range for k_p with at least one positive real root in Example 2.4

Since $n + m$ is odd, we see from (2.123) and (2.125) that $\nu_{\text{odd}}(-\omega^2, k_p)$ must have at least one positive real root of odd multiplicity. Rearranging terms in (2.123), we can represent k_p in terms of ω:

$$k_p(\omega) := \frac{-6.6\omega^2 + 1}{\omega^2 + 25}. \tag{2.135}$$

The range of $k_p(\omega)$ such that (2.135) has at least one real, positive, distinct, finite zero with odd multiplicity is determined to be $k_p \in (-6.5, 0.04)$ as shown in Fig. 2.3. This range is called the *allowable range* for k_p.

For illustration purposes, let us fix a value of $k_p^* = -1$. This value is contained in the allowable range for k_p. The corresponding root is $\omega = 2.1547$ and it is shown in Fig. 2.4. We also define $\omega_0 := 0,\ \omega_\infty := \infty$.

We calculate j from Eq. (2.124). For 0^+, we substitute 0.001. Then,

$$j = \text{sgn}[(-1 + 6.6)(0.001)^3 + (25(-1) - 1)(0.001)] = -1. \tag{2.136}$$

Then, the signature in (2.125) is given by

$$-(i_0 - 2i_1 + i_2) = 4. \tag{2.137}$$

Let $I_1 = \{-1, +1, -1\}$ be the string $\{i_0, i_1, i_2\}$ that satisfies the signature condition in (2.137). The stabilizing set in the space of (k_p, k_i) for a fixed $k_p^* = -1$ is given by the intersection of the set of inequalities

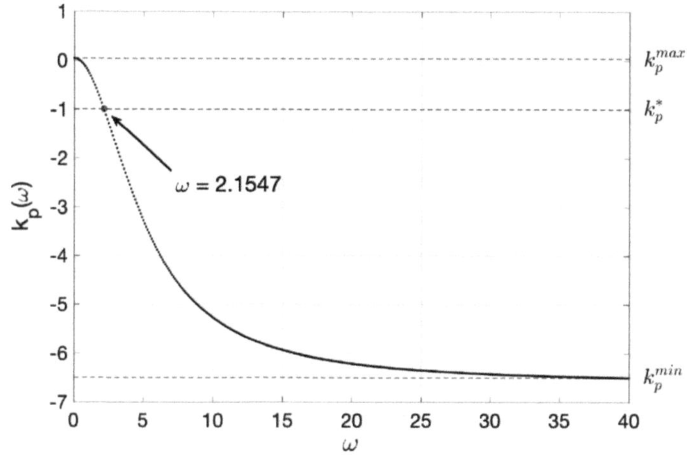

Fig. 2.4 Root for $k_p(\omega)$ with a fixed $k_p^* = -1$ in Example 2.4

$$\begin{aligned} k_i &< 0, \\ 29.6427k_i + 16.5154 &> 0, \\ (k_i + 8.2) \cdot \infty + 25k_i &< \infty. \end{aligned} \tag{2.138}$$

Therefore, the stabilizing set for a fixed $k_p^* = -1$ is

$$-0.5571 < k_i < 0. \tag{2.139}$$

By sweeping over different k_p values within the interval $(-6.5, 0.04)$, we can generate the set of stabilizing gains in (k_p, k_i) space, shown in Fig. 2.5.

2.6 Computation of the Stabilizing Set for Delay-Free Systems with PID Controllers

Consider the system configuration in Fig. 2.1 with a plant $P(s)$ and a PID controller

$$C(s) = k_p + \frac{k_i}{s} + k_d s. \tag{2.140}$$

The procedure to compute the stabilizing set is the following:

1. Calculate the characteristic equation derived from Fig. 2.1

$$\delta(s) = sD(s) + (k_d s^2 + k_p s + k_i)N(s). \tag{2.141}$$

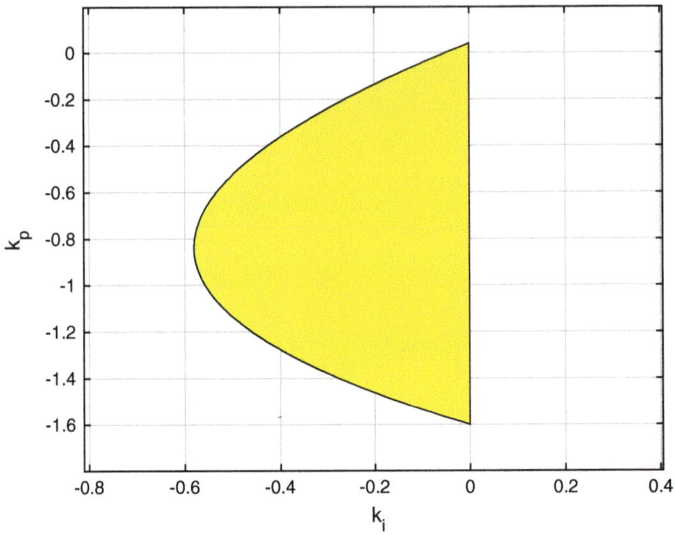

Fig. 2.5 Stabilizing set in Example 2.4

2. Form the new polynomial

$$\nu(s) := \delta(s)N(-s). \tag{2.142}$$

3. The even and odd decomposition of the polynomial $\nu(s)$ in (2.142) is denoted:

$$\nu(s) = \nu_{\text{even}}(s^2, k_i, k_d) + s\nu_{\text{odd}}(s^2, k_p). \tag{2.143}$$

4. Fix $k_p = k_p^*$ and let $0 < \omega_1 < \omega_2 < \cdots < \omega_{l-1}$ be the finite frequencies that are real and positive roots of

$$\nu_{\text{odd}}(-\omega^2, k_p^*) = 0 \tag{2.144}$$

of odd multiplicities. Define $\omega_0 := 0$ and $\omega_l := \infty$.
5. Write

$$j = \text{sgn}[\nu_{\text{odd}}(0^+, k_p)]. \tag{2.145}$$

Let $\deg[D(s)] := n$, $\deg[N(s)] := m \leq n$, and let z^+ and z^- be the number of zeros in $\mathbb{C}^+$ and $\mathbb{C}^-$ of the plant, respectively. Let $i_0, i_1, \ldots$ be the sequence of numbers such that $i_t \in \{+1, -1\}$, $\forall t \in \{0, \ldots, l\}$. If $n + m$ is odd, the signature requirement for stability is

$$j(i_0 - 2i_1 + 2i_2 + \cdots + (-1)^{l-1}2i_{l-1} + (-1)^l i_l) = n - m + 1 + 2z^+, \tag{2.146}$$

if $n+m$ is even, the signature requirement for stability is

$$j(i_0 - 2i_1 + 2i_2 + \cdots + (-1)^{l-1}2i_{l-1}) = n - m + 1 + 2z^+. \tag{2.147}$$

6. Let $I_1, I_2, I_3, \ldots$ be the distinct strings of $i_0, i_1, \ldots$ that satisfy the signature condition in (2.146) or (2.147). For a fixed $k_p = k_p^*$, the stabilizing set in the space of (k_i, k_d) can be computed by solving the set of linear inequalities:

$$\nu_{\text{even}}(-\omega_t^2, k_i, k_d)i_t > 0 \tag{2.148}$$

for all $t \in \{0, \ldots, l\}$ if $n+m$ is odd or for all $t \in \{0, \ldots, l-1\}$ if $n+m$ is even.
7. Every string I_j in the previous step creates a stability region $\mathcal{S}_j^o(k_p^*)$ and the region is a convex set since it is the intersection of solutions of linear (or affine) inequalities. Thus, the stabilizing region for the fixed $k_p = k_p^*$ is a union of convex sets:

$$\mathcal{S}^o(k_p^*) = \cup_j \mathcal{S}_j^o(k_p^*). \tag{2.149}$$

8. The stabilizing set in the space of (k_p, k_i, k_d) can be computed by sweeping k_p over the real axis and by repeating the steps above.

Example 2.5 Let us consider the plant

$$P(s) = \frac{s-3}{s^3 + 4s^2 + 5s + 2} \tag{2.150}$$

and the controller

$$C(s) = \frac{k_d s^2 + k_p s + k_i}{s}. \tag{2.151}$$

The closed-loop characteristic polynomial is

$$\begin{aligned}\delta(s, k_p, k_i, k_d) =& s^4 + (k_d + 4)s^3 + (k_p - 3k_d + 5)s^2 \\ &+ (k_i - 3k_p + 2)s - 3k_i.\end{aligned} \tag{2.152}$$

Here, $n = 3$, $m = 1$, and $N(-s) = -s - 3$. Therefore, we obtain

$$\begin{aligned}\nu(s) &= \delta(s, k_p, k_i, k_d)N(-s) \\ &= -s^5 + (-k_d - 7)s^4 + (-k_p - 17)s^3 + (9k_d - k_i - 17)s^2 \\ &\quad + (9k_p - 6)s + 9k_i\end{aligned} \tag{2.153}$$

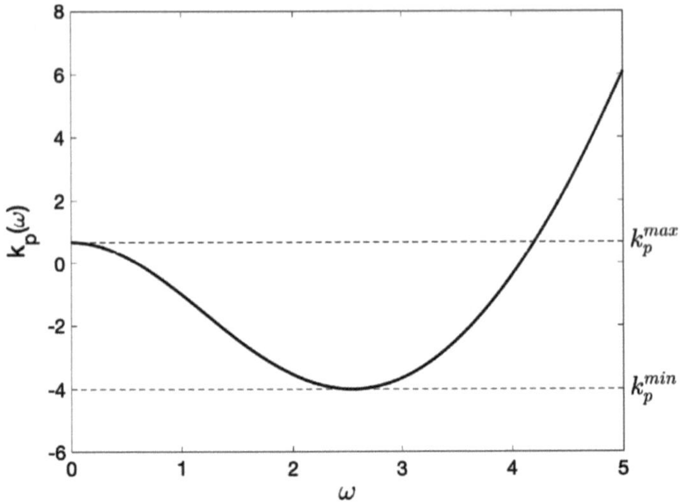

Fig. 2.6 Range for k_p with at least two positive real roots in Example 2.5

with the even and odd decomposition:

$$\begin{aligned}\nu(j\omega, k_p, k_i, k_d) &= (-k_d - 7)\omega^4 + (k_i - 9k_d + 17)\omega^2 + 9k_i \\ &\quad + j[-\omega^5 + (k_p + 17)\omega^3 + (9k_p - 6)\omega] \\ &= p(\omega) + jq(\omega).\end{aligned} \tag{2.154}$$

We find that $z^+ = 1$ so that the signature requirement (2.13) on $\nu(s)$ for stability is

$$n - m + 1 + 2z^+ = 5. \tag{2.155}$$

Since the degree of $\nu(s)$ is odd we use Eq. (2.147). We see from the signature formulas that $q(\omega)$ must have at least two positive real roots of odd multiplicity. The range of k_p such that $q(\omega, k_p)$ has at least two real, positive, distinct, finite zeros with odd multiplicities roots was determined to be $k_p \in (-4, 0.65)$ which is the allowable range for k_p shown in Fig. 2.6. From the expression for $q(\omega)$ in (2.154), we can represent k_p in terms of ω as

$$k_p(\omega) = \frac{\omega^4 - 17\omega^2 + 6}{\omega^2 + 9}. \tag{2.156}$$

For illustration purposes, let us fix a value of $k_p^* = -1$, which is contained in the allowable range for k_p. Now, we can find the roots from Fig. 2.6. The roots are $\omega_1 = 1$ and $\omega_2 = 3.873$ as shown in Fig. 2.7. Consider $\omega_0 = 0$, $\omega_1 = 1$, and $\omega_2 = 3.873$. For a small value of 0^+, we consider 0.001. Then,

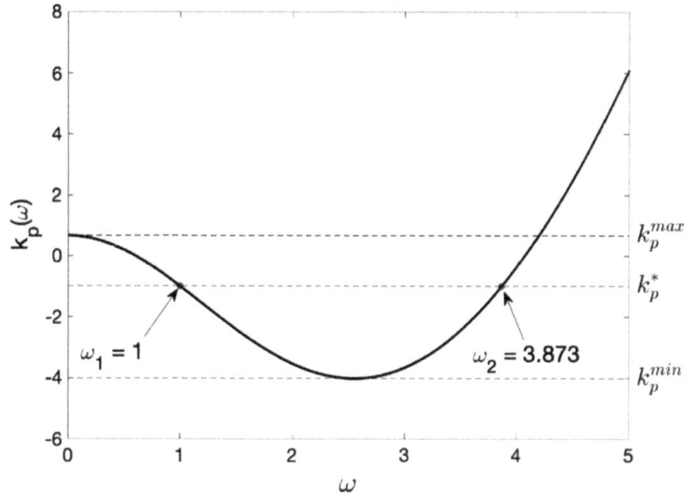

Fig. 2.7 Roots for $k_p(\omega)$ with a fixed $k_p^* = -1$ in Example 2.5

$$j = \text{sgn}[-(0.001)^5 + (-1 + 17)(0.001)^3 + (9(-1) - 6)] = -1. \tag{2.157}$$

Then, the signature expression (2.147) is given by

$$-(i_0 - 2i_1 + 2i_1) = 5, \tag{2.158}$$

and $I_1 = \{-1, +1, -1\}$ is the string of i_0, i_1, i_2 that satisfies the signature condition in (2.158). The stabilizing set in the space of (k_p, k_i, k_d) for a fixed $k_p^* = -1$ is thus given by the set of inequalities

$$\begin{aligned} k_i &< 0 \\ -k_d + k_i + 1 &> 0 \\ -15k_d + k_i - 55 &< 0. \end{aligned} \tag{2.159}$$

Therefore, the stabilizing set for $k_p^* = -1$ is shown in Fig. 2.8.

By sweeping over different k_p values within the interval $(-4, 0.65)$, we can generate the set of stabilizing (k_p, k_i, k_d) values. This set is shown in Fig. 2.9.

Remark 2.3 The signature method for calculating the PID stabilizing set $\mathcal{S}$ can be easily adapted to compute the stabilizing sets for controllers of the types

$$C_1(s) = \frac{k_2 s^2 + k_1 s + k_0}{s^2},$$

$$C_2(s) = \frac{k_2 s^2 + k_1 s + k_0}{s^2 + \omega_r^2}.$$

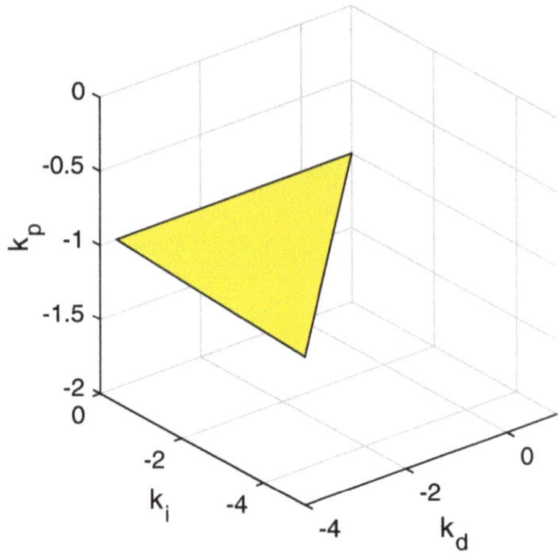

Fig. 2.8 2-D stabilizing set for Example 2.5 with $k_p^* = -1$

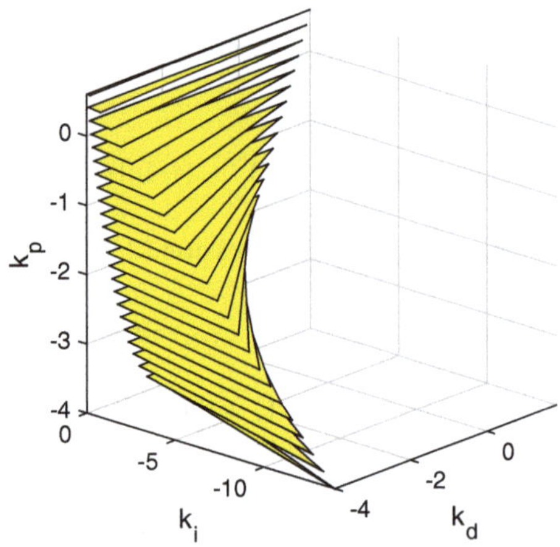

Fig. 2.9 Complete 3-D stabilizing set for Example 2.5

$C_1(s)$ makes the output $y(t)$ track step and ramp inputs $r(t)$ and reject step and ramp disturbances in the closed loop. $C_2(s)$ is called a *resonant controller* and makes the output track a sinusoidal input $r(t)$ and reject a sinusoidal disturbance at the radian frequency ω_r.

The main difference in the computation of the stabilizing set is in the characteristic polynomial $\delta(s)$. $sD(s)$ for the PID controller is replaced by $s^2D(s)$ and $(s^2 + \omega_r^2)D(s)$ for $C_1(s)$ and $C_2(s)$, respectively.

The signature requirement for stability with $C_1(s)$ or $C_2(s)$ is

$$\sigma(\nu) = n - m + 2 + 2z^+.$$

The plant $P(s)$, controlled by the resonant controller $C_2(s)$, is assumed to have no $j\omega$-axis zeros at $\omega = \omega_r$ since the closed loop cannot be stabilized otherwise.

2.7 σ-Hurwitz Stability[1]

In some applications, it may be necessary to push the closed-loop characteristic roots to the left of a line $s = -\sigma$ to speed up the time response. In this section, we present a constructive determination of $\mathcal{S}(\sigma)$, the subset of $\mathcal{S}$ for which the closed-loop poles have real parts less than $-\sigma$. By this means, we can also determine the maximum achievable σ for a given plant. It is simply the smallest σ for which $\mathcal{S}(\sigma) = \varnothing$.

Consider a monic polynomial $\delta(s)$ of degree n with real coefficients. Write

$$\delta(s) = (s - \lambda_1) \cdots (s - \lambda_n), \tag{2.160}$$

where $\lambda_i \in \mathbb{C}$, $\forall i \in \{1, \ldots, n\}$.

Definition 2.6 For $\sigma \geq 0$, $\delta(s)$ is *σ-Hurwitz* stable if all its roots have real part less than $-\sigma$, that is, $\text{Re}\{\lambda_i\} < -\sigma \ \forall i$.

Define s' such that $s = s' - \sigma$ and $\delta'(s') := \delta(s' - \sigma)$. Then we have the following fact.

Fact 3 *$\delta'(s')$ is Hurwitz if and only if $\delta(s)$ is σ-Hurwitz.*

Remark 2.4 A Hurwitz polynomial is a special case of a σ-Hurwitz polynomial with $\sigma = 0$.

2.7.1 σ-Hurwitz PID Stabilizing Set

2.7.1.1 Problem Formulation

Consider a unity feedback control loop with a PID controller and a plant in Fig. 2.1, where

$$P(s) = \frac{N(s)}{D(s)}, \tag{2.161}$$

$$C(s) = k_p + \frac{k_i}{s} + k_d s. \tag{2.162}$$

Let

$$\mathcal{S}(\sigma) := \left\{(k_p, k_i, k_d) : \delta(s, k_p, k_i, k_d) \text{ is } \sigma\text{-Hurwitz}\right\} \tag{2.163}$$

denote the set of PID controllers that stabilize $P(s)$ for which the characteristic polynomial $\delta(s, k_p, k_i, k_d)$ is σ-Hurwitz.

---

[1] In this section, σ is a positive real number instead of the Hurwitz signature. We temporarily denote the Hurwitz signature as simply "signature".

Fact 4 *If* $0 \le \sigma_1 \le \sigma_2$, *then* $\mathcal{S}(\sigma_1) \supseteq \mathcal{S}(\sigma_2)$.

We find below $\mathcal{S}(\sigma)$ for a prescribed σ. Moreover, we constructively determine the maximum achievable $\sigma < \infty$ for which $\mathcal{S}(\sigma)$ just becomes empty.

2.7.1.2 Signature Method for σ-Hurwitz Polynomials

In order for $\delta(s)$ to be σ-Hurwitz, we examine whether or not $\delta'(s')$ is Hurwitz using Fact 3. Observe that

$$\delta'(s') = \delta(s' - \sigma) \tag{2.164}$$

$$= (s' - \sigma)\underbrace{D(s' - \sigma)}_{=:D'(s')} + (k_i + k_p(s' - \sigma) + k_d(s' - \sigma)^2)\underbrace{N(s' - \sigma)}_{=:N'(s')}. \tag{2.165}$$

Define $N'_{\text{even}}(s'^2)$, $N'_{\text{odd}}(s'^2)$, $D'_{\text{even}}(s'^2)$, and $D'_{\text{odd}}(s'^2)$ through

$$N'(s') = N'_{\text{even}}(s'^2) + s' N'_{\text{odd}}(s'^2), \tag{2.166}$$

$$D'(s') = D'_{\text{even}}(s'^2) + s' D'_{\text{odd}}(s'^2). \tag{2.167}$$

Then,

$$N'(-s') = N'_{\text{even}}(s'^2) - s' N'_{\text{odd}}(s'^2). \tag{2.168}$$

Let $\mathbb{C}_\sigma^-$ denote the open half-plane to the left of the line $\text{Re}\{s\} = -\sigma$, $\mathbb{C}_\sigma^+$ the closed half-plane to the right of $\text{Re}\{s\} = -\sigma$, and z_σ^- and z_σ^+ the numbers of roots of $N'(s')$ in $\mathbb{C}_\sigma^-$ and $\mathbb{C}_\sigma^+$, respectively. Then, $N'(-s')$ has z_σ^+ roots in $\mathbb{C}_\sigma^-$ and z_σ^- roots in $\mathbb{C}_\sigma^+$.

Assume $N(s)$ has no roots on $\text{Re}\{s\} = -\sigma$. Consider the net change in the phase of $N(s)|_{s=-\sigma+j\omega}$ from $\omega = 0$ to $\omega = \infty$. This is equivalent to the net change in the phase of $N'(s')|_{s'=j\omega}$ and

$$\Delta_{\omega=0}^{\infty} \angle N'(j\omega) = \frac{\pi}{2}(z_\sigma^- - z_\sigma^+). \tag{2.169}$$

We call $z_\sigma^- - z_\sigma^+$ the *σ-signature* of $N'(s')$ and denote it as

$$\sigma\text{-signature}(N') := z_\sigma^- - z_\sigma^+. \tag{2.170}$$

We form the new polynomial

$$\nu'(s') := \delta'(s')N'(-s') \tag{2.171}$$

and state the condition for σ-Hurwitz stability of the closed-loop system in terms of $\nu'(s')$ in the following.

Theorem 2.3 *The closed-loop system is σ-Hurwitz stable if and only if*

$$\sigma\text{-}signature(\nu') = n + 1 - m + 2z_\sigma^+. \tag{2.172}$$

Proof By Fact 3, it suffices to show that $\delta'(s')$ is Hurwitz stable. This is equivalent to

$$\begin{aligned}\sigma\text{-signature}(\nu') &= n + 1 - z_\sigma^- + z_\sigma^+ \\ &= n + 1 - m + 2z_\sigma^+.\end{aligned}$$ □

We substitute (2.166)–(2.168) into (2.171), to get

$$\begin{aligned}\nu'(s') &= \delta'(s')N'(-s') \\ &= Q_0(s') + (k_2 + k_3 s'^2)Q_1(s') + k_1 s' Q_1(s'),\end{aligned} \tag{2.173}$$

where

$$\begin{aligned}Q_0(s') &= (s' - \sigma)(D'_{\text{even}}(s') + s' D'_{\text{odd}}(s'))(N'_{\text{even}}(s') - s' N'_{\text{odd}}(s')), \\ Q_1(s') &= N'^2_{\text{even}}(s') - s'^2 N'^2_{\text{odd}}(s'),\end{aligned}$$

$$k_1 = k_p - 2\sigma k_d, \quad k_2 = -\sigma k_p + k_i + \sigma^2 k_d, \quad k_3 = k_d.$$

It is easy to see that the transformation from k_p, k_i, k_d to k_1, k_2, k_3 is a linear map and is invertible for all σ. We have

$$\begin{bmatrix} k_p \\ k_i \\ k_d \end{bmatrix} = \begin{bmatrix} 1 & 0 & 2\sigma \\ \sigma & 1 & \sigma^2 \\ 0 & 0 & 1 \end{bmatrix} \begin{bmatrix} k_1 \\ k_2 \\ k_3 \end{bmatrix}. \tag{2.174}$$

We can see that $\nu'(s')$ exhibits the parameter separation property in terms of k_1, k_2, k_3 just as $\nu(s)$ does with respect to k_p, k_i, k_d.

By fixing $k_1 = k_1^*$, we can determine the zeros of the imaginary part of $\nu'(j\omega)$. There exist sets of linear inequalities in terms of k_2 and k_3 for such fixed k_1^*. By intersecting the half-planes from the inequalities in each set, we find the stabilizing set in (k_2, k_3) space. By sweeping over different k_1 values, we obtain the stabilizing set in (k_1, k_2, k_3) space. By the transformation in (2.174), we obtain $\mathcal{S}(\sigma)$ in (k_p, k_i, k_d) space for a prescribed σ. The maximum achievable σ can be found by increasing σ until $\mathcal{S}(\sigma)$ just becomes empty.

2.7.2 Computation of Achievable σ

The constructive procedure of the algorithm that finds $\mathcal{S}(\sigma)$, the subset of the stabilizing set $\mathcal{S}$ for a prescribed σ, is demonstrated by three examples.

Example 2.6 Consider the plant and controller:

$$P(s) = \frac{s-2}{s^2+4s+3}, \quad C(s) = k_p + \frac{k_i}{s}. \tag{2.175}$$

Substitute (2.175) into (2.3) and (2.1) with $k_d = 0$. We have

$$N(s) = s-2, \quad N'(s') = s' - \sigma - 2, \tag{2.176}$$
$$N'_{\text{even}}(s'^2) = -\sigma - 2, \quad N'_{\text{odd}}(s'^2) = 1, \tag{2.177}$$

$$D(s) = s^2 + 4s + 3, \tag{2.178}$$
$$\begin{aligned} D'(s') &= (s'-\sigma)^2 + 4(s'-\sigma) + 3, \\ &= s'^2 + (4-2\sigma)s' + \sigma^2 - 4\sigma + 3, \end{aligned} \tag{2.179}$$
$$D'_{\text{even}}(s'^2) = s'^2 + \sigma^2 - 4\sigma + 3, \quad D'_{\text{odd}}(s'^2) = -2\sigma + 4. \tag{2.180}$$

Substituting (2.177), (2.180) into (2.173) and evaluating $\nu'(s')$ at $s' = j\omega$, we get

$$\nu'(j\omega) = p(\omega, \sigma, k_2) + jq(\omega, \sigma, k_1), \tag{2.181}$$

where

$$p(\omega, \sigma, k_2) = p_1(\omega, \sigma) + k_2 p_2(\omega, \sigma), \tag{2.182}$$
$$q(\omega, \sigma, k_1) = q_1(\omega, \sigma) + k_1 q_2(\omega, \sigma), \tag{2.183}$$

and

$$\begin{aligned} p_1(\omega, \sigma) &= -\omega^4 + (11 - 10\sigma)\omega^2 + (\sigma^4 - 2\sigma^3 - 5\sigma^2 + 6\sigma), \\ p_2(\omega, \sigma) &= \omega^2 + \sigma^2 + 4\sigma + 4, \\ q_1(\omega, \sigma) &= (6 - 2\sigma)\omega^3 - (2\sigma^3 + 2\sigma^2 - 16\sigma + 6)\omega, \\ q_2(\omega, \sigma) &= \omega\left[\omega^2 + \sigma^2 + 4\sigma + 4\right]. \end{aligned}$$

Suppose we fix $\sigma = 0.5$. Then, by Theorem 2.3, we need for σ-Hurwitz stability

$$\sigma\text{-signature}(\nu') = n + 1 - m + 2z_\sigma^+ \tag{2.184}$$
$$= 2 + 1 - 1 + 2 \cdot 1 \tag{2.185}$$
$$= 4. \tag{2.186}$$

The admissible range for k_1 is such that $q(\omega, \sigma = 0.5, k_1)$ must have at least one positive zero with odd multiplicity. This was graphically determined to be $(-5, -0.2)$.

For a fixed $k_1 \in (-5, -0.2)$, for instance, $k_1 = -1$, we have

$$q(\omega, \sigma = 0.5, k_1 = -1) = 4\omega^3 - 5\omega, \tag{2.187}$$

and the nonnegative distinct zeros of $q(\omega, \sigma = 0.5, k_1 = -1)$ with odd multiplicities are

$$\omega_0 = 0, \ \omega_1 = \frac{\sqrt{5}}{2}. \tag{2.188}$$

We also define $\omega_2 := \infty$. Since

$$\mathrm{sgn}[q(0^+, 0.5, -1)] = -1,$$

it follows that the stabilizing k_2 values corresponding to $k_1 = -1$ must satisfy the string of inequalities:

$$p_1(\omega_0, \sigma = 0.5) + k_2 p_2(\omega_0, \sigma = 0.5) < 0, \tag{2.189}$$
$$p_1(\omega_1, \sigma = 0.5) + k_2 p_2(\omega_1, \sigma = 0.5) > 0, \tag{2.190}$$
$$p_1(\omega_2, \sigma = 0.5) + k_2 p_2(\omega_2, \sigma = 0.5) < 0. \tag{2.191}$$

Substituting for ω_0, ω_1 and ω_2, we have

$$1.5625 + 6.25k_2 < 0, \tag{2.192}$$
$$7.499 + 7.499k_2 > 0, \tag{2.193}$$

and this yields $-1 < k_2 < -0.25$.

Using (2.174), the line segment in (k_1, k_2) space

$$\{(k_1, k_2) : k_1 = -1, -1 < k_2 < -0.25\} \tag{2.194}$$

is mapped back to the line segment in (k_p, k_i) space

$$\left\{(k_p, k_i) : k_p = -1, -1.5 < k_i < -0.75\right\}. \tag{2.195}$$

By sweeping over different k_1 values within the interval $(-5, -0.2)$ and repeating the procedure, we get the stabilizing set in (k_1, k_2) space and finally mapping the set back to (k_p, k_i) space we get $\mathcal{S}(\sigma)$ for $\sigma = 0.5$.

The sets $\mathcal{S}(\sigma)$ are displayed for different σ values in Fig. 2.10. The set colored in yellow is the Hurwitz stabilizing set and it is consistent with $\mathcal{S}(\sigma)$ when $\sigma = 0$. Observe that the sets, $\mathcal{S}(\sigma)$, are telescoping as σ increases, which is consistent with Fact 4. It vanishes at around $\sigma = 1.1$. Thus, the maximum achievable σ can be determined to be approximately 1.1.

In order to demonstrate the movement of the closed-loop poles, the controllers are selected from the boundary points in the $\mathcal{S}(\sigma)$ sets for different σ values and the closed-loop poles are depicted in Fig. 2.11. It is clear that the poles are pushed to

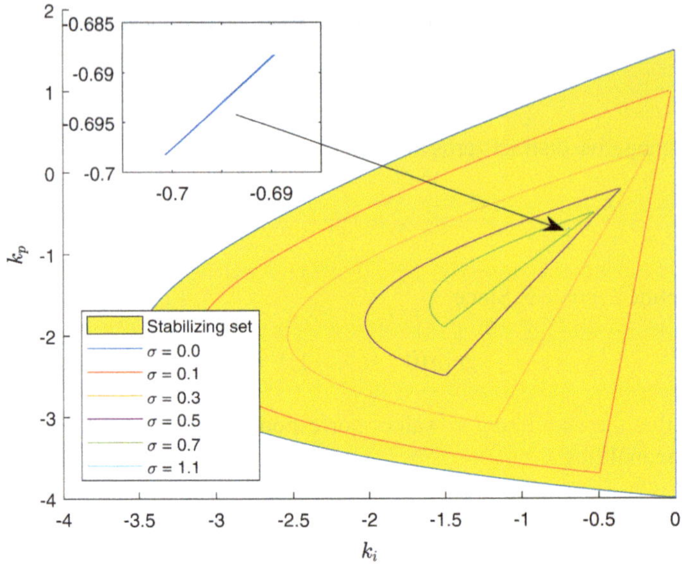

Fig. 2.10 $\mathcal{S}(\sigma)$ sets for various σ values. © 2018 IEEE. Reproduced from [7] with permission

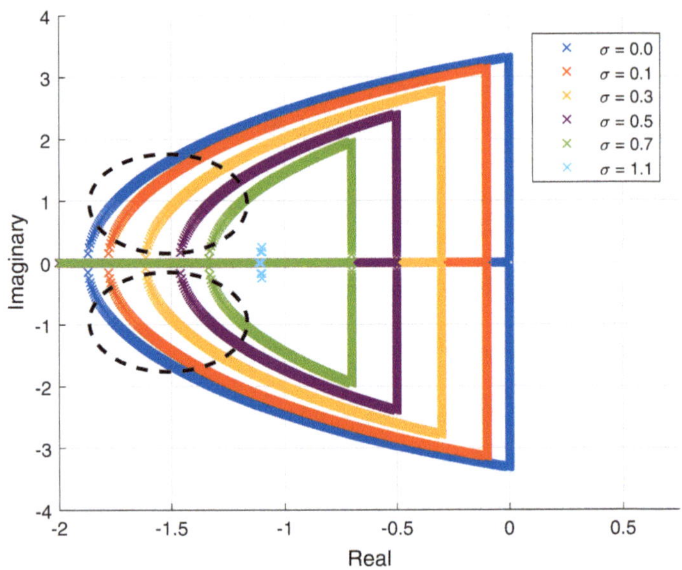

Fig. 2.11 The closed-loop poles for various σ values. © 2018 IEEE. Reproduced from [7] with permission

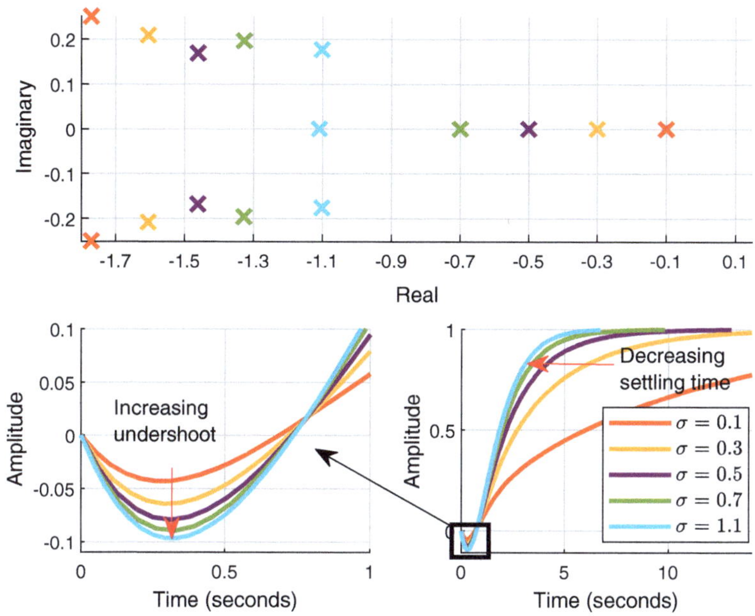

Fig. 2.12 Closed-loop poles (top) and step responses (bottom) for σ values. © 2018 IEEE. Reproduced from [7] with permission

the left as σ increases. The closed-loop poles appear to be telescoping similar to the $\mathcal{S}(\sigma)$ sets in this example.

Since the plant is of second order and the controller is first order, the resulting closed-loop system has three poles. Some complex conjugate poles appear to be pushed more than the prescribed σ, as marked in dashed ellipses in Fig. 2.11, but the third root is located at $s = -\sigma$. It should be noted that the complex conjugate poles are not the rightmost poles but the real root at $s = -\sigma$ is the rightmost pole.

Location of the closed-loop poles for several PI controllers and the corresponding step responses are plotted in Fig. 2.12. We chose PI gains so that each closed-loop system had a pair of complex poles whose imaginary part was around 0.2. Observe that settling time decreases as σ increases as expected but undershoot also increases. The closed-loop poles are lined up on the line $\text{Re}\{s\} = -\sigma$ when $\sigma = 1.1$.

Example 2.7 Consider the plant and the controller:

$$P(s) = \frac{N(s)}{D(s)}, \quad C(s) = k_p + \frac{k_i}{s} + k_d s, \tag{2.196}$$

where

$$N(s) = s^3 - 2s^2 - s - 1, \tag{2.197}$$
$$D(s) = s^6 + 2s^5 + 32s^4 + 26s^3 + 65s^2 - 8s + 1. \tag{2.198}$$

From (2.197) and (2.198), we have

$$N'_{\text{even}}(s'^2) = -3\sigma s'^2 - \sigma^3 - 2\sigma^2 + \sigma - 1, \tag{2.199}$$

$$N'_{\text{odd}}(s'^2) = s'^2 + 3\sigma^2 + 4\sigma - 1, \tag{2.200}$$

$$\begin{aligned} D'_{\text{even}}(s'^2) = &\, s'^6 + (15\sigma^2 - 10\sigma + 32)s'^4 \\ &+ (15\sigma^4 - 20\sigma^3 + 192\sigma^2 - 78\sigma + 65)s'^2 \\ &+ \sigma^6 - 2\sigma^5 + 32\sigma^4 - 26\sigma^3 + 65\sigma^2 + 8\sigma + 1, \end{aligned} \tag{2.201}$$

$$\begin{aligned} D'_{\text{odd}}(s'^2) = &\,(2 - 6\sigma)s'^4 - (20\sigma^3 - 20\sigma^2 + 128\sigma - 26)s'^2 \\ &- 6\sigma^5 + 10\sigma^4 - 128\sigma^3 + 78\sigma^2 - 130\sigma - 8. \end{aligned} \tag{2.202}$$

Substituting the above equations into (2.173) and evaluating $\nu'(s')$ at $s' = j\omega$, we get

$$\nu'(j\omega) = p(\omega, \sigma, k_2, k_3) + jq(\omega, \sigma, k_1), \tag{2.203}$$

where

$$p(\omega, \sigma, k_2, k_3) = p_1(\omega, \sigma) + (k_2 - \omega^2 k_3)p_2(\omega, \sigma) \tag{2.204}$$
$$q(\omega, \sigma, k_1) = q_1(\omega, \sigma) + k_1 q_2(\omega, \sigma). \tag{2.205}$$

and

$$\begin{aligned} p_1(\omega, \sigma) = &\, N'_{\text{odd}}(-\omega^2)(D'_{\text{even}}(-\omega^2) - \sigma D'_{\text{odd}}(-\omega^2))\omega^2 \\ &- D'_{\text{odd}}(-\omega^2)N'_{\text{even}}(-\omega^2)\omega^2 - \sigma D'_{\text{even}}(-\omega^2)N'_{\text{even}}(-\omega^2), \end{aligned} \tag{2.206}$$
$$p_2(\omega, \sigma) = \{N'_{\text{odd}}(-\omega^2)\}^2\omega^2 + \{N'_{\text{even}}(-\omega^2)\}^2, \tag{2.207}$$
$$\begin{aligned} q_1(\omega, \sigma) = &\, D'_{\text{odd}}(-\omega^2)N'_{\text{odd}}(-\omega^2)\omega^3 + \sigma D'_{\text{even}}(-\omega^2)N'_{\text{odd}}(-\omega^2)\omega \\ &+ N'_{\text{even}}(-\omega^2)(D'_{\text{even}}(-\omega^2) - \sigma D'_{\text{odd}}(-\omega^2))\omega, \end{aligned} \tag{2.208}$$
$$q_2(\omega, \sigma) = \omega\left[\{N'_{\text{odd}}(-\omega^2)\}^2\omega^2 + \{N'_{\text{even}}(-\omega^2)\}^2\right]. \tag{2.209}$$

The sets $\mathcal{S}(\sigma)$ are displayed in Figs. 2.13, 2.14, and 2.15. The maximum achievable σ is found to be 0.1655. The algorithm allows us to observe how these sets shrink in a telescoping manner as σ increases. Of course, the complexity of computation of the stabilizing sets increases with the order of the system because the number of linear inequalities increases.

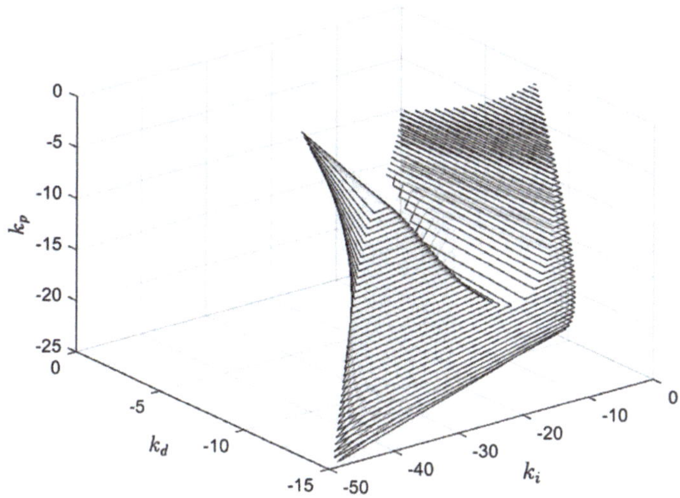

Fig. 2.13 $\mathcal{S}(\sigma)$ set for $\sigma = 0$. © 2018 IEEE. Reproduced from [7] with permission

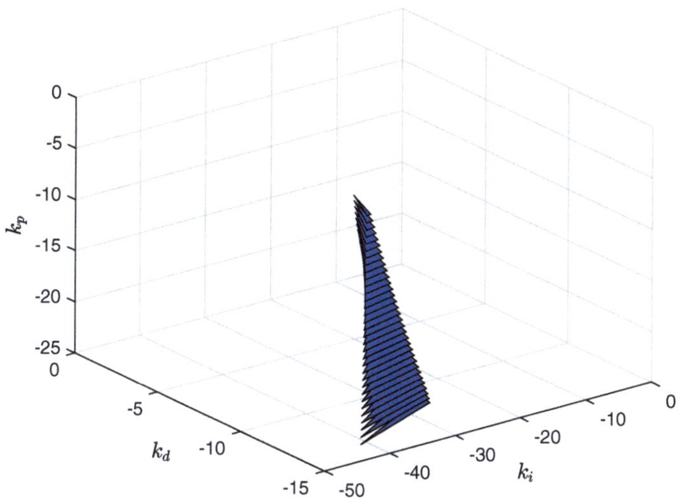

Fig. 2.14 $\mathcal{S}(\sigma)$ set for $\sigma = 0.1$. © 2018 IEEE. Reproduced from [7] with permission

2.8 Computation of the Stabilizing Set of Delay-Free Systems with First-Order Controllers

Lead and Lag controllers are often used in control engineering to reshape the loop frequency response to provide improved stability margins. However, the design methods used are often based on trial and error, but nevertheless, useful because of the

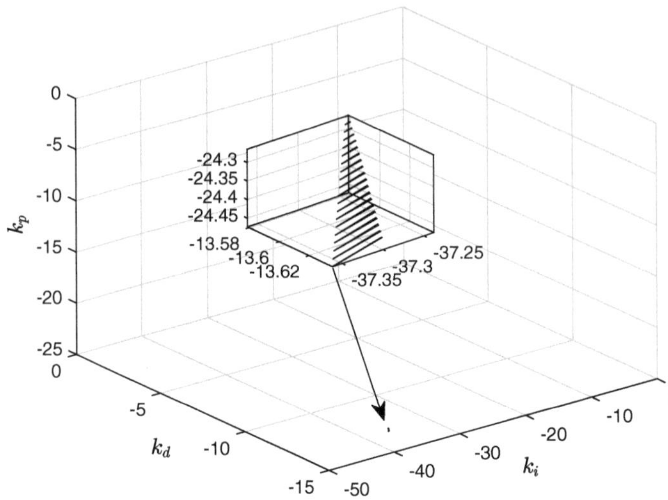

Fig. 2.15 $\mathcal{S}(\sigma)$ set for $\sigma = 0.1655$. © 2018 IEEE. Reproduced from [7] with permission

widespread use of first-order controllers. Indeed, they are next to PID controllers in importance and usage.

In this section, we provide the procedure to calculate all first-order controllers that stabilize a given linear time-invariant continuous-time system. These sets of controllers can be displayed in the 2-D or 3-D parameter spaces of the controllers. The main tool used is Neimark's D-decomposition, which was originally introduced in 1949.

2.8.1 *Root Distribution Invariant Regions*

The Neimark's D-decomposition technique determines root distribution invariant regions of a polynomial with respect to a region in the complex plane, by mapping the boundary of the region onto the coefficient or parameter space. For Hurwitz stability problems, the boundary in question is the imaginary axis including the points at infinity. This concept can be applied to the characteristic polynomial to determine the stabilizing set for the first-order controller

$$C(s) := \frac{x_1 s + x_2}{s + x_3} \tag{2.210}$$

connected in a unity feedback control loop to the rational proper plant

$$P(s) = \frac{N(s)}{D(s)}, \tag{2.211}$$

with $N(s)$ and $D(s)$ being coprime. The closed-loop characteristic polynomial is

$$\begin{aligned}\delta(s) &= D(s)(s + x_3) + N(s)(x_1 s + x_2)\\ &= [D_{\text{even}}(s^2) + s D_{\text{odd}}(s^2)](s + x_3) + [N_{\text{even}}(s^2) + s N_{\text{odd}}(s^2)](x_1 s + x_2)\\ &= [s^2 D_{\text{odd}}(s^2) + x_3 D_{\text{even}}(s^2) + x_2 N_{\text{even}}(s^2) + x_1 s^2 N_{\text{odd}}(s^2)]\\ &\qquad + s[D_{\text{even}}(s^2) + x_3 D_{\text{odd}}(s^2) + x_2 N_{\text{odd}}(s^2) + x_1 N_{\text{even}}(s^2)].\end{aligned} \tag{2.212}$$

With $s = j\omega$, we have

$$\begin{aligned}\delta(j\omega) = &[-\omega^2 N_{\text{odd}}(-\omega^2)x_1 + N_{\text{even}}(-\omega^2)x_2 + D_{\text{even}}(-\omega^2)x_3 - \omega^2 D_{\text{odd}}(-\omega^2)]\\ &+ j\omega[N_{\text{even}}(-\omega^2)x_1 + N_{\text{odd}}(-\omega^2)x_2 + D_{\text{odd}}(-\omega^2)x_3 + D_{\text{even}}(-\omega^2)].\end{aligned} \tag{2.213}$$

The complex root boundary in the space x_1, x_2, x_3 is given by the equation

$$\delta(j\omega) = 0, \qquad \omega \in (0, +\infty) \tag{2.214}$$

and the real root boundary is given by

$$\delta(0) = 0, \qquad \delta_{n+1} = 0, \tag{2.215}$$

where δ_{n+1} denotes the leading coefficient of $\delta(s)$. Thus, with usual notation

$$-\omega^2 N_{\text{odd}}(-\omega^2)x_1 + N_{\text{even}}(-\omega^2)x_2 + D_{\text{even}}(-\omega^2)x_3 - \omega^2 D_{\text{odd}}(-\omega^2) = 0 \tag{2.216}$$

$$\omega[N_{\text{even}}(-\omega^2)x_1 + N_{\text{odd}}(-\omega^2)x_2 + D_{\text{odd}}(-\omega^2)x_3 + D_{\text{even}}(-\omega^2)] = 0. \tag{2.217}$$

These lead to

$$N_{\text{even}}(0)x_2 + D_{\text{even}}(0)x_3 = 0 \tag{2.218}$$

$$d_n + x_1 n_n = 0, \tag{2.219}$$

where d_n and n_n denote the coefficients of s^n in $D(s)$ and $N(s)$, respectively. In matrix form, (2.216) and (2.217) are

$$\underbrace{\begin{bmatrix} \omega^2 N_{\text{odd}}(-\omega^2) & -N_{\text{even}}(-\omega^2) \\ N_{\text{even}}(-\omega^2) & N_{\text{odd}}(-\omega^2) \end{bmatrix}}_{A(\omega)} \begin{bmatrix} x_1 \\ x_2 \end{bmatrix} = \begin{bmatrix} D_{\text{even}}(-\omega^2)x_3 - \omega^2 D_{\text{odd}}(-\omega^2) \\ -D_{\text{odd}}(-\omega^2)x_3 - D_{\text{even}}(-\omega^2) \end{bmatrix}. \tag{2.220}$$

Consider first the case when $|A(\omega)| \neq 0$ for each $\omega > 0$. Then

$$|A(\omega)| = \omega^2 N_{\text{odd}}^2(-\omega^2) + N_{\text{even}}^2(-\omega^2) > 0, \quad \forall\ \omega > 0, \tag{2.221}$$

and, for every x_3 (2.220) has a unique solution $x_1(\omega)$ and $x_2(\omega)$ at each $\omega > 0$ given by

$$\begin{aligned} x_1(\omega) = \frac{1}{|A(\omega)|} &\left(\left[N_{\text{odd}}(-\omega^2) D_{\text{even}}(-\omega^2) - N_{\text{even}}(-\omega^2) D_{\text{odd}}(-\omega^2) \right] x_3 \right. \\ &\left. -\omega^2 N_{\text{odd}}(-\omega^2) D_{\text{odd}}(-\omega^2) - N_{\text{even}}(-\omega^2) D_{\text{even}}(-\omega^2) \right) \quad (2.222) \\ x_2(\omega) = \frac{1}{|A(\omega)|} &\left(\left[-N_{\text{even}}(-\omega^2) D_{\text{even}}(-\omega^2) - \omega^2 N_{\text{odd}}(-\omega^2) D_{\text{odd}}(-\omega^2) \right] x_3 \right. \\ &\left. +\omega^2 N_{\text{even}}(-\omega^2) D_{\text{odd}}(-\omega^2) - \omega^2 N_{\text{odd}}(-\omega^2) D_{\text{even}}(-\omega^2) \right). \quad (2.223) \end{aligned}$$

For a fixed value of x_3, let ω run from 0 to ∞. The above equations trace out a curve in the (x_1, x_2) plane corresponding to the complex root space boundary, which along with the straight lines (2.218) and (2.219) partition the parameter space into a set of open root distribution invariant regions. Finally sweeping over x_3, we find these regions.

The case $|A(\omega)| = 0$ for some $\omega \neq 0$ can be discarded as it can be easily shown that in this case $D(s)$ and $N(s)$ have a common root at $s = \pm j\omega$.

2.8.2 *First-Order Stabilizing Set Computation Procedure*

The procedure to compute the first-order controller stabilizing set is the following:

1. Calculate the characteristic polynomial derived from Fig. 2.1 as in (2.212).
2. Substitute $s = j\omega$ in the characteristic equation (2.212) as in (2.213).
3. Find the stability boundary for complex roots at $\omega = 0$. This boundary is given by setting $\delta(j0) = 0$ as in (2.215).
4. Find the stability boundary for complex roots at $\omega > 0$ using (2.222) and (2.223). For a fixed value of x_3 and $\omega > 0$, we have the curve in the (x_1, x_2) plane that represents the stability boundary for the complex roots. Each region formed corresponds to a set of characteristic polynomials with a fixed number of RHP roots.
5. For a fixed x_3, pick a point inside each region and calculate the roots of the characteristic equation. Select the regions with no RHP roots. By sweeping over x_3, it is possible to see the stability regions in three dimensions for a given plant, if one exists.

The following example illustrates these computations.

Example 2.8 Let us consider the continuous-time system represented in Fig. 2.1 using the plant

$$P(s) = \frac{s-2}{s^2+0.6s-0.1} \tag{2.224}$$

and the controller

$$C(s) = \frac{x_1 s + x_2}{s + x_3}. \tag{2.225}$$

Following the stabilizing set computation procedure presented in Sect. 2.8.2, the first step is given by

$$\begin{aligned} N_{\text{even}}(s^2) &= -2 \\ N_{\text{odd}}(s^2) &= 1 \\ D_{\text{even}}(s^2) &= s^2 - 0.1 \\ D_{\text{odd}}(s^2) &= 0.6. \end{aligned} \tag{2.226}$$

Then, the characteristic polynomial is given by (2.212)

$$\delta(s) = \left[(x_1 + x_3 + 0.6)\, s^2 - 0.1x_3 - 2x_2\right] + s\left[s^2 - 2x_1 + x_2 + 0.6x_3 - 0.1\right]. \tag{2.227}$$

Substituting $s = j\omega$ in the characteristic equation (2.227), we get

$$\begin{aligned} \delta(j\omega) = &\left[-\,(x_1 + x_3 + 0.6)\,\omega^2 - 2x_2 - 0.1x_3\right] \\ &+ j\omega\left[-\omega^2 - 2x_1 + x_2 + 0.6x_3 - 0.1\right]. \end{aligned} \tag{2.228}$$

For the stability boundaries, the two conditions to be satisfied are described in (2.218) and (2.219). For $\omega = 0$, we have

$$-2x_2 - 0.1x_3 = 0. \tag{2.229}$$

Then, there exist a real root boundary at

$$x_2 = -0.05x_3. \tag{2.230}$$

For $\omega > 0$, using (2.222) and (2.223) we have

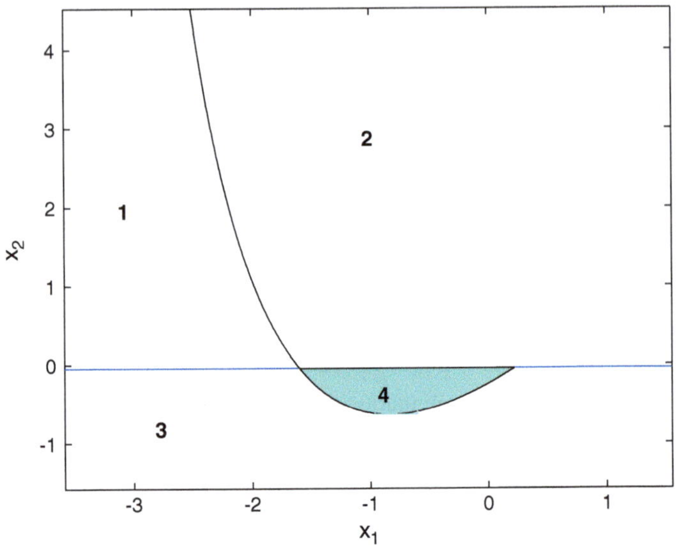

Fig. 2.16 Root-invariant regions for $x_3 = 1$ in Example 2.8. © 2015 IEEE. Reproduced from [5] with permission

$$
\begin{aligned}
x_1(\omega) &= \frac{(-\omega^2 - 0.1)x_3 - 2.6\omega^2 - 0.2}{\omega^2 + 4} \\
x_2(\omega) &= \frac{(-2.6\omega^2 - 0.2)x_3 + \omega^4 - 1.1\omega^2}{\omega^2 + 4}.
\end{aligned} \tag{2.231}
$$

For illustration purposes, let $x_3 = 1$. Then, the stability region for $x_3 = 1$ is shown in Fig. 2.16. The regions are numbered to represent signature-invariant root regions. Then, we can pick any value contained in the regions and check the root distribution. It was found that the region numbered 4 is the stabilizing region. By sweeping x_3 from -0.4 to 8, we obtain the following three-dimensional figure shown in Fig. 2.17, which represents the stabilizing set for our plant $P(s)$ in (2.224).

2.9 Notes and References

The characterization of all stabilizing PID controllers for a given delay-free linear time-invariant plant was developed by Ho, Datta, and Bhattacharyya in [8]. For an extensive description, the reader can refer to the book Structure and Synthesis of PID Controllers by Datta, Ho, and Bhattacharyya in [4].

The results in Sects. 2.2, 2.3 are taken from the book Linear Control Theory: Structure, Robustness, and Optimization by Bhattacharyya, Datta, and Keel in [3]. Also, more information about these results is available in [9] for the first-order controller case and for continuous-time PI/PID controller.

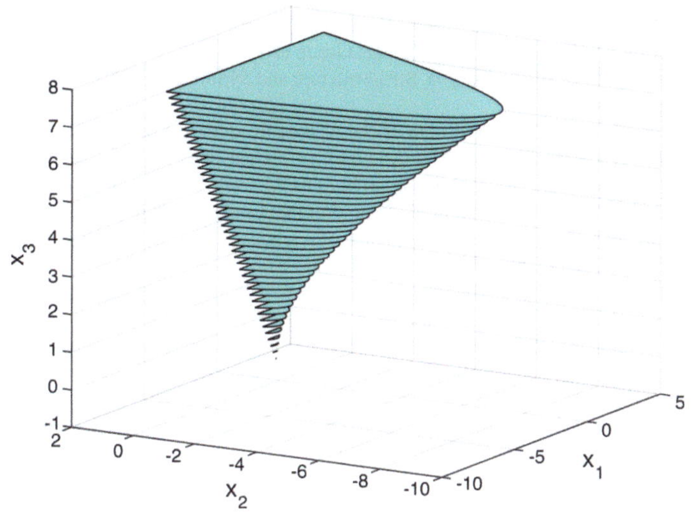

Fig. 2.17 Stability region for $-0.4 \leq x_3 \leq 8$ in Example 2.8. © 2015 IEEE. Reproduced from [5] with permission

In [1], a linear quadratic regulator (LQR) design with a prescribed σ, called the degree of stability, was proposed using state feedback. The control law minimized a quadratic performance index and the closed-loop poles had real parts less than $-\sigma$. In [2], a branch and bound algorithm for computing the stability degree was proposed when the system was subject to parametric perturbations. State feedback controller design with prescribed damping ratio as well as degree of stability was proposed in [11]. The algorithm was based on the control law by which the closed-loop poles were inside a prescribed disk.

A procedure for calculating PID gains for "optimal" stability degree was proposed in [13]. The plant model was assumed to be all-pole, i.e., the plant had no zeros. PID controller design for dominant pole placement was proposed in [6, 14]. In dominant pole placement, the desired locations of a pair of poles are chosen by a designer. The PID gains are selected by the desired dominant pole pair. The gains are further tuned so that the rest of the poles have real part less than that of the dominant poles. A similar approach was taken for first-order controllers in [10] and for Proportional–Integral–Retarded (PIR) controllers in [12].

References

1. Anderson, B.D., Moore, J.B.: Linear system optimisation with prescribed degree of stability. Proc. Inst. Electr. Eng. **116**(12), 2083–2087 (1969)
2. Balakrishnan, V., Boyd, S., Balemi, S.: Branch and bound algorithm for computing the minimum stability degree of parameter-dependent linear systems. Int. J. Robust Nonlinear Control **1**(4), 295–317 (1991)

3. Bhattacharyya, S.P., Datta, A., Keel, L.H.: Linear Control Theory: Structure, Robustness, and Optimization. CRC Press Taylor & Francis Group (2009)
4. Datta, A., Ho, M.-T., Bhattacharyya, S.P.: Structure and Synthesis of PID Controllers. Springer, Berlin (2000)
5. Diaz-Rodriguez, I.D.: Modern design of classical controllers: continuous-time first order controllers. In: Proceedings of the 41st Annual Conference of the IEEE Industrial Electronics Society, Student Forum. IECON, pp. 000070–000075 (2015)
6. Dincel, E., Söylemez, M.T.: Limitations on dominant pole pair selection with continuous PI and PID controllers. In: 2016 International Conference on Control, Decision and Information Technologies (CoDIT), pp. 741–745 (2016)
7. Han, S., Bhattacharyya, S.: PID controller synthesis using a σ-Hurwitz stability criterion. IEEE Control Syst. Lett. **2**(3), 525–530 (2018)
8. Ho, M.-T., Datta, A., Bhattacharyya, S.P.: A linear programming characterization of all stabilizing PID controllers. In: Proceedings of the 1997 American Control Conference, pp. 3922–3928 (1997)
9. Keel, L.H., Bhattacharyya, S.P.: Controller synthesis free of analytical models: three term controllers. IEEE Trans. Autom. Control **53**(6), 1353–1369 (2008)
10. Madady, A., Reza-Alikhani, H.-R.: First-order controllers design employing dominant pole placement. In: 2011 19th Mediterranean Conference on Control & Automation (MED), pp. 1498–1503 (2011)
11. Misra, P.: LQR design with prescribed damping and degree of stability. In: 1996 Proceedings of the 1996 IEEE International Symposium on Computer-Aided Control System Design, pp. 68–70. IEEE (1996)
12. Ramírez, A., Mondié, S., Garrido, R.: Proportional integral retarded control of second order linear systems. In: 2013 IEEE 52nd Annual Conference on Decision and Control (CDC), pp. 2239–2244 (2013)
13. Shubladze, A.: A procedure for calculating the optimal stability of PID controls. 2. Autom. Remote Control **48**(6), 748–756 (1987)
14. Srivastava, S., Pandit, V.: A PI/PID controller for time delay systems with desired closed loop time response and guaranteed gain and phase margins. J. Process Control **37**, 70–77 (2016)

Chapter 3
Stabilizing Sets for Ziegler–Nichols Plants

Abstract In this chapter, we consider the PID stabilization of first-order plants cascaded with a time-delay. Such plants are called Ziegler–Nichols plants and there is a considerable interest in PID controller design for such plants. A complete solution to the problem of determining the stabilizing set for P, PI, and PID controllers was obtained in 2002. This solution is based on an extension of the classical Hermite–Biehler theorem for polynomials to quasipolynomials. Here, we summarize the results in the form of algorithms omitting the details of the extensive proofs. These appear in sources cited in the Notes and References section for the benefit of the interested reader.

3.1 Introduction

Linear time-invariant systems often involve time delays. These can be due to communication, transportation, transmission delays, or inertia effects involved in the control process. Time delays occur in electrical, chemical, hydraulic, and pneumatic systems and transmission lines, robotics, and industrial processes. One of the most studied cases is the Ziegler–Nichols plant because many of the industrial processes are approximated by this type of system. The study of this kind of systems involving time delays gives rise to characteristic equations known as quasipolynomials. Pontryagin was among the first researchers to study these quasipolynomials. He provided necessary and sufficient conditions for the roots of a given quasipolynomial to have negative real parts. Based on the work of Pontryagin, a suitable extension of the Hermite–Biehler Theorem was developed to study the stability of certain classes of quasipolynomials. The complete solutions to compute the stabilizing sets of Ziegler–Nichols plants were obtained by Guillermo, Datta, and Bhattacharyya in 2001.

I. D. Díaz-Rodríguez et al., *Analytical Design of PID Controllers*,
https://doi.org/10.1007/978-3-030-18228-1_3

3.2 PI Controller Stabilizing Sets for Ziegler–Nichols Plants

Consider a linear time-invariant system given by

$$P(s) := \frac{k}{1+Ts}e^{-Ls}, \tag{3.1}$$

referred to as a Ziegler–Nichols plant and a PI controller

$$C(s) = \frac{k_p s + k_i}{s}, \tag{3.2}$$

where k is the steady-state gain, L is the time-delay, T is the time constant of the plant, and k_p, k_i are the controller gains. The objective is to analytically determine the region in the (k_p, k_i) parameter space for which the closed-loop system is stable. When the time-delay L of the plant model is zero, the characteristic equation of the closed-loop system is given by

$$\delta(s) = Ts^2 + (kk_p + 1)s + kk_i. \tag{3.3}$$

From Eq. (3.3), we conclude that for the closed-loop stability of the delay-free system, we must have

$$kk_i > 0, \quad kk_p + 1 > 0, \quad T > 0 \tag{3.4}$$

or

$$kk_i < 0, \quad kk_p + 1 < 0, \quad T < 0. \tag{3.5}$$

Clearly, (3.4) must be satisfied for an open-loop stable plant while (3.5) must be satisfied for an open-loop unstable plant. Assuming that the steady-state gain k of the plant is positive, we obtain the following conditions for closed-loop stability of the delay-free system:

$$k_p > -\frac{1}{k}, \quad k_i > 0 \qquad \text{open-loop stable plant, } T > 0 \tag{3.6}$$

$$k_p < -\frac{1}{k}, \quad k_i < 0 \qquad \text{open-loop unstable plant, } T < 0. \tag{3.7}$$

For $L > 0$, the closed-loop characteristic equation of the system is given by

$$\delta(s) = (kk_i + kk_p)e^{-Ls} + (1 + Ts)s. \tag{3.8}$$

In the following subsections, we summarize the procedure to compute the complete PI stabilizing set obtained by Silva, Datta, and Bhattacharyya in 2001 by considering the two different cases: open-loop stable Ziegler–Nichols plants and open-loop unstable Ziegler–Nichols plants.

3.2.1 Open-Loop Stable Ziegler–Nichols Plants

For, the case of an open-loop stable Ziegler–Nichols plant with a PI controller, we have $T > 0$. Furthermore, let us assume that $k > 0$, and $L > 0$. The procedure to compute the PI stabilizing set is the following:

1. For $L = 0$, the characteristic equation is

$$\delta(s) = Ts^2 + (kk_p + 1)s + kk_i. \tag{3.9}$$

   For stability, it is required

$$k_p > -\frac{1}{k}, \quad k_i > 0. \tag{3.10}$$

2. For $L > 0$, the characteristic equation is

$$\delta(s) = (kk_i + kk_p)e^{-Ls} + (1 + Ts)s. \tag{3.11}$$

   Considering $\delta^*(s) = e^{Ls}\delta(s)$ we have

$$\delta^*(s) = (kk_i + kk_p s) + (1 + Ts)se^{Ls}. \tag{3.12}$$

3. Calculate

$$\delta^*(j\omega) = \delta_r(\omega) + j\delta_i(j\omega), \tag{3.13}$$

   where

$$\delta_r(\omega) = kk_i - \omega\sin(L\omega) - T\omega^2\cos(L\omega) \tag{3.14}$$

$$\delta_i(\omega) = \omega\left[kk_p + \cos(L\omega) - T\omega\sin(L\omega)\right]. \tag{3.15}$$

4. Make a change in variable $z = L\omega$ and calculate the new real and imaginary parts of $\delta^*(z)$

$$\delta_r(z) = k\left[k_i - a(z)\right] \tag{3.16}$$

$$\delta_i(z) = \frac{z}{L}\left[kk_p + \cos(z) - \frac{T}{L}z\sin(z)\right], \tag{3.17}$$

   where

$$a(z) = \frac{z}{kL}\left[\sin(z) + \frac{T}{L}z\cos(z)\right]. \tag{3.18}$$

5. Pick a value for $k_p = k_p^*$ in the range

$$-\frac{1}{k} < k_p < \frac{T}{kL}\sqrt{\alpha_1^2 + \frac{L^2}{T^2}}, \tag{3.19}$$

where $\alpha = \alpha_1$ is the solution of

$$\tan(\alpha) = -\frac{T}{L}\alpha \tag{3.20}$$

in the interval $(\frac{\pi}{2}, \pi)$.

6. Let z_j denote the jth real root of $\delta_i(z)$, namely, the jth real root of

$$\left[kk_p^* + \cos(z) - \frac{T}{L}z\sin(z)\right] = 0. \tag{3.21}$$

The roots can be found graphically for the following cases:

- $-\frac{1}{k} < k_p^* < \frac{1}{k}$: in this case, determine the intersection of the functions $\frac{kk_p^*+\cos(z)}{\sin(z)}$ and $\frac{T}{L}z$.
- $k_p^* = \frac{1}{k}$: in this case, determine the intersection of the functions $kk_p + cos(z)$ and $\frac{T}{L}z\sin(z)$.
- $\frac{1}{k} < k_p^* < \frac{T}{kL}\sqrt{\alpha_1^2 + \frac{L^2}{T^2}}$, in this case determine the intersection of the functions $\frac{kk_p^*+\cos(z)}{\sin(z)}$ and $\frac{T}{L}z$.

7. Compute the parameters $a_j := a(z)|_{z=z_j}$ using (3.18).
8. If $\cos(z_j) > 0$ go to the next step. If not, set $j = j + 2$ and go to step 6.
9. Determine the lower and upper bounds for k_i from

$$0 < k_i < \min_{l=1,3,5,\ldots j}\{a_l\}. \tag{3.22}$$

10. Update k_p^* from (3.19) and repeat steps 6 through 10 until the range of k_p in (3.19) is exhausted.

We illustrate the previous procedure in the following example.

Example 3.1 Consider an open-loop stable Ziegler–Nichols plant given by

$$P(s) = \frac{1}{2s+1}e^{-0.3s}, \tag{3.23}$$

and the PI controller in (3.2). The characteristic equation is given by

$$\delta(s) = (2s+1)s + (k_p s + k_i)e^{-0.3s}. \tag{3.24}$$

By (3.12),

$$\delta^*(s) = e^{0.3s}(2s+1)s + (k_p s + k_i). \tag{3.25}$$

For $L = 0$, we have

$$\delta(s) = 2s^2 + (k_p + 1)s + k_i. \tag{3.26}$$

For stability, it is required $k_p > -1$, $k_i > 0$. For $L > 0$ and by (3.8),

$$\delta^*(j\omega) = \delta_r(\omega) + j\delta_i(j\omega), \tag{3.27}$$

where

$$\delta_r(\omega) = k_i - \omega\sin(0.3\omega) - 2\omega^2\cos(0.5\omega) \tag{3.28}$$

$$\delta_i(\omega) = \omega\left[5k_p + \cos(0.5\omega) + 12\omega\sin(0.5\omega)\right]. \tag{3.29}$$

By (3.18), we can calculate the range for k_p for stability

$$-1 < k_p < 6.6667\sqrt{\alpha_1^2 + 0.0225}. \tag{3.30}$$

Following all the steps of the procedure, we get the stabilizing set shown in Fig. 3.1.

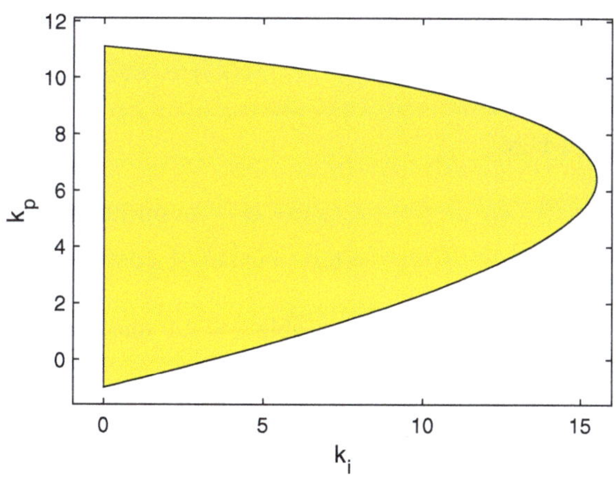

Fig. 3.1 Stabilizing set for an open-loop stable Ziegler–Nichols plant using a PI controller design in Example 3.1

3.2.2 Open-Loop Unstable Ziegler–Nichols Plants

For the case of an open-loop unstable Ziegler–Nichols plant with a PI controller, we have $T < 0$, $k > 0$, and $L > 0$. Now, for the closed-loop stability of the delay-free system, we require

$$k_p < -\frac{1}{k}, k_i < 0. \tag{3.31}$$

The procedure to compute the PI stabilizing set is the following:

1. For $L = 0$, the characteristic equation is

$$\delta(s) = Ts^2 + (kk_p + 1)s + kk_i. \tag{3.32}$$

For stability, it is required

$$k_p < -\frac{1}{k}, \quad k_i < 0. \tag{3.33}$$

2. Calculate the characteristic equation

$$\delta(s) = (kk_i + kk_p)e^{-Ls} + (1 + Ts)s, \tag{3.34}$$

and set $\delta^*(s) = e^{Ls}\delta(s)$ to get

$$\delta^*(s) = (kk_i + kk_p s) + (1 + Ts)se^{Ls}. \tag{3.35}$$

3. Calculate

$$\delta^*(j\omega) = \delta_r(\omega) + j\delta_i(j\omega), \tag{3.36}$$

where

$$\delta_r(\omega) = kk_i - \omega\sin(L\omega) - T\omega^2\cos(L\omega) \tag{3.37}$$

$$\delta_i(\omega) = \omega\left[kk_p + \cos(L\omega) - T\omega\sin(L\omega)\right]. \tag{3.38}$$

4. Make a change in variable $z = L\omega$ and calculate the new real and imaginary parts of $\delta^*(z)$

$$\delta_r(z) = k\,[k_i - a(z)] \tag{3.39}$$

$$\delta_i(z) = \frac{z}{L}\left[kk_p + \cos(z) - \frac{T}{L}z\sin(z)\right], \tag{3.40}$$

where

$$a(z) = \frac{z}{kL}\left[\sin(z) + \frac{T}{L}z\cos(z)\right]. \tag{3.41}$$

5. Pick a value for $k_p = k_p^*$ in the range

$$\frac{T}{kL}\sqrt{\alpha_1^2 + \frac{L^2}{T^2}} < k_p < -\frac{1}{k}, \tag{3.42}$$

where α_1 is the solution of

$$\tan(\alpha) = -\frac{T}{L}\alpha \tag{3.43}$$

in the interval $(0, \frac{\pi}{2})$.

6. Let z_j denote the jth real root of $\delta_i(z)$, namely, the jth real root of:

$$\left[kk_p^* + \cos(z) - \frac{T}{L}z\sin(z)\right] = 0. \tag{3.44}$$

The roots can be found graphically for the following case:

- $\frac{T}{kL}\sqrt{\alpha_1^2 + \frac{L^2}{T^2}} < k_p^* < -\frac{1}{k}$: in this case take the intersection of the functions $\frac{kk_p^* + \cos(z)}{\sin(z)}$ and $\frac{T}{L}z$.

7. Compute the parameters $a_j := a(z)|_{z=z_j}$ using (3.41).
8. If $\cos(z_j) > 0$ go to the next step. If not, set $j = j + 2$ and go to step 6.
9. Determine the lower and upper bounds for k_i from:

$$\max_{l=1,3,5,\ldots j}\{a_l\} < k_i < 0. \tag{3.45}$$

10. Update k_p^* from (3.42) and repeat steps 6 through 10 until the range of k_p in (3.42) is exhausted.

We illustrate the previous procedure in the following example:

Example 3.2 Consider an open-loop unstable Ziegler–Nichols plant given by

$$P(s) = \frac{5}{-12s + 1}e^{-0.5s} \tag{3.46}$$

and the PI controller (3.2). The characteristic equation is given by

$$\delta(s) = (-12s + 1)s + 5(k_p s + k_i)e^{-0.5s}. \tag{3.47}$$

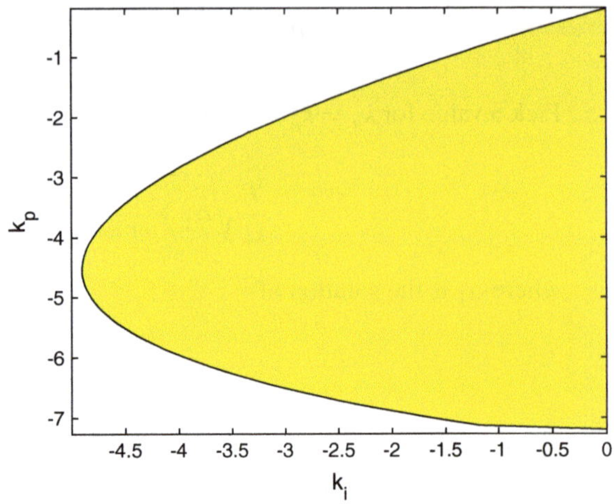

Fig. 3.2 Stabilizing set for an open-loop unstable Ziegler–Nichols plant using a PI controller: Example 3.2

By (3.35),

$$\delta^*(s) = e^{0.5s}(-12s+1)s + (k_p s + k_i)5. \tag{3.48}$$

For $L = 0$, we have

$$\delta(s) = -12s^2 + (5k_p + 1)s + 5k_i. \tag{3.49}$$

For stability, it is required $k_p < -\frac{1}{5}$, $k_i < 0$. For $L > 0$ and by (3.37) and (3.38),

$$\delta^*(j\omega) = \delta_r(\omega) + j\delta_i(j\omega) \tag{3.50}$$

where

$$\delta_r(\omega) = 5k_i - \omega\sin(0.5\omega) + 12\omega^2\cos(0.5\omega) \tag{3.51}$$

$$\delta_i(\omega) = \omega\left[5k_p + \cos(0.5\omega) + 12\omega\sin(0.5\omega)\right]. \tag{3.52}$$

By (3.42), we can calculate the range fo k_p for stability

$$-4.8\sqrt{\alpha_1^2 + 0.0017} < k_p < -\frac{1}{5}. \tag{3.53}$$

Following all the steps of the previous procedure, we get the stabilizing set in Fig. 3.2.

3.3 PID Controller Stabilizing Sets for Ziegler–Nichols Plants

Let us consider a linear time-invariant plant

$$P(s) := \frac{k}{1+Ts}e^{-Ls}, \tag{3.54}$$

where k is the steady-state gain, L is the time-delay, and T is the time constant. Consider a PID controller of the form

$$C(s) = \frac{k_d s^2 + k_p s + k_i}{s}, \tag{3.55}$$

where k_p is the proportional gain, k_i is the integral gain, and k_d is the derivative gain. Our objective is to analytically determine the set of controller parameters (k_p, k_i, k_d) for which the closed-loop system is stable.

We first analyze the system without the time-delay, i.e., $L = 0$. In this case, the closed-loop characteristic equation of the system is given by

$$\delta(s) = (T + kk_d)s^2 + (1 + kk_p)s + kk_i. \tag{3.56}$$

Since this is a second-order polynomial, closed-loop stability is equivalent to all the coefficients having the same sign. Assuming that the steady-state gain k of the plant is positive, these conditions are

$$k_p > -\frac{1}{k}, \qquad k_i > 0 \qquad \text{and} \qquad k_d > -\frac{T}{k} \tag{3.57}$$

or

$$k_p < -\frac{1}{k}, \qquad k_i < 0 \qquad \text{and} \qquad k_d < -\frac{T}{k}. \tag{3.58}$$

A minimal requirement for any control design is that the delay-free closed-loop system be stable. Consequently, it will be henceforth assumed in this section that the PID gains used to stabilize the plant with delay always satisfy one of the conditions (3.57) or (3.58).

Next, consider the case where the time-delay of the plant model is different from zero. The closed-loop characteristic equation of the system is then

$$\delta(s) = (kk_i + kk_p s + kk_d s^2)e^{-Ls} + (1 + Ts)s. \tag{3.59}$$

In the following subsections, we summarize the procedure to compute the complete PID stabilizing set considering the two different cases: open-loop stable Ziegler–Nichols plants and open-loop unstable Ziegler–Nichols plants.

3.3.1 Open-Loop Stable Ziegler–Nichols Plants

For open-loop stable Ziegler–Nichols plants, we have $T > 0$, $k > 0$, and $L > 0$. The procedure to compute the PID stabilizing set is the following:

1. For $L = 0$, the characteristic equation is

$$\delta(s) = (T + kk_d)s^2 + (kk_p + 1)s + kk_i. \tag{3.60}$$

   For stability, it is required that

$$k_p > -\frac{1}{k}, \quad k_i > 0 \ \ , k_d > -\frac{T}{k}. \tag{3.61}$$

2. For $L > 0$, the characteristic equation is

$$\delta(s) = (kk_i + kk_p s + kk_d s^2)e^{-Ls} + (1 + Ts)s. \tag{3.62}$$

   Setting $\delta^*(s) = e^{Ls}\delta(s)$, we have

$$\delta^*(s) = (kk_i + kk_p s + kk_d s^2) + (1 + Ts)se^{Ls}. \tag{3.63}$$

3. Calculate

$$\delta^*(j\omega) = \delta_r(\omega) + j\delta_i(j\omega), \tag{3.64}$$

   where

$$\delta_r(\omega) = kk_i - kk_d\omega^2 - \omega\sin(L\omega) - T\omega^2\cos(L\omega) \tag{3.65}$$

$$\delta_i(\omega) = \omega\left[kk_p + \cos(L\omega) - T\omega\sin(L\omega)\right]. \tag{3.66}$$

4. Make a change in variable $z = L\omega$ and calculate the new real and imaginary parts of $\delta^*(z)$

$$\delta_r(z) = kk_i - \frac{kk_d}{L^2}z^2 - \frac{1}{L}z\sin(z) - \frac{T}{L^2}z^2cos(z) \tag{3.67}$$

$$\delta_i(z) = \frac{z}{L}\left[kk_p + \cos(z) - \frac{T}{L}z\sin(z)\right]. \tag{3.68}$$

5. Pick a value for $k_p = k_p^*$ in the range

$$-\frac{1}{k} < k_p < \frac{1}{k}\left[\frac{T}{L}\alpha_1 \sin(\alpha_1) - \cos(\alpha_1)\right], \tag{3.69}$$

where α_1 is the solution of

$$\tan(\alpha) = -\frac{T}{T+L}\alpha \tag{3.70}$$

in the interval $(0, \pi)$.

6. Find the roots z_1 and z_2 of $\delta_i(z)$ from

$$\left[kk_p^* + \cos(z) - \frac{T}{L}z\sin(z)\right] = 0. \tag{3.71}$$

The roots can be found graphically for the following cases:

- $-\frac{1}{k} < k_p^* < \frac{1}{k}$: in this case, find the intersection of the functions $\frac{kk_p^*+\cos(z)}{\sin(z)}$ and $\frac{T}{L}z$.
- $k_p^* = \frac{1}{k}$: in this case, find the intersection of the functions $kk_p + cos(z)$ and $\frac{T}{L}z\sin(z)$.
- $\frac{1}{k} < k_p^* < \frac{T}{kL}\left[\frac{T}{L}\alpha_1\sin(\alpha_1) - \cos(\alpha_1)\right]$: in this case find the intersection of the functions $\frac{kk_p^*+\cos(z)}{\sin(z)}$ and $\frac{T}{L}z$.

7. Compute the parameters $m_j(z_j)$ and $b_j(z_j)$ for $j = 1, 2$ where

$$m(z) = \frac{L^2}{z^2} \tag{3.72}$$

$$b(z) = -\frac{L}{kz}\left[\sin(z) + \frac{T}{L}z\cos(z)\right]. \tag{3.73}$$

8. Calculate the (k_i, k_d) stabilizing set using Fig. 3.3.
9. Go to step 5 and update k_p^* within the range in (3.69) and repeat the steps 6 through 9 until the range in (3.69) is exhausted.

The following example illustrates this procedure.

Example 3.3 Consider an open-loop stable continuous-time Ziegler–Nichols plant

$$P(s) = \frac{1}{2s+1}e^{-2s}, \tag{3.74}$$

and the PID controller (3.55). The characteristic equation is given by

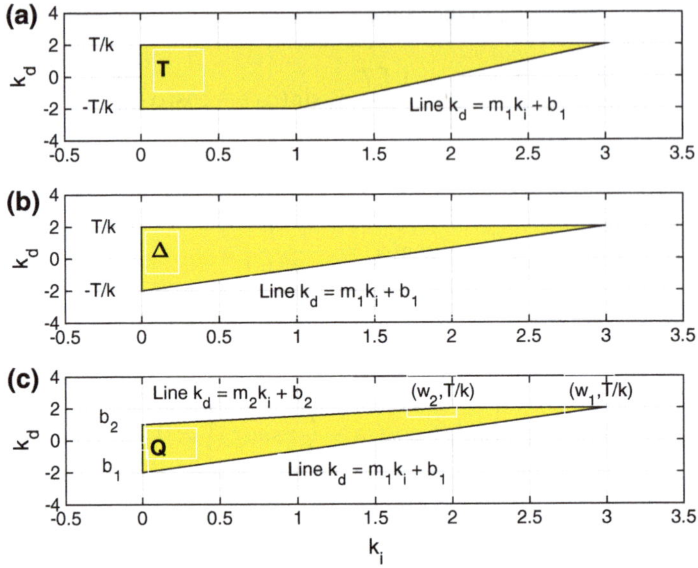

Fig. 3.3 Stabilizing region of (k_i, k_d) for: **a** $-\frac{1}{k} < k_p < \frac{1}{k}$, **b** $k_p = \frac{1}{k}$, **c** $\frac{1}{k} < k_p < \frac{T}{kL}\left[\frac{T}{L}\alpha_1 \sin(\alpha_1) - \cos(\alpha_1)\right]$. © 2002 IEEE. Reproduced from [9] with permission

$$\delta(s) = (2s + 1)s + (k_d s^2 + k_p s + k_i)e^{-2s}, \tag{3.75}$$

and by (3.63)

$$\delta^*(s) = e^{2s}(2s + 1)s + (k_d s^2 + k_p s + k_i). \tag{3.76}$$

For $L = 0$, we have

$$\delta(s) = (k_d + 2)s^2 + (k_p + 1)s + k_i. \tag{3.77}$$

For stability, it is required

$$k_p > -1, \quad k_i > 0, \quad k_d > -2. \tag{3.78}$$

For $L > 0$,

$$\delta^*(j\omega) = \delta_r(\omega) + j\delta_i(j\omega), \tag{3.79}$$

where

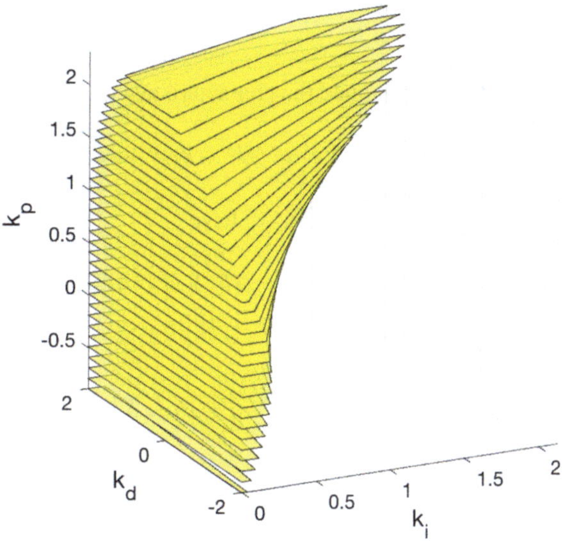

Fig. 3.4 Stabilizing Set for an open-loop stable Ziegler–Nichols plant using a PID controller design in Example 3.3

$$\delta_r(\omega) = k_i - k_d\omega^2 - \omega\sin(2\omega) - 2\omega^2\cos(2\omega) \tag{3.80}$$

$$\delta_i(\omega) = \omega\left[k_p + \cos(2\omega) - 2\omega\sin(2\omega)\right]. \tag{3.81}$$

By (3.69), we can calculate the range of k_p for stability

$$-1 < k_p < [\alpha_1\sin(\alpha_1) - \cos(\alpha_1)]. \tag{3.82}$$

Following all the steps of the previous procedure, we get the stabilizing set shown in Fig. 3.4.

3.3.2 *Open-Loop Unstable Ziegler–Nichols Plants*

We consider here the PID stabilization of an open-loop unstable Ziegler–Nichols plant. In this case, $T < 0$ in (3.54). Furthermore, let us assume that $k > 0$ and $L > 0$. Then it can be shown that $\left|\frac{T}{L}\right| > 0.5$ is necessary for the existence of a stabilizing controller. The procedure to compute the PID stabilizing set when $\left|\frac{T}{L}\right| > 0.5$ is the following:

1. For $L = 0$, calculate the characteristic equation

$$\delta(s) = (T + kk_d)s^2 + (kk_p + 1)s + kk_i. \tag{3.83}$$

For stability, it is required

$$k_p < -\frac{1}{k}, \quad k_i < 0, \quad k_d < -\frac{T}{k}. \tag{3.84}$$

2. For $L > 0$, calculate the characteristic equation

$$\delta(s) = (kk_i + kk_p s + kk_d s^2)e^{-Ls} + (1 + Ts)s. \tag{3.85}$$

Setting $\delta^*(s) = e^{Ls}\delta(s)$, we have

$$\delta^*(s) = (kk_i + kk_p s + kk_d s^2) + (1 + Ts)se^{Ls}. \tag{3.86}$$

3. Calculate

$$\delta^*(j\omega) = \delta_r(\omega) + j\delta_i(j\omega), \tag{3.87}$$

where

$$\delta_r(\omega) = kk_i - kk_d\omega^2 - \omega\sin(L\omega) - T\omega^2\cos(L\omega) \tag{3.88}$$
$$\delta_i(\omega) = \omega\left[kk_p + \cos(L\omega) - T\omega\sin(L\omega)\right]. \tag{3.89}$$

4. Make a change in variable $z = L\omega$ and calculate the new real and imaginary parts of $\delta^*(z)$

$$\delta_r(z) = kk_i - \frac{kk_d}{L^2}z^2 - \frac{1}{L}z\sin(z) - \frac{T}{L^2}z^2 cos(z) \tag{3.90}$$
$$\delta_i(z) = \frac{z}{L}\left[kk_p + \cos(z) - \frac{T}{L}z\sin(z)\right]. \tag{3.91}$$

5. For $\left|\frac{T}{L}\right| > 0.5$, pick a value for $k_p = k_p^*$ in the range

$$\frac{1}{k}\left[\frac{T}{L}\alpha_1\sin(\alpha_1) - \cos(\alpha_1)\right] < k_p < -\frac{1}{k}, \tag{3.92}$$

where α_1 is the solution of

$$\tan(\alpha) = -\frac{T}{T+L}\alpha \tag{3.93}$$

in the $(0, \pi)$ interval. For the case $\left|\frac{T}{L}\right| = 1$, $\alpha_1 = \frac{\pi}{2}$.

6. Find the real roots z_1 and z_2 of

$$\left[kk_p^* + \cos(z) - \frac{T}{L}z\sin(z)\right] = 0. \tag{3.94}$$

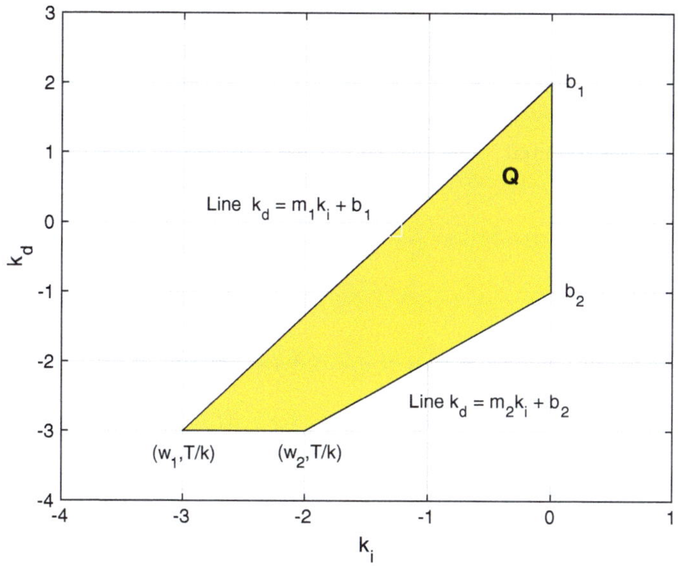

Fig. 3.5 Stabilizing region of (k_i, k_d) for: $\frac{1}{k}\left[\frac{T}{L}\alpha_1\sin(\alpha_1)-\cos(\alpha_1)\right]<k_p<-\frac{1}{k}$. w_1 and w_2 are defined by the intersections of the lines $k_d=\frac{T}{k}$ and $k_d=m_1k_i+b_1$ and $k_d=m_2k_i+b_2$, respectively. © 2002 IEEE. Reproduced from [9] with permission

The roots can be found graphically for the following case:

- $\frac{T}{kL}\left[\frac{T}{L}\alpha_1\sin(\alpha_1)-\cos(\alpha_1)\right]<k_p^*<-\frac{1}{k}$: in this case, determine the intersection of the functions $\frac{kk_p^*+\cos(z)}{\sin(z)}$ and $\frac{T}{L}z$.

7. Compute the parameters $m_j := m(z)|_{z=z_j}$ and $b_j := b(z)|_{z=z_j}$ for $j = 1,\ 2$, where

$$m(z)=\frac{L^2}{z^2} \tag{3.95}$$

$$b(z)=-\frac{L}{kz}\left[\sin(z)+\frac{T}{L}z\cos(z)\right]. \tag{3.96}$$

8. Calculate the $(k_i,\ k_d)$ stabilizing set as described in Fig. 3.5.
9. Go to step 5 and update k_p^* within the range in (3.92) and repeat the steps 6 through 9 until the range in (3.92) is exhausted.

We illustrate the previous procedure in the following example.

Example 3.4 Consider an open-loop unstable Ziegler–Nichols plant

$$P(s)=\frac{2}{-3s+1}e^{-0.5s} \tag{3.97}$$

and the PID controller (3.55). The characteristic equation, for $L = 0$ is given by

$$\delta(s) = (-3 + 2k_d)s^2 + (2k_p + 1)s + 2k_i. \tag{3.98}$$

For stability, it is required

$$k_p < -\frac{1}{2}, \quad k_i < 0, \quad k_d < \frac{3}{2}. \tag{3.99}$$

For $L > 0$

$$\delta^*(j\omega) = \delta_r(\omega) + j\delta_i(j\omega), \tag{3.100}$$

where

$$\delta_r(\omega) = 2k_i - 2k_d\omega^2 - \omega\sin(0.5\omega) + 3\omega^2\cos(0.5\omega), \tag{3.101}$$

$$\delta_i(\omega) = \omega\left[2k_p + \cos(0.5\omega) + 3\omega\sin(0.5\omega)\right]. \tag{3.102}$$

By (3.92), we can calculate the range of k_p for stability

$$\frac{1}{2}\left[-\frac{3}{2}\alpha_1\sin(\alpha_1) - \cos(\alpha_1)\right] < k_p < -\frac{1}{2}. \tag{3.103}$$

Following all the steps of the previous procedure, we get the stabilizing set in Fig. 3.6.

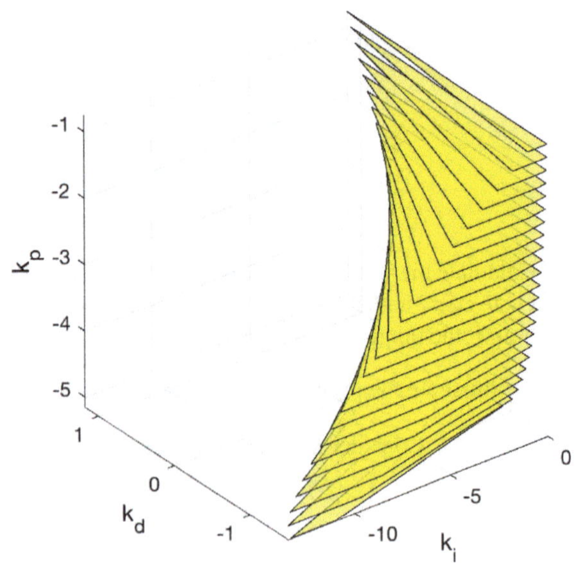

Fig. 3.6 Stabilizing Set for an open-loop unstable Ziegler–Nichols plant using a PID controller design in Example 3.4

3.4 Notes and References

The complete description of Pontryagin's results can be found in his original paper [7]. For further results on quasipolynomials, the reader is referred to the book by Bellman and Cooke [1]. As an application of Pontryagin's results, the stability of time-delay systems has been an active area of research and early results can be found in [4–6].

The characterization of all stabilizing PI and PID controllers for a first-order plant with time delay was initially developed by Silva, Datta, and Bhattacharyya [8, 9]. The constant gain and pure integral controllers were reported in the book by Datta, Ho, and Bhattacharyya [3]. The mathematical derivation of all stabilizing PI and PID controllers for a given Ziegler–Nichols plant can be found in [2] and [10] where complete derivations were developed with detailed proofs followed by numerical examples. An extension of these results to PI and PID controller synthesis with prescribed gain and phase margins is discussed in Chap. 6 of this book.

References

1. Bellman, R.E., Cooke, K.L.: Differential-Difference Equations
2. Bhattacharyya, S.P., Datta, A., Keel, L.H.: Linear Control Theory Structure, Robustness, and Optimization. CRC Press Taylor and Francis Group, Boca Raton (2009)
3. Datta, A., Ho, M.-T., Bhattacharyya, S.P.: Structure and Synthesis of PID Controllers. Springer, Berlin (2000)
4. Karmarkar, J., Siljak, D.: Stability analysis of systems with time delay. In: Proceedings of the Institution of Electrical Engineers (IET), vol. 117, pp. 1421–1424 (1970)
5. Malek-Zavarei, M., Jamshidi, M.: Time-delay Systems: Analysis, Optimization and Applications. Elsevier Science Inc., Amsterdam (1987)
6. Marshall, J.E.: Control of Time-Delay Systems. IEE Control Engineering Series, vol. 10. The Institution of Electrical Engineers, London, U.K. and Peter Peregrinus Ltd. Stevenage, U.K. (1979)
7. Pontryagin, L.S.: On the zeros of some elementary transcendental functions. Amer. Math. Soc. Transl **2**(1), 95–110 (1955)
8. Silva, G.J., Datta, A., Bhattacharyya, S.P.: PI stabilization of first-order systems with time delay. Automatica, 2025–2031 (2001)
9. Silva, G.J., Datta, A., Bhattacharyya, S.P.: New results on the synthesis of PID controllers. IEEE Trans. Autom. Control **47**, 241–252 (2002)
10. Silva, G.J., Datta, A., Bhattacharyya, S.P.: PID Controllers for Time Delay Systems. Birkhuser, Basel (2004)

Chapter 4
Stabilizing Sets for Linear Time-Invariant Discrete-Time Plants

Abstract In this chapter, we consider the synthesis and design of digital PID controllers which can provide tracking and disturbance rejection of constant reference and disturbance discrete-time signals. We develop procedures and algorithms to determine the set of PID controllers that stabilize a given linear time-invariant discrete-time plant. This algorithm determines the three parameter stabilizing set via a linear programming problem in two variables with a single sweeping parameter.

4.1 Introduction

Digital control is ubiquitous in motion control, process control, robotics, and myriad other applications. Many, if not most, servomechanisms employ digital control. In these systems, signals are processed in discrete time with a prescribed sampling period. The most common problem is to make the plant output asymptotically track a discrete-time step input in the presence of a discrete-time constant disturbance. This can be accomplished by a digital PI or PID controller, which contains digital integral action and which stabilizes the discrete-time plant in question. This stabilization problem is solved in this chapter.

The main results are developed based on some preliminary mathematics related to root counting. First, we develop the Tchebyshev representation of the image of a polynomial or rational function over a circle in the complex plane. Then, a Hermite–Biehler-like theorem is developed for Schur stability, that is, for the roots of a polynomial to lie within the unit disk. This is followed by signature formulas that can compute the number of roots within the unit circle from an evaluation of the polynomial or rational function over the unit circle. These results are then combined to develop our algorithm for determining the PID stabilizing set. The computation

Sections 4.1, 4.2, 4.3, 4.4, and 4.8 are reproduced from S. P. Bhattacharyya, A. Datta, L. H. Keel, *Linear System Theory: Structure, Robustness, and Optimization. Taylor & Francis LLC Books*, with permission © 2008 Taylor & Francis LLC Books.

I. D. Díaz-Rodríguez et al., *Analytical Design of PID Controllers*,
https://doi.org/10.1007/978-3-030-18228-1_4

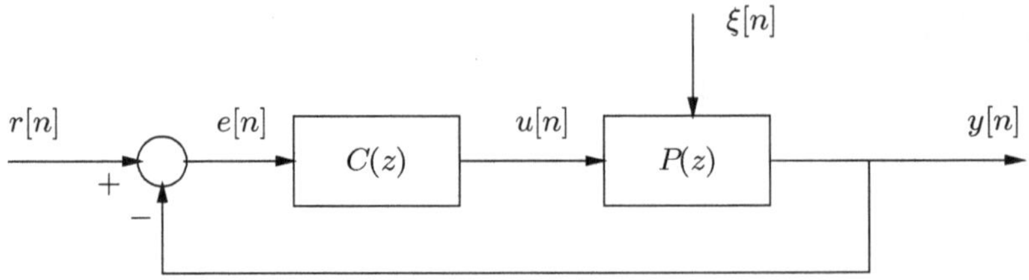

Fig. 4.1 A discrete-time feedback control system. © Taylor & Francis LLC Books. Reproduced from [1] with permission

amounts to a linear programming problem in two variables with a sweeping parameter. Examples of this computation and some illustrative design applications are provided.

Consider the sampled data or discrete-time feedback control system shown in Fig. 4.1. When the reference and disturbance signals are arbitrary discrete-time step inputs, it is possible to show that the tracking error $e[n]$ converges asymptotically to zero provided the digital controller $C(z)$ includes (a) a discrete-time integrator and (b) the closed-loop is stable. Condition (a) is achieved by including $z = 1$ as a pole of $C(z)$. The condition (b) reduces to the requirement that the closed-loop characteristic polynomial be Schur stable, that is, have all its roots within the unit disk. We rule out plants whose control transfer functions have a zero at $z = 1$, since such plants cannot be stabilized by digital PID controllers. The set of controller gains rendering the closed-loop characteristic polynomial Schur stable is called the *stabilizing set*. Our objective is to develop constructive methods to compute this set. With the stabilizing set in hand, one can search for subsets achieving various design objectives, as shown in later chapters.

We first determine the complex plane image of a real polynomial or rational function over a circle of radius ρ in the complex plane centered at the origin. It is determined and expressed in terms of Tchebyshev polynomials of the first and second kinds. A formula is developed for root counting with respect to circular regions in terms of this Tchebyshev representation. This formula is a Hermite–Biehler-type result for Schur stability. Using these results, we show how the PID controller can be reparametrized so that the stabilizing set is obtained as the solution of sets of linear inequalities in two variables for a fixed value of the third variable. By sweeping or gridding over the third variable, the complete stabilizing set can be determined constructively. The solution shows that the stabilizing set for any Discrete-time Linear Time-invariant (DTLTI) plant, when it is nonempty, consists of unions of convex polygons in the space of the PID gains.

Using the above computation, we further solve two design problems. The first problem is related to deadbeat control, wherein one places all closed-loop characteristic roots at the origin so that the transients are zeroed out in a finite number of steps. In general, deadbeat control is not possible using PID controllers and a reasonable goal is to place the closed-loop characteristic roots as close to the origin as possible so that the transient error decays quickly. Such designs have been advocated in the literature on sampled data control systems. We show how the stabilization solution obtained by us can be exploited to give a constructive determination of such "maximally" deadbeat designs. The second problem involves the determination of the maximum delay in the loop that a given plant under PID control can be made to tolerate. We show how our solution can also be extended to determine this maximum delay for a given DTLTI plant.

4.2 Preliminaries

We consider a discrete-time control system consisting of a Single-input Single-output (SISO) plant described by its z-domain transfer function $P(z)$ and the unity feedback controller $C(z)$ in Fig. 4.2.

The transfer functions shown in Fig. 4.2 are rational functions and we write

$$P(z) = \frac{N(z)}{D(z)}, \qquad C(z) = \frac{N_C(z)}{D_C(z)}.$$

The *characteristic polynomial* of the closed-loop system is

$$\delta(z) := D_C(z)D(z) + N_C(z)N(z)$$

and a necessary and sufficient condition for stability of the closed-loop control system is that the characteristic roots, namely the zeros of $\delta(z)$, have magnitude less than unity. This condition is commonly referred to as *Schur stability* of $\delta(z)$.

The stabilization problem can be stated as the problem of determining $C(z)$ so that for the given $P(z)$, the closed-loop characteristic polynomial $\delta(z)$ is Schur. For a fixed structure controller, such as a PID controller, $C(z)$ is characterized by a set of gains $\mathbf{K}$ (three gains in case of PID) and these gains must be chosen to stabilize $\delta(z)$ if possible. An important problem in multiobjective design is the determination of the entire set $\mathcal{S}$ of stabilizing gains in a constructive way. A useful characterization of $\mathcal{S}$ should allow the designer to test the feasibility of imposing various performance constraints and checking their attainability with the controller parameters ranging over the stabilizing set.

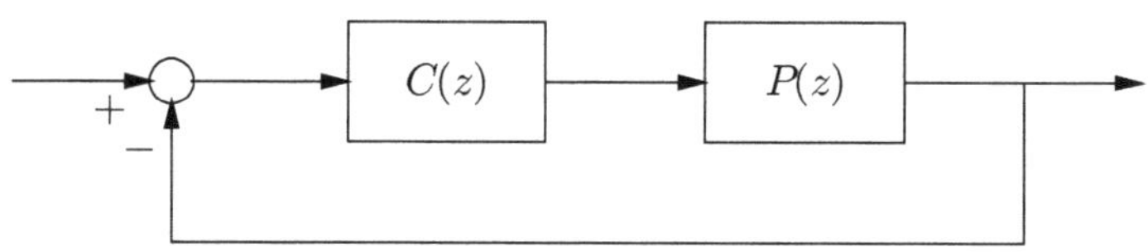

Fig. 4.2 A unity feedback system

4.3 Tchebyshev Representation and Root Clustering

The stabilization results require us to determine the complex plane image of polynomials and rational functions on a circle of radius ρ centered at the origin. The image is used for certain root counting formulas which we need to develop. These formulas, in turn, depend on the Tchebyshev representation discussed here.

4.3.1 Tchebyshev Representation of Real Polynomials

Let us consider a polynomial $P(z) = a_n z^n + \cdots + a_0$ with real coefficients. The image of $P(z)$ evaluated on the circle $\mathcal{C}_\rho$ of radius ρ, centered at the origin is:

$$\left\{ P(z) \ : \ z = \rho e^{j\theta}, \quad 0 \leq \theta \leq 2\pi \right\}. \tag{4.1}$$

Since the a_i's are real for all $i \in 0, \ldots, n$ $P\left(\rho e^{j\theta}\right)$ and $P\left(\rho e^{-j\theta}\right)$ are complex conjugate. Thus, it suffices to determine the image of the upper half of the circle:

$$\left\{ P(z) \ : \ z = \rho e^{j\theta}, \quad 0 \leq \theta \leq \pi \right\}. \tag{4.2}$$

Since

$$z^k \Big|_{z=\rho e^{j\theta}} = \rho^k (\cos k\theta + j \sin k\theta), \tag{4.3}$$

we have

$$\begin{aligned} & P\left(\rho e^{j\theta}\right) \\ &= \underbrace{\left(a_n \rho^n \cos n\theta + \cdots + a_1 \rho \cos\theta + a_0\right)}_{\bar{R}(\rho,\theta)} + j \underbrace{\left(a_n \rho^n \sin n\theta + \cdots + a_1 \rho \sin\theta\right)}_{\bar{I}(\rho,\theta)} \\ &= \bar{R}(\rho, \theta) + j \bar{I}(\rho, \theta). \end{aligned} \tag{4.4}$$

It is well known that $\cos k\theta$, and $\frac{\sin k\theta}{\sin\theta}$ can be written as polynomials in $\cos\theta$ using Tchebyshev polynomials. Write $u := -\cos\theta$. Then as θ runs from $0 \rightarrow \pi$, u runs from -1 to $+1$. Now

$$e^{j\theta} = \cos\theta + j \sin\theta = -u + j\sqrt{1 - u^2} \tag{4.5}$$

and we have

$$c_k(u) := \cos k\theta \quad \text{and} \quad s_k(u) := \frac{\sin k\theta}{\sin\theta} \tag{4.6}$$

where $c_k(u)$ and $s_k(u)$ are real polynomials in u and are known as the Tchebyshev polynomials of the first and second kind, respectively. It is easy to show that

Table 4.1 Tchebyshev polynomials of the first and second Kind. © Taylor & Francis LLC Books. Reproduced from [1] with permission

k	$c_k(u,\rho)$	$s_k(u,\rho)$
1	$-\rho u$	ρ
2	$\rho^2\left(2u^2-1\right)$	$-2\rho^2 u$
3	$\rho^3\left(-4u^3+3u\right)$	$\rho^3\left(4u^2-1\right)$
4	$\rho^4\left(8u^4-8u^2+1\right)$	$\rho^4\left(-8u^3+4u\right)$
5	$\rho^5\left(-16u^5+20u^3-5u\right)$	$\rho^5\left(16u^4-12u^2+1\right)$
$\vdots$	$\vdots$	$\vdots$

$$s_k(u) = -\frac{c_k'(u)}{k}, \quad k = 1, 2, \ldots \tag{4.7}$$

and that the Tchebyshev polynomials satisfy the recursive relation:

$$c_{k+1}(u) = -uc_k(u) - \left(1-u^2\right)s_k(u), \quad k = 1, 2, \ldots \tag{4.8}$$

From (4.6), (4.7), and (4.8), we can determine $c_k(u)$ and $s_k(u)$ for $k = 1, 2, 3, \ldots$.

Now

$$(\rho e^{j\theta})^k = \rho^k \cos k\theta + j\rho^k \sin k\theta \tag{4.9}$$

and so we define the generalized Tchebyshev polynomials as follows:

$$c_k(u,\rho) := \rho^k c_k(u), \quad s_k(u,\rho) := \rho^k s_k(u), \quad k = 0, 1, 2\ldots \tag{4.10}$$

and note that

$$s_k(u,\rho) = -\frac{1}{k}\cdot\frac{d\left[c_k(u,\rho)\right]}{du}, \qquad k = 1, 2, \ldots \tag{4.11}$$

$$c_{k+1}(u,\rho) = -\rho u c_k(u,\rho) - \left(1-u^2\right)\rho s_k(u,\rho), \qquad k = 1, 2, \ldots \tag{4.12}$$

The generalized Tchebyshev polynomials are displayed in Table 4.1 for $k = 1, \ldots, 5$.

Using the notation in (4.10),

$$P\left(\rho e^{j\theta}\right) = R(u,\rho) + j\sqrt{1-u^2}T(u,\rho) =: P_c(u,\rho), \tag{4.13}$$

where

$$R(u,\rho) = a_n c_n(u,\rho) + a_{n-1}c_{n-1}(u,\rho) + \cdots + a_1 c_1(u,\rho) + a_0, \tag{4.14}$$

$$T(u,\rho) = a_n s_n(u,\rho) + a_{n-1}s_{n-1}(u,\rho) + \cdots + a_1 s_1(u,\rho). \tag{4.15}$$

$R(u, \rho)$ and $T(u, \rho)$ are polynomials in u and ρ. The complex plane image of $P(z)$ as z traverses the upper half of the circle $\mathcal{C}_\rho$ can be obtained by evaluating $P_c(u, \rho)$ as u runs from -1 to $+1$.

Lemma 4.1 *For a fixed* $\rho > 0$,

(a) if $P(z)$ *has no roots on the circle of radius* $\rho > 0$, $(R(u, \rho), T(u, \rho))$ *have no common roots for* $u \in [-1, 1]$ *and* $R(\pm 1, \rho) \neq 0$.
(b) if $P(z)$ *has* $2m$ *roots at* $z = -\rho\,(z = +\rho)$, *then* $R(u, \rho)$ *and* $T(u, \rho)$ *have* m *roots each at* $u = +1$ $(u = -1)$.
(c) if $P(z)$ *has* $2m - 1$ *roots at* $z = -\rho\,(z = +\rho)$, *then* $R(u, \rho)$ *and* $T(u, \rho)$ *have* m *and* $m - 1$ *roots, respectively at* $u = +1$ $(u = -1)$.
(d) if $P(z)$ *has* q_i *pairs of complex conjugate roots at* $z = -\rho u_i \pm j\rho\sqrt{1 - u_i^2}$, *for* $u_i \neq \pm 1$, *then* $R(u, \rho)$ *and* $T(u, \rho)$ *each have* q_i *real roots at* $u = u_i$.

Proof To prove (a) note that $P(\rho e^{j\theta}) \neq 0$ for $\theta \in [0, \pi]$ and therefore $P_c(u, \rho) \neq 0$ for $u \in [-1, +1]$; hence, the result. The statements in (b)–(d) above may be verified by direct calculation. □

When the circle of interest is the unit circle, that is $\rho = 1$, we will write $P_c(u, 1) = P_c(u)$ and also

$$R(u, 1) =: R(u), \qquad T(u, 1) =: T(u)$$

for notational simplicity.

4.3.2 *Interlacing Conditions for Root Clustering and Schur Stability*

The formulas of the last section can be used to derive conditions for root clustering in circular regions, that is for the roots to lie strictly within a circle of radius ρ. For Schur stability, we simply take $\rho = 1$. As before, let $P(z)$ be a real polynomial of degree n and

$$\begin{aligned} P\left(\rho e^{j\theta}\right) &= \bar{R}(\theta, \rho) + j\bar{I}(\theta, \rho), \qquad \text{where } u = -\cos\theta \\ &= R(u, \rho) + j\sqrt{1 - u^2}\,T(u, \rho) \end{aligned} \tag{4.16}$$

where $R(u, \rho)$ and $T(u, \rho)$ are real polynomials of degree n and $n - 1$, respectively, in u, for fixed ρ.

Theorem 4.1 $P(z)$ *has all its zeros strictly within* $\mathcal{C}_\rho$ *if and only if*

(a) $R(u, \rho)$ *has* n *real distinct zeros* r_i, $i = 1, 2, \ldots, n$ *in* $(-1, 1)$.
(b) $T(u, \rho)$ *has* $n - 1$ *real distinct zeros* t_j, $j = 1, 2, \ldots, n - 1$ *in* $(-1, 1)$.

(c) The zeros r_i and t_j interlace:

$$-1 < r_1 < t_1 < r_2 < t_2 < \cdots < t_{n-1} < r_n < +1.$$

Proof Let

$$t_j = -\cos\alpha_j, \qquad \alpha_j \in (0, \pi), \qquad j = 1, 2, \ldots, n-1$$

or

$$\begin{aligned} \alpha_j &= -\cos^{-1} t_j, \qquad j = 1, 2, \ldots, n-1 \\ \alpha_0 &= 0, \\ \alpha_n &= \pi \end{aligned}$$

and let

$$\beta_i = -\cos^{-1} r_i, \qquad i = 1, 2, \ldots, n, \qquad \beta_i \in (0, \pi).$$

Then, $(\alpha_0, \alpha_1, \ldots, \alpha_n)$ are the $n+1$ zeros of $\bar{I}(\theta, \rho) = 0$ and $(\beta_1, \beta_2, \ldots, \beta_{n-1})$ are the n zeros of $\bar{R}(\theta, \rho) = 0$. The condition (c) means that α_i and β_j satisfy:

$$0 = \alpha_0 < \beta_1 < \alpha_1 < \beta_2 < \cdots < \beta_{n-1} < \alpha_n = \pi. \qquad (4.17)$$

The conditions (a)–(c) imply that the plot of $P\left(\rho e^{j\theta}\right)$ for $\theta \in [0, \pi]$ turns counter-clockwise through exactly $2n$ quadrants, and this condition is equivalent to $P(z)$ having n zeros inside the circle $\mathcal{C}_\rho$. □

Remark 4.1 The conditions (a), (b), and (c) given in Theorem 4.1 may be referred to as Interlacing Conditions on $R(u, \rho)$ and $T(u, \rho)$. By setting $\rho = 1$, we obtain conditions for Schur stability in terms of interlacing of the zeros of $R(u)$ and $T(u)$. This constitutes a Hermite–Biehler-like theorem for Schur stability.

4.3.3 Tchebyshev Representation of Rational Functions

Let $Q(z)$ be a ratio of two real polynomials $P_1(z)$ and $P_2(z)$. We compute the image of $Q(z)$ on $\mathcal{C}_\rho$ and write it as the corresponding Tchebyshev representation $Q_c(u, \rho)$. Let

$$P_i(z)|_{z=-\rho u + j\rho\sqrt{1-u^2}} = R_i(u, \rho) + j\sqrt{1-u^2}\,T_i(u, \rho), \qquad \text{for } i = 1, 2. \qquad (4.18)$$

Then,

$$
\begin{aligned}
&Q(z)|_{z=-\rho u+j\rho\sqrt{1-u^2}} \\
&= \left.\frac{P_1(z)}{P_2(z)}\right|_{z=-\rho u+j\rho\sqrt{1-u^2}} \\
&= \left.\frac{P_1(z)P_2\left(z^{-1}\right)}{P_2(z)P_2\left(z^{-1}\right)}\right|_{z=-\rho u+j\rho\sqrt{1-u^2}} \\
&= \frac{\left(R_1(u,\rho)+j\sqrt{1-u^2}T_1(u,\rho)\right)\left(R_2(u,\rho)-j\sqrt{1-u^2}T_2(u,\rho)\right)}{\left(R_2(u,\rho)+j\sqrt{1-u^2}T_2(u,\rho)\right)\left(R_2(u,\rho)-j\sqrt{1-u^2}T_2(u,\rho)\right)} \\
&= \underbrace{\left(\frac{R_1(u,\rho)R_2(u,\rho)+\left(1-u^2\right)T_1(u,\rho)T_2(u,\rho)}{R_2^2(u,\rho)+\left(1-u^2\right)T_2^2(u,\rho)}\right)}_{=:R(u,\rho)} \\
&\qquad + j\sqrt{1-u^2}\underbrace{\left(\frac{T_1(u,\rho)R_2(u,\rho)-R_1(u,\rho)T_2(u,\rho)}{R_2^2(u,\rho)+\left(1-u^2\right)T_2^2(u,\rho)}\right)}_{=:T(u,\rho)}
\end{aligned}
\tag{4.19}
$$

and we denote

$$
Q_c(u,\rho) := R(u,\rho) + j\sqrt{1-u^2}T(u,\rho). \tag{4.20}
$$

Note that, $R(u,\rho)$, $T(u,\rho)$ are rational functions of the real variable u which runs from -1 to $+1$. This representation will be needed in a later section on the solution of the maximally deadbeat problem.

4.4 Root Counting Formulas

In this section, we develop some formulas for counting the root distribution with respect to the circle $\mathcal{C}_\rho$, for real polynomials and real rational functions. These formulas will be necessary for our solution of the stabilization problem but are also of independent interest. We begin by relating root distribution to phase unwrapping.

4.4.1 Phase Unwrapping and Root Distribution

Let $\phi_P(\theta) := \angle P(\rho e^{j\theta})$ denote the *phase* of the polynomial $P(z)$ evaluated at $z = \rho e^{j\theta}$ and let $\Delta_{\theta_1}^{\theta_2}[\phi_P(\theta)]$ denote the net change in or *unwrapped phase* of $P(\rho e^{j\theta})$ as θ increases from θ_1 to θ_2. Similar notation applies to the rational function $Q(z)$ with

Tchebyshev representation $Q_C(u, \rho)$. Let $\phi_{Q_C}(u) = \angle Q_C(u, \rho)$ denote the phase of $Q_C(u, \rho)$ and $\Delta_{u_1}^{u_2}\left[\phi_{Q_c}(u)\right]$ the net change in or unwrapped phase of $Q_C(u, \rho)$ as u increases from u_1 to u_2.

Lemma 4.2 *Let the real polynomial $P(z)$ have i roots in the interior of the circle $\mathcal{C}_\rho$ and no roots on the circle. Then,*

$$\Delta_0^{\pi}[\phi_P(\theta)] = \pi i.$$

Proof From geometric considerations, it is easily seen that each interior root contributes 2π to $\Delta_0^{2\pi}[\phi_P(\theta)]$ and therefore by the symmetry of roots about the real axis, the interior roots contribute πi to $\Delta_0^{\pi}[\phi_P(\theta)]$. □

We state the corresponding result for a rational function. The proof is similar to the previous lemma and is omitted.

Lemma 4.3 *Let $Q(z) = \frac{P_1(z)}{P_2(z)}$ where the real polynomials $P_1(z)$ and $P_2(z)$ have i_1 and i_2 roots, respectively, in the interior of the circle $\mathcal{C}_\rho$ and no roots on the circle. Then*

$$\Delta_0^{\pi}[\phi_Q(\theta)] = \pi\,(i_1 - i_2) = \Delta_{-1}^{+1}[\phi_{Q_C}(u)].$$

4.4.2 Root Counting and Tchebyshev Representation

In this section, we first develop formulas to determine the number of roots of a real polynomial, inside a circle $\mathcal{C}_\rho$, to its Tchebyshev representation. Let us begin with a real polynomial $P(z)$ and its Tchebyshev representation

$$P_C(u, \rho) = R(u, \rho) + j\sqrt{1 - u^2}\,T(u, \rho)$$

as developed before. Henceforth, let $t_1, \ldots, t_k$ denote the real distinct zeros of $T(u, \rho)$ of odd multiplicity, for $u \in (-1, 1)$, ordered as follows:

$$-1 < t_1 < t_2 < \cdots < t_k < +1.$$

Suppose also that $T(u, \rho)$ has p zeros at $u = -1$ and let $f^{(i)}(x_0)$ denote the ith derivative to $f(x)$ evaluated at $x = x_0$. Let us also define

$$\operatorname{sgn}[x] = \begin{cases} -1 & \text{if } x < 0, \\ 0 & \text{if } x = 0, \\ 1 & \text{if } x > 0. \end{cases}$$

Theorem 4.2 *Let $P(z)$ be a real polynomial with no roots on the circle $\mathcal{C}_\rho$ and suppose that $T(u, \rho)$ has p zeros at $u = -1$. Then, the number of roots i of $P(z)$ in the interior of the circle $\mathcal{C}_\rho$ is given by*

$$i = \frac{1}{2}\text{sgn}\left[T^{(p)}(-1,\rho)\right]\left(\text{sgn}\left[R(-1,\rho)\right] + 2\sum_{j=1}^{k}(-1)^j \text{sgn}\left[R\left(t_j,\rho\right)\right]\right.$$
$$\left. +(-1)^{k+1}\text{sgn}\left[R(+1,\rho)\right]\right). \qquad (4.21)$$

Proof Recall that

$$P(\rho e^{j\theta}) = \bar{R}(\theta,\rho) + j\bar{I}(\theta,\rho)$$

and define $\theta_i, i = 1, \ldots, k$ such that

$$t_i = -\cos\theta_i, \qquad \text{for } \theta_i \in [0,\pi],$$

where t_i's are the zeros of $T(u,\rho)$ of odd multiplicity, for $u \in (-1,1)$. Let $\theta_0 := 0$, $t_0 := -1$ and $\theta_{k+1} := \pi$, and note that the θ_i, $i = 0, 1 \ldots, k+1$ are zeros of $\bar{I}(\theta,\rho)$. The proof depends on the following elementary and easily verified facts which are first stated. (In the following, θ_i^+ denotes the point immediately to the right of θ_i).

(a)

$$\Delta_0^{\pi}[\phi(\theta)] = \pi i,$$

(b)

$$\Delta_0^{\pi}[\phi(\theta)] = \Delta_0^{\theta_1}[\phi(\theta)] + \Delta_{\theta_1}^{\theta_2}[\phi(\theta)] + \cdots + \Delta_{\theta_k}^{\pi}[\phi(\theta)],$$

(c)

$$\Delta_{\theta_i}^{\theta_{i+1}}[\phi(\theta)] = \frac{\pi}{2}\text{sgn}\left[\bar{I}\left(\theta_i^+,\rho\right)\right]\left(\text{sgn}\left[\bar{R}\left(\theta_i,\rho\right)\right) - \text{sgn}\left[\bar{R}\left(\theta_{i+1},\rho\right)\right]\right),$$
$$\text{for } i = 0, 1, \ldots, k,$$

(d)

$$\text{sgn}\left[\bar{I}\left(\theta_i^+,\rho\right)\right] = -\text{sgn}\left[\bar{I}\left(\theta_{i+1}^+,\rho\right)\right], \quad i = 0, 1, \ldots, k,$$

(e)

$$\text{sgn}\left[\bar{I}(0^+,\rho)\right] = \text{sgn}\left[T^{(p)}(-1,\rho)\right],$$

(f)

$$\text{sgn}\left[\bar{R}\left(\theta_i,\rho\right)\right] = \text{sgn}\left[R\left(t_i,\rho\right)\right], \quad i = 0, 1, \ldots, k.$$

Using (a)–(f), we have

$$
\begin{aligned}
\pi i &= \Delta_0^{\pi}[\phi(\theta)] \\
&= \Delta_0^{\theta_1}[\phi(\theta)] + \cdots + \Delta_{\theta_k}^{\pi}[\phi(\theta)], && \text{by (a) and (b)} \\
&= \frac{\pi}{2}\Big\{\operatorname{sgn}\left[\bar{I}(0^+,\rho)\right]\left(\operatorname{sgn}\left[\bar{R}(0,\rho)\right] - \operatorname{sgn}\left[\bar{R}(\theta_1,\rho)\right]\right) + \cdots \\
&\qquad \cdots + \operatorname{sgn}\left[\bar{I}\left(\theta_k^+,\rho\right)\right]\left(\operatorname{sgn}\left[\bar{R}(\theta_k,\rho)\right] - \operatorname{sgn}\left[\bar{R}(\pi,\rho)\right]\right)\Big\}, && \text{by (c)} \\
&= \frac{\pi}{2}\operatorname{sgn}\left[\bar{I}\left(0^+,\rho\right)\right]\Big\{\left(\operatorname{sgn}\left[\bar{R}(0,\rho)\right] - \operatorname{sgn}\left[\bar{R}(\theta_1,\rho)\right]\right) \\
&\qquad - \left(\operatorname{sgn}\left[\bar{R}(\theta_1,\rho)\right] - \operatorname{sgn}\left[\bar{R}(\theta_2,\rho)\right]\right) + \\
&\qquad \cdots + (-1)^k\left(\operatorname{sgn}\left[\bar{R}(\theta_k,\rho)\right] - \operatorname{sgn}\left[\bar{R}(\pi,\rho)\right]\right)\Big\}, && \text{by (d)} \\
&= \frac{\pi}{2}\operatorname{sgn}[T^{(p)}(-1,\rho)]\Big\{\operatorname{sgn}\left[\bar{R}(0,\rho)\right] - 2\operatorname{sgn}\left[\bar{R}(\theta_1,\rho)\right] + 2\operatorname{sgn}\left[\bar{R}(\theta_2,\rho)\right] + \\
&\qquad \cdots + (-1)^k\operatorname{sgn}\left[\bar{R}(\theta_k,\rho)\right] + (-1)^{k+1}\operatorname{sgn}\left[\bar{R}(\pi,\rho)\right]\Big\}, && \text{by (e)} \\
&= \frac{\pi}{2}\operatorname{sgn}[T^{(p)}(-1,\rho)]\Big\{\operatorname{sgn}[R(-1,\rho)] - 2\operatorname{sgn}[R(t_1,\rho)] + 2\operatorname{sgn}[R(t_2,\rho)] + \\
&\qquad \cdots + (-1)^k 2\operatorname{sgn}[R(t_k,\rho)] + (-1)^{k+1}\operatorname{sgn}[R(+1,\rho)]\Big\}, && \text{by (f)}
\end{aligned}
$$

from which the result (4.21) follows. □

The result derived above can now be extended to the case of rational functions. Let $Q(z) = \frac{P_1(z)}{P_2(z)}$ where $P_i(z), i = 1, 2$ are polynomials with real coefficients. Let

$$R_i(u,\rho) + j\sqrt{1-u^2}T_i(u,\rho), i = 1, 2$$

denote the Tchebyshev representations of $P_i(z), i = 1, 2$ and $Q_C(u,\rho)$ denote the Tchebyshev representation of $Q(z)$ on the circle $\mathcal{C}_\rho$. Let $R(u,\rho)$, $T(u,\rho)$ be defined by

$$
\begin{aligned}
R(u,\rho) &= R_1(u,\rho)R_2(u,\rho) + (1-u^2)T_1(u,\rho)T_2(u,\rho), \\
T(u,\rho) &= T_1(u,\rho)R_2(u,\rho) - R_1(u,\rho)T_2(u,\rho).
\end{aligned}
$$

Suppose that $T(u,\rho)$ has p zeros at $u = -1$ and let $t_1 \dots t_k$ denote the real distinct zeros of $T(u,\rho)$ of odd multiplicity ordered as follows:

$$-1 < t_1 < t_2 < \cdots < t_k < +1.$$

Theorem 4.3 *Let $Q(z) = \frac{P_1(z)}{P_2(z)}$ where $P_i(z), i = 1, 2$ are real polynomials with i_1 and i_2 zeros, respectively, inside the circle C_ρ and no zeros on it. Then,*

$$i_1 - i_2 = \frac{1}{2}\operatorname{sgn}\left[T^{(\rho)}(-1,\rho)\right]\left(\operatorname{sgn}\left[R(-1,\rho)\right] + 2\sum_{j=1}^{k}(-1)^j \operatorname{sgn}\left[R\left(t_j,\rho\right)\right]\right.$$
$$\left. +(-1)^{k+1}\operatorname{sgn}\left[R(+1,\rho)\right]\right). \tag{4.22}$$

Proof The proof is based on the representation of $Q_C(u,\rho)$ developed in (4.20). Since the denominator of (4.20) is strictly positive for $u \in [-1,+1]$, it follows that the phase unwrapping can be computed from the numerator. The rest of the proof is similar to the proof for the polynomial case and is omitted. □

4.5 Digital PI, PD, and PID Controllers

In this section, we give general parametrizations of PI, PD, and PID controllers, respectively, in terms of parameters K_0, K_1, K_2. These will be used in the sequel to compute the Stabilizing Sets. T denotes the sampling period.

1. For PI controllers, we have

$$C(z) = k_p + k_i T \cdot \frac{z}{z-1} = \frac{\left(k_p + k_i T\right) z - k_p}{z-1} = \frac{\left(k_p + k_i T\right)\left(z - \frac{k_p}{k_i T + k_p}\right)}{z-1}.$$

Thus, we rewrite

$$C(z) = \frac{K_1 z + K_0}{z-1}, \tag{4.23}$$

where

$$k_p = -K_0 \quad \text{and} \quad k_i = \frac{K_1 + K_0}{T}. \tag{4.24}$$

2. For PD controllers, we have

$$C(z) = k_p + \frac{k_d}{T}\cdot\frac{z-1}{z} = \frac{\left(k_p T + k_d\right) z - k_d}{Tz} = \frac{\left(k_p + \frac{k_d}{T}\right)\left(z - \frac{\frac{k_d}{T}}{k_p + \frac{k_d}{T}}\right)}{z}.$$

We rewrite

$$C(z) = \frac{K_1 z + K_0}{z}, \tag{4.25}$$

where

$$k_p = K_0 + K_1 \quad \text{and} \quad k_d = -K_0 T. \tag{4.26}$$

3. The general formula of a discrete-time PID controller, using backward differences for differentiation to preserve causality, is

$$\begin{aligned} C(z) &= k_p + k_i T \cdot \frac{z}{z-1} + \frac{k_d}{T} \cdot \frac{z-1}{z} \\ &= \frac{\left(k_p + k_i T + \frac{k_d}{T}\right) z^2 + \left(-k_p - \frac{2k_d}{T}\right) z + \frac{k_d}{T}}{z(z-1)}. \end{aligned}$$

We use the general representation

$$C(z) = \frac{K_2 z^2 + K_1 z + K_0}{z(z-1)}, \tag{4.27}$$

where

$$k_p = -K_1 - 2K_0, \quad k_i = \frac{K_0 + K_1 + K_2}{T}, \tag{4.28}$$

and

$$k_d = K_0 T. \tag{4.29}$$

4.6 Computation of the Stabilizing Set

The results of the previous sections on Tchebyshev representations, root counting and clustering, and the representations of digital PID controllers are used here to develop constructive techniques for computing the stabilizing set. The main idea is to construct a polynomial or rational function such that the controller parameters are *separated* as much as possible into the real and imaginary parts. By applying the root counting formulas to this function, we can often "linearize" the problem. We emphasize that other root counting formulas such as Jury's test applied to these problems result in difficult nonlinear problems, which are often impossible to solve. In this section, we outline how the technique works for PI and PID controllers, and present a complete development along with an example in the next section.

4.7 PI Controllers

Consider the unity feedback control loop in Fig 4.2 with the plant $P(z)$ as

$$P(z) := \frac{N(z)}{D(z)} \tag{4.30}$$

with $\deg[D(z)] = n$ and $\deg[N(z)] \leq n$. The PI controller is represented as:

$$C(z) = \frac{K_1 z + K_0}{z - 1}. \tag{4.31}$$

The procedure to compute the PI stabilizing set is summarized in the following:

1. Represent the polynomials $N(z)$ and $D(z)$ from (4.30) as

$$D(e^{j\theta}) := R_D(u) + j\sqrt{1-u^2}T_D(u), \tag{4.32}$$

$$N(e^{j\theta}) := R_N(u) + j\sqrt{1-u^2}T_N(u), \tag{4.33}$$

where

$$N(z) = a_n z^n + a_{n-1}z^{n-1} + \cdots + a_1 z + a_0, \tag{4.34}$$

$$D(z) = b_n z^n + b_{n-1}z^{n-1} + \cdots + b_1 z + b_0 \tag{4.35}$$

with $a_0, a_1, \ldots, a_n$ and $b_0, b_1, \ldots, b_n$ real, and

$$\begin{aligned} R_N(u) &= a_n c_n(u) + a_{n-1}c_{n-1}(u) + \cdots + a_1 c_1(u) + a_0, \\ T_N(u) &= a_n s_n(u) + a_{n-1}s_{n-1}(u) + \cdots + a_1 s_1(u), \\ R_D(u) &= b_n c_n(u) + b_{n-1}c_{n-1}(u) + \cdots + b_1 c_1(u) + b_0, \\ T_D(u) &= b_n s_n(u) + b_{n-1}s_{n-1}(u) + \cdots + b_1 s_1(u), \end{aligned} \tag{4.36}$$

where

$$c_k(u) = \cos k\theta \quad \text{and} \quad s_k(u) = \frac{\sin k\theta}{\sin\theta}, \quad k = 1, 2, 3, \ldots, n \tag{4.37}$$

are the Tchebyshev polynomials of the first and second kind. With

$$u := -\cos\theta \quad \text{and} \quad z = e^j\theta = -u + j\sqrt{1-u^2}. \tag{4.38}$$

The generalized Tchebyshev polynomials are presented in Table 4.1, where $s_k(u)$ and $c_k(u)$ are obtained recursively as

$$s_k(u) = -\frac{1}{k}\frac{d[c_k(u)]}{du}, \tag{4.39}$$

$$c_{k+1}(u) = -uc_k(u) - (1-u^2)s_k(u). \tag{4.40}$$

2. Calculate the characteristic polynomial from the closed-loop system in Fig. 4.2

$$\delta(z) = (z-1)D(z) + (K_0 + K_1 z)N(z). \tag{4.41}$$

3. Determine

$$\delta(z)N(z^{-1}) = (z-1)D(z)N(z^{-1}) + (K_0 + K_1 z)N(z)N(z^{-1}). \tag{4.42}$$

4. Use the Tchebyshev representations to calculate

$$\begin{aligned}\delta(z)N(z^{-1})|_{z=e^{j\theta},u=-\cos\theta}\\ &= (-u-1+j\sqrt{1-u^2})(P_1(u)+j\sqrt{1-u^2}P_2(u))\\ &\quad + jK_1\sqrt{1-u^2}P_3(u) - K_1uP_3(u) + K_0P_3(u),\end{aligned} \tag{4.43}$$

where

$$\begin{aligned}P_1(u) &= R_D(u)R_N(u) + (1-u^2)T_D(u)T_N(u),\\ P_2(u) &= R_N(u)T_D(u) - T_N(u)R_D(u),\\ P_3(u) &= R_N^2(u) + (1-u^2)T_N^2(u),\end{aligned} \tag{4.44}$$

where $R_N(u)$, $R_D(u)$, $T_N(u)$, and $T_D(u)$ are calculated as (4.36). Let $N_r(z)$ denote the reverse polynomial of $N(z)$,

$$\begin{aligned}\delta(z)N(z^{-1})|_{z=e^{j\theta},u=-\cos\theta} &= \left.\frac{\delta(z)N_r(z)}{z^l}\right|_{z=e^{j\theta},u=-\cos\theta}\\ &= R(u,K_0,K_1) + \sqrt{1-u^2}T(u,K_1),\end{aligned} \tag{4.45}$$

where

$$\begin{aligned}R(u,K_0,K_1) &= -(u+1)P_1(u) - (1-u^2)P_2(u) - (K_1u - K_0)P_3(u),\\ T(u,K_1) &= P_1(u) - (u+1)P_2(u) + K_1P_3(u).\end{aligned} \tag{4.46}$$

5. Fixing a specific value of K_1, we can calculate the zeros of t_i of $T(u, K_1)$ which are real, distinct, and of odd multiplicity for $u \in (-1, +1)$:

$$-1 < t_1 < t_2 < \cdots < t_k < +1. \tag{4.47}$$

6. For this fixed K_1, calculate the set of strings of signs for the real part $R(t_j, K_0, K_1)$, corresponding to stability satisfying

$$(n+1)+i_{N_r}-l=\frac{1}{2}\mathrm{sgn}\left[T^{(p)}(-1)\right]\Big(\mathrm{sgn}[R(-1,K_0,K_1)]$$
$$+2\sum_{j=1}^{k}(-1)^j\mathrm{sgn}[R(t_j,K_0,K_1)]+(-1)^{k+1}\mathrm{sgn}[R(+1,K_0,K_1)]\Big). \quad (4.48)$$

   where i_{N_r} is the number of zeros of $N_r(z)$ inside the unit circle. For fixed K_1, this leads to linear inequalities in K_0.
7. Sweep over the K_1 range for which an adequate number k of real roots t_k exist in $(-1, 1)$ to satisfy (4.48). This number is lower bounded by $n+1+i_{N_r}-l$.

Example 4.1 Consider the discrete-time system represented in Fig. 4.2, with

$$P(z)=\frac{z-0.1}{z^3+0.1z-0.25} \quad \text{and} \quad C(z)=\frac{K_0+K_1z}{z-1}. \quad (4.49)$$

Using the Tchebyshev representation with $\rho=1$, we have

$$\begin{aligned}
R_N(u)&=-u-0.1,\\
T_N(u)&=1,\\
R_D(u)&=-4u^3+2.9u-0.25,\\
T_D(u)&=4u^2-0.9,\\
P_1(u)&=0.4u^3+2u^2-0.04u-0.875,\\
P_2(u)&=0.34-0.4u^2-2u,\\
P_3(u)&=0.2u+1.01.
\end{aligned} \quad (4.50)$$

Then, we have

$$\begin{aligned}
R(u,K_0,K_1)&=-0.8u^4-4.4u^3-(1.22+0.2K_1)u^2\\
&\quad+(2.915-1.01K_1+0.2K_0)u+(0.535+1.01K_0),\\
T(u,K_1)&=0.8u^3+4.4u^2+(1.62+0.5K_1)u+(-1.215+1.01K_1).
\end{aligned} \quad (4.51)$$

Since $P(z)$ is of order 3 and $C(z)$, the PI controller, is of order 1, the number of roots of $\delta(z)$ inside the unit circle is required to be 4, for stability. Then,

$$i_1-i_2=\underbrace{(i_\delta+i_{N_r})}_{i_1}-\underbrace{l}_{i_2}=3, \quad (4.52)$$

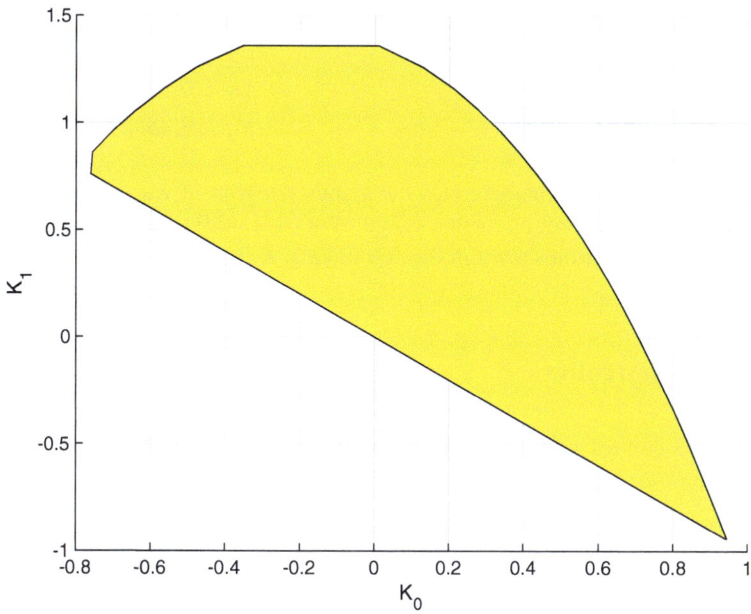

Fig. 4.3 Stabilizing set in Example 4.1

where i_δ and i_{N_r} are the number of roots of $\delta(z)$ and the reverse polynomial of $N(z)$, respectively, and l is the degree of $N(z)$. Since the required i_δ is 4, $i_{N_r} = 0$, and $l = 1$, $i_1 - i_2$ is required to be 3. Therefore, we require two real roots of $T(u, K_1)$ to satisfy the stability condition. We can find the feasible range for K_1, so there exist at least two real roots $\in (-1, 1)$. For this example, this range is $K_1 \in [-0.94, 1.415]$. Following the stabilizing set procedure for the range of K_1, we obtain the stability region shown in Fig. 4.3 in (K_0, K_1) space. To illustrate the example in detail, we fix $K_1 = 1$.Then, the real roots of $T(u, K_1)$ in $(-1, +1)$ are -0.5535 and 0.0919. Furthermore, $\text{sgn}[T(-1)] = +1$, and $i_1 - i_2 = 3$ requires that

$$\begin{aligned}
&\frac{1}{2}\text{sgn}[T(-1)]\big\{\text{sgn}[R(-1, K_0)] - 2\text{sgn}[R(-0.5535, K_0)] \\
&+ 2\text{sgn}[R(0.0919, K_0)] - \text{sgn}[R(1, K_0)]\big\} \\
&= 3.
\end{aligned} \tag{4.53}$$

Here, the only valid sign sequence satisfying the last equation is

$$\begin{aligned}
\text{sgn}[R(-1, K_0)] = +1, \quad \text{sgn}[R(-0.5535, K_0)] = -1, \\
\text{sgn}[R(0.0919, K_0)] = +1, \qquad \text{sgn}[R(1, K_0)] = -1.
\end{aligned} \tag{4.54}$$

Corresponding to this sequence, we have the following set of linear inequalities:

$$\begin{aligned} K_0 &> -1, & K_0 &< 0.3151, \\ K_0 &> -0.6754, & K_0 &< 3.4545. \end{aligned} \tag{4.55}$$

This set of inequalities characterizes the stability region in K_0 space for $K_1 = 1$. By repeating this procedure for the range of $K_1 \in [-0.94, 1.415]$, we obtain the complete stability region shown in Fig. 4.3 in (K_0, K_1) space.

4.8 PID Controllers

Consider the control system in Fig 4.2 with a rational and proper plant transfer function

$$P(z) := \frac{N(z)}{D(z)} \tag{4.56}$$

with deg $D(z) = n$ and deg $N(z) \leq n$. The PID controller is of the form

$$C(z) = \frac{K_0 + K_1 z + K_2 z^2}{z(z-1)}. \tag{4.57}$$

The procedure to compute the PID stabilizing set is the following:

1. Form the characteristic polynomial of the closed-loop system in Fig. 4.2 of degree $n+2$:

$$\delta(z) = z(z-1)D(z) + (K_0 + K_1 z + K_2 z^2)N(z). \tag{4.58}$$

2. Determine

$$\begin{aligned} z^{-1}\delta(z)N(z^{-1}) &= (z-1)D(z)N(z^{-1}) \\ &\quad + (K_0 z^{-1} + K_1 + K_2 z)N(z)N(z^{-1}). \end{aligned} \tag{4.59}$$

3. Use the Tchebyshev representations to calculate

$$\begin{aligned} z^{-1}\delta(z)N(z^{-1}) &= -(u+1)P_1(u) - (1-u^2)P_2(u) \\ &\quad - [(K_0 + K_2)u - K_1]P_3(u) \\ &\quad + j\sqrt{1-u^2}\cdot[-(u+1)P_2(u) + P_1(u) + (K_2 - K_0)P_3(u)], \end{aligned} \tag{4.60}$$

where

$$
\begin{aligned}
P_1(u) &= R_D(u)R_N(u) + (1-u^2)T_D(u)T_N(u),\\
P_2(u) &= R_N(u)T_D(u) - T_N(u)R_D(u),\\
P_3(u) &= R_N^2(u) + (1-u^2)T_N^2(u)
\end{aligned}
\tag{4.61}
$$

and $R_N(u)$, $R_D(u)$, $T_N(u)$, and $T_D(u)$ are calculated as in (4.36). Now, let $K_3 := K_2 - K_0$. Rewriting (4.60), we have

$$
\begin{aligned}
z^{-1}\delta(z)N(z^{-1}) &= -(u+1)P_1(u) - (1-u^2)P_2\\
&\quad - [(2K_2 - K_3)u) - K_1]P_3(u)\\
&\quad + j\sqrt{1-u^2}\cdot[-(u+1)P_2(u) + P_1(u) + K_3P_3(u)]\\
&= R(u, K_1, K_2, K_3) + j\sqrt{1-u^2}T(u, K_3).
\end{aligned}
\tag{4.62}
$$

4. Fixing a specific value of K_3, we can calculate the zeros of t_i of $T(u, K_3)$ which are real, distinct, and of odd multiplicity for $u \in (-1, +1)$:

$$
-1 < t_1 < t_2 < \cdots < t_k < +1. \tag{4.63}
$$

5. For this fixed K_3, calculate the set of strings of sign patterns for the real part $R(t_j, K_1, K_2, K_3)$, corresponding to stability, using

$$
\begin{aligned}
(n+2) + i_{N_r} - (l+1) &= \frac{1}{2}\mathrm{sgn}\left[T^{(p)}(-1)\right]\Big(\mathrm{sgn}[R(-1, K_1, K_2, K_3)]\\
&\quad +2\sum_{j=1}^{k}(-1)^j\mathrm{sgn}[R(t_j, K_1, K_2, K_3)] + (-1)^{k+1}\mathrm{sgn}[R(+1, K_1, K_2, K_3)]\Big),
\end{aligned}
\tag{4.64}
$$

where i_{N_r} is the number of zeros of $N_r(z)$ inside the unit circle. For fixed K_3, this leads to linear inequalities in (K_1, K_2).
6. Sweep over the K_3 range for which an adequate number of real roots t_k exist in $(-1, 1)$ for $T(u, K_3) = 0$.

Example 4.2 Consider the plant $P(z)$.

$$
P(z) = \frac{1}{z^2 - 0.25}.
$$

Then,

$$
\begin{aligned}
R_D(u) &= 2u^2 - 1.25,\\
T_D(u) &= -2u,\\
R_N(u) &= 1,\\
T_N(u) &= 0
\end{aligned}
$$

and

$$\begin{aligned} P_1(u) &= 2u^2 - 1.25, \\ P_2(u) &= -2u, \\ P_3(u) &= 1. \end{aligned}$$

Since $P(z)$ is of order 2 and $C(z)$, the PID controller, is of order 2, the number of roots of $\delta(z)$ inside the unit circle is required to be 4, for stability. From Theorem 4.2,

$$i_i - i_2 = \underbrace{(i_\delta + i_{N_r})}_{i_1} - \underbrace{(l+1)}_{i_2},$$

where i_δ and i_{N_r} are the numbers of roots of $\delta(z)$ and the reverse polynomial of $N(z)$ inside the unit circle, respectively, and l is the degree of $N(z)$. Since the required i_δ is 4, $i_{N_r} = 0$, and $l = 0$, $i_1 - i_2$ is required to be 3. To illustrate the example in detail, we first fix $K_3 = 1.3$. Then, the real roots of $T(u, K_3)$ in $(-1, 1)$ are -0.4736 and -0.0264. Furthermore, $\text{sgn}[T(-1)] = 1$, and from Theorem 4.2, $i_1 - i_2 = 3$ requires that

$$\begin{aligned} \frac{1}{2}\text{sgn}[T(-1)]\Big(& \text{sgn}[R(-1, K_1, K_2)] - 2\text{sgn}[R(-0.4736, K_1, K_2)] \\ & + 2\text{sgn}[R(-0.0264, K_1, K_2)] - \text{sgn}[R(1, K_1, K_2)]\Big) = 3. \end{aligned}$$

We have only one admissible string satisfying the above equation, namely

sgn[R(-1, K_1, K_2)]	1
sgn[R(-0.4736, K_1, K_2)]	-1
sgn[R(-0.0264, K_1, K_2)]	1
sgn[R(1, K_1, K_2)]	-1
$2(i_1 - i_2)$	6

Corresponding to this string, we have the following set of linear inequalities:

$$\begin{aligned} -1.3 + K_1 + 2K_2 &> 0, \\ -0.9286 + K_1 + 0.9472 &< 0, \\ 1.1286 + K_1 + 0.0528K_2 &> 0, \\ -0.2 + K_1 - 2K_2 &< 0. \end{aligned}$$

This set of inequalities characterizes the stability region in (K_1, K_2) space for the fixed $K_3 = 1.3$. By repeating this procedure for the range of K_3 for which $T(u, K_3)$ has at least two real roots, we obtain the stability region shown in the left of Fig. 4.4. Using

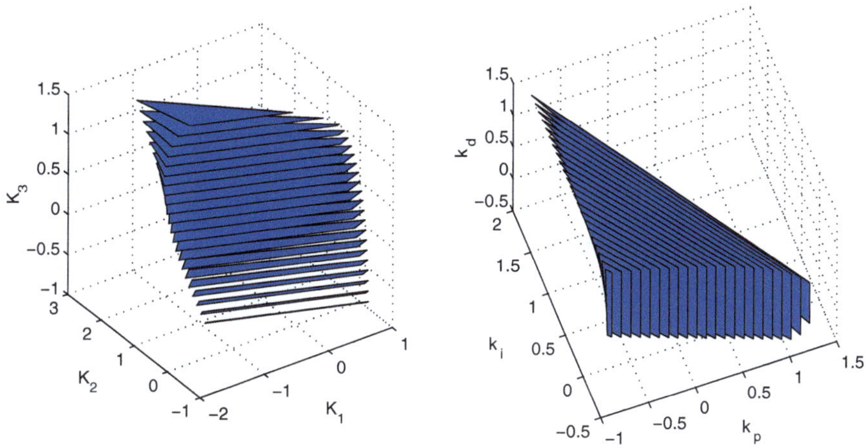

Fig. 4.4 Stability regions in (K_1, K_2, K_3) space (left) and (k_p, k_i, k_d) space (right). © Taylor & Francis LLC Books. Reproduced from [1] with permission

$$\begin{bmatrix} k_p \\ k_i \\ k_d \end{bmatrix} = \begin{bmatrix} -2 & -1 & 0 \\ \frac{1}{T} & \frac{1}{T} & \frac{1}{T} \\ T & 0 & 0 \end{bmatrix} \begin{bmatrix} K_0 \\ K_1 \\ K_2 \end{bmatrix} = \begin{bmatrix} -2 & -1 & 0 \\ \frac{1}{T} & \frac{1}{T} & \frac{1}{T} \\ T & 0 & 0 \end{bmatrix} \begin{bmatrix} 0 & 1 & -1 \\ 1 & 0 & 0 \\ 0 & 1 & 0 \end{bmatrix} \begin{bmatrix} K_1 \\ K_2 \\ K_3 \end{bmatrix}$$
$$= \begin{bmatrix} -1 & -2 & 2 \\ \frac{1}{T} & \frac{2}{T} & -\frac{1}{T} \\ 0 & T & -T \end{bmatrix} \begin{bmatrix} K_1 \\ K_2 \\ K_3 \end{bmatrix}.$$

We can determine, for a fixed sampling period T, the stabilizing region in (k_p, k_i, k_d) space as shown in the right of Fig. 4.4.

4.8.1 Maximally Deadbeat Control

An important design technique in digital control is deadbeat control, wherein one places all closed-loop poles at the origin. If this is used in conjunction with integral control the tracking error is zeroed out in a finite number of sampling steps. Deadbeat control requires in general that we be able to control all the poles of the system. However, such a pole placement design is in general not possible when a lower order controller is used. Thus, we are motivated to design a PID controller that places the closed-loop characteristic roots as close to the origin as possible. The transient response of such a system will decay out faster than any other design and therefore, the fastest possible convergence of the error under PID control will be achieved.

The design scheme to be developed will attempt to place the closed-loop poles within a circle of *minimum* radius ρ. Let $\mathcal{S}_\rho$ denote the set of PID controllers achieving such a closed-loop root cluster. We show below how $\mathcal{S}_\rho$ can be computed for fixed

ρ. The minimum value of ρ can be found by determining the value ρ^* for which $\mathcal{S}_{\rho^*} = \varnothing$ but $\mathcal{S}_\rho \neq \varnothing$, $\rho > \rho^*$.

Now, let us again consider the PID controller

$$C(z) = \frac{K_2 z^2 + K_1 z + K_0}{z(z-1)} \tag{4.65}$$

and the characteristic polynomial

$$\delta(z) = z(z-1)D(z) + \left(K_2 z^2 + K_1 z + K_0\right) N(z). \tag{4.66}$$

Note that

$$D(z)|_{z=-\rho u + j\rho\sqrt{1-u^2}} = R_D(u,\rho) + j\sqrt{1-u^2}T_D(u,\rho), \tag{4.67}$$

$$N(z)|_{z=-\rho u + j\rho\sqrt{1-u^2}} = R_N(u,\rho) + j\sqrt{1-u^2}T_N(u,\rho), \tag{4.68}$$

and

$$\begin{aligned} N\left(\rho^2 z^{-1}\right)\big|_{z=-\rho u + j\rho\sqrt{1-u^2}} &= N(z)|_{z=-\rho u - j\rho\sqrt{1-u^2}} \\ &= R_N(u,\rho) - j\sqrt{1-u^2}T_N(u,\rho). \end{aligned} \tag{4.69}$$

We now evaluate

$$\begin{aligned} &\rho^2 z^{-1}\delta(z)N\left(\rho^2 z^{-1}\right) \\ &= \rho^2 z^{-1} \underbrace{\left[z(z-1)D(z) + \left(K_2 z^2 + K_1 z + K_0\right)N(z)\right]}_{\delta(z)} N\left(\rho^2 z^{-1}\right) \end{aligned} \tag{4.70}$$

over the circle $\mathcal{C}_\rho$

$$\rho^2 z^{-1}\delta(z)N\left(\rho^2 z^{-1}\right)\big|_{z=-\rho u + j\rho\sqrt{1-u^2}}$$

$$\begin{aligned} &= -\rho^2(\rho u + 1)P_1(u,\rho) - \rho^3\left(1-u^2\right)P_2(u,\rho) \\ &\quad - \left[\left(K_0 + K_2\rho^2\right)\rho u - K_1\rho^2\right]P_3(u,\rho) \\ &\quad + j\sqrt{1-u^2}\left[\rho^3 P_1(u,\rho) - \rho^2(\rho u + 1)P_2(u,\rho) + \left(K_2\rho^2 - K_0\right)\rho P_3(u,\rho)\right], \end{aligned} \tag{4.71}$$

where

$$P_1(u,\rho) = R_D(u,\rho)R_N(u,\rho) + \left(1-u^2\right)T_D(u,\rho)T_N(u,\rho), \tag{4.72}$$

$$P_2(u,\rho) = R_N(u,\rho)T_D(u,\rho) - T_N(u,\rho)R_D(u,\rho), \tag{4.73}$$

$$P_3(u,\rho) = R_N^2(u,\rho) + \left(1-u^2\right)T_N^2(u,\rho). \tag{4.74}$$

By letting

$$K_3 := K_2\rho^2 - K_0, \tag{4.75}$$

we have

$$\begin{aligned}
&\rho^2 z^{-1}\delta(z)N\left(\rho^2 z^{-1}\right)\big|_{z=-\rho u+j\rho\sqrt{1-u^2}} \\
&= -\rho^2(\rho u+1)P_1(u,\rho) - \rho^3\left(1-u^2\right)P_2(u,\rho) \\
&\quad -\left[\left(2K_2\rho^2 - K_3\right)\rho u - K_1\rho^2\right]P_3(u,\rho) \\
&\quad +j\sqrt{1-u^2}\left[\rho^3 P_1(u,\rho) - \rho^2(\rho u+1)P_2(u,\rho) + K_3\rho P_3(u,\rho)\right].
\end{aligned} \tag{4.76}$$

To determine the set of controllers achieving root clustering inside a circle of radius ρ we proceed as before: fix K_3, use the root counting formulas of Sect. 4.4, develop linear inequalities in K_2, K_3 and sweep over the requisite range of K_3. This procedure is then performed as ρ decreases until the set of stabilizing PID parameters just disappears. The following example illustrates this procedure.

Example 4.3 We consider the same plant used in Example 4.2. Figure 4.5 (left) shows the stabilizing set in the PID gain space at $\rho = 0.275$. For a smaller value of ρ, the stabilizing region in PID parameter space disappears. This means that there is no PID controller available to push all closed-loop poles inside a circle of radius smaller than 0.275. From this, we select a point inside the region that is

$$K_0 = 0.0048, \quad K_1 = -0.3195, \quad K_2 = 0.6390$$

and $K_3 = 0.0435$. From the relationship in (4.75), we have

$$\begin{bmatrix} k_p \\ k_i \\ k_d \end{bmatrix} = \begin{bmatrix} -1 & -2\rho^2 & 2 \\ \frac{1}{T} & \frac{\rho^2}{T}+\frac{1}{T} & -\frac{1}{T} \\ 0 & \rho^2 T & -T \end{bmatrix} \begin{bmatrix} K_1 \\ K_2 \\ K_3 \end{bmatrix} = \begin{bmatrix} 0.3099 \\ 0.3243 \\ 0.0048 \end{bmatrix}$$

Figure 4.5 (right) shows the closed-loop poles that lie inside the circle of radius $\rho = 0.275$. The roots are

$$0.2500 \pm j0.1118 \quad \text{and} \quad 0.2500 \pm j0.0387.$$

We select several sets of stabilizing PID parameters from the set obtained in Example 4.2 ($\rho = 1$) and compare the step responses between them. Figure 4.6 shows that the maximally deadbeat design produces nearly deadbeat response.

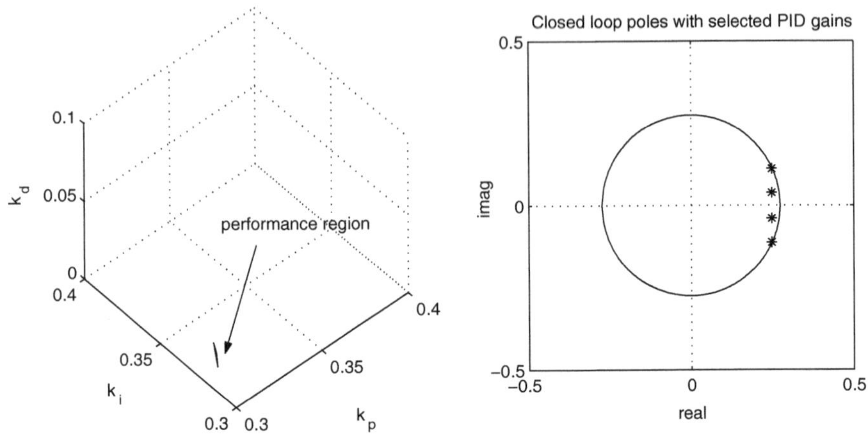

Fig. 4.5 Stability regions with $\rho = 0.275$ (left). Closed-loop poles of the selected PID gains (right). © Taylor & Francis LLC Books. Reproduced from [1] with permission

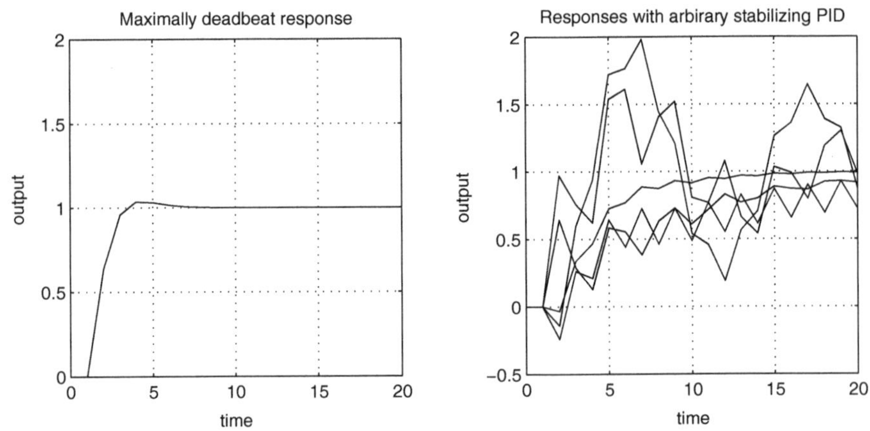

Fig. 4.6 Maximally deadbeat design (left). Arbitrary stabilization (right). © Taylor & Francis LLC Books. Reproduced from [1] with permission

4.8.2 Maximal Delay Tolerance Design

In some control systems, an important design parameter is the delay tolerance of the loop, that is the maximum delay that can be inserted into the loop without destabilizing it. In digital control, a delay of k sampling instants is represented by z^{-k}. We use this to determine the maximum delay that a control loop under PID control can be designed to tolerate. This gives the limit of delay tolerance achievable for the given plant under PID control.

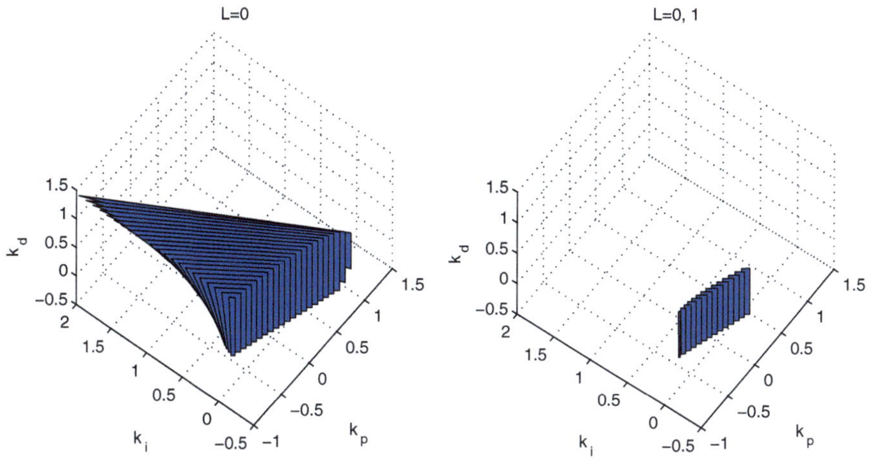

Fig. 4.7 Stability region for delayed systems. © Taylor & Francis LLC Books. Reproduced from [1] with permission

Let the plant be

$$P(z) = \frac{N(z)}{D(z)}. \tag{4.77}$$

We consider the problem of finding the maximum delay L^* such that the plant can be stabilized by a PID controller. In other words, we find the maximum value L^* such that the stabilizing PID gain set that simultaneously stabilizes the set of plants

$$z^{-L}P(z) = \frac{N(z)}{z^L D(z)}, \qquad \text{for } L = 0, 1, \ldots, L^* \tag{4.78}$$

is not empty. Let $\mathcal{S}_i$ be the set of PID gains that stabilizes the plant $z^{-i}P(z)$. Then it is clear that

$$\cap_{i=0}^{L} \mathcal{S}_i \text{ stabilizes } z^i P(z) \text{ for all } i = 0, 1, \ldots, L. \tag{4.79}$$

We illustrate this computation with an example.

Example 4.4 Consider the plant used in Example 4.2. Figure 4.7 (left) shows the stabilizing PID gains when there is no delay (i.e., $L = 0$).

The figure on the right shows the stabilizing PID gains when $L = 0, 1$. As seen in the figure, the size of the set is reduced as the delay increases.

In many systems, the set disappears for a sufficiently large value of L^*. This is the maximum delay that can be stabilized by any PID controller. To illustrate, we fix $K_3 = 1$. Figure 4.8 shows that the stability region reduces when the required time delay increases and for this system with delay $L \in (0, 3)$ the region vanishes.

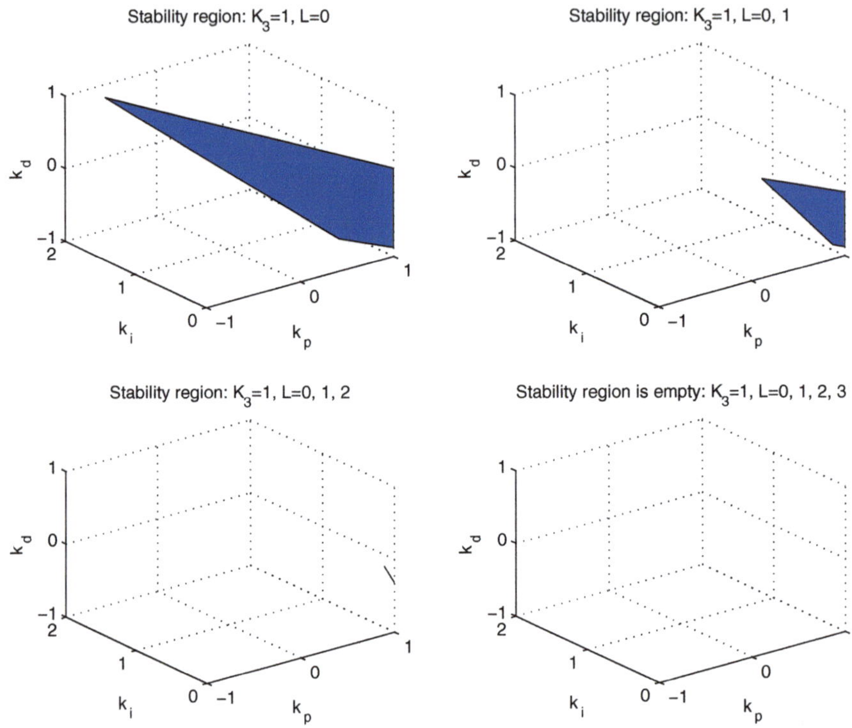

Fig. 4.8 Stability region for delayed systems. © Taylor & Francis LLC Books. Reproduced from [1] with permission

4.9 Notes and References

The procedures presented in this chapter are summarized from [1]. Stabilization sets for discrete-time PID controllers were calculated in [3]. For applications of digital systems to Power Electronics see [2]. Tchebyshev polynomials of the first and second kind are treated in [4].

References

1. Bhattacharyya, S.P., Datta, A., Keel, L.H.: Linear Control Theory Structure, Robustness, and Optimization. CRC Press/Taylor & Francis Group, Boca Raton (2009)
2. Buso, S., Mattavelli, P.: Digital Control in Power Electronics. Morgan and Claypool Publishers, San Rafael (2006)
3. Keel, L.H., Rego, J.I., Bhattacharyya, S.P.: A new approach to digital PID controller design. IEEE Trans. Autom. Control **48**(4), 687–692 (2003)
4. Mason, J.C., Handscomb, D.C.: Chebyshev Polynomials. Chapman and Hall/CRC, Boca Raton (2002)

Chapter 5
Computation of Stabilizing Sets from Frequency Response Data

Abstract The focus of this chapter is on direct data-driven synthesis and design of controllers. We show that the complete set of stabilizing PID and first-order controllers for a finite-dimensional linear time-invariant (LTI) plant, possibly cascaded with a delay, can be calculated directly from the frequency response (Nyquist/Bode) data $P(j\omega)$ for $\omega \in [0, \infty)$ without the need of producing an identified analytical model. The solutions have important features. For example, it is not necessary to know the order of the plant or even the number of left half-plane or right half-plane poles or zeros. The solution also identifies, in the case of PID controllers an exact low-frequency band over which the plant data must be known with accuracy and beyond which the plant information may be rough or approximate. These constitute important guidelines for identification when the latter is to be used for control design. The model-free approach to control synthesis and design developed here is an attractive complement to modern and postmodern model-based design methods, which require complete information on the plant and generally, produce a single controller. A discussion is included, with an illustrative example, of the sharp differences that can occur between model-free and model-based approaches when computing sets of stabilizing controllers. For example, it is shown, that the identified model of a high-order system can be non-PID stabilizable whereas the original data indicates it is PID stabilizable. The results given here could be a significant improvement over classical control loop-shaping approaches since we obtain complete sets of controllers achieving the design specifications. It can enhance fuzzy and neural approaches which are model-free but cannot guarantee stability and performance. Finally, these results open the door to adaptive, model-free, fixed order designs of real-world systems.

I. D. Díaz-Rodríguez et al., *Analytical Design of PID Controllers*,
https://doi.org/10.1007/978-3-030-18228-1_5

5.1 Introduction

In the previous chapters, we have described results on the design of PID controllers which are generically crucial in many industries and are implemented in electrical, mechanical, hydraulic, fluidic, and pneumatic systems. It is important to note that these approaches are *model-based.*

The purpose of this chapter is to show that at least for three-term controllers, synthesis and design can be carried out directly from frequency response measurements on the plant without constructing a state space or transfer function model. The main results show that complete sets achieving stability and various performance specifications can be obtained from Nyquist/Bode data without constructing an identified model. It is emphasized that our solution does not require knowledge of the order of the system nor the numbers of LHP or RHP poles or zeros, and no identification of the plant is needed. It will be seen in the sequel that, in the case of PID controllers, the solution specifies an exact "low-frequency band" where the plant frequency response must be known accurately and beyond which rough data or measurements suffice. These features have important implications in real-world control engineering, where models are often unavailable, measurements can be made only over a restricted range of frequencies and where guarantees of various performance specifications must still be made.

The results given are valid for stable and unstable LTI systems, possibly containing a delay, and specifically deal with PID and first-order controllers. They can be extended to general three-term controllers with minor modifications. In general, a higher order controller would have more than three adjustable parameters. In such cases, the three-term theory can be applied by fixing various sets of parameters and leaving three adjustable terms. An important advantage of dealing with three design terms is that 2D and 3D graphics can be used to display the resulting sets of controllers in parameter space.

In practice, the frequency response data can be readily obtained for stable plants by direct measurements and the theory given here can, therefore, be applied to stable plants without constructing a transfer function or state space model. For unstable plants, frequency response data could be obtained if a feedback compensator that stabilizes the plant is known. In the latter case, measurements can be made on the stable closed-loop system and the plant frequency response data extracted from it by "dividing out" the known compensator. This kind of procedure is also necessary for the identification of unstable systems and also for determining the Nyquist plot of unstable systems. By *synthesis*, we mean that the *complete* set of controllers of the prescribed type (PID or first-order) achieving stability can be computed. This is the first essential step to enabling the design of systems achieving multiple performance specifications. We may also consider this as the model-free fixed order version of the well-known YJBK parametrization of all stabilizing controllers.

We show in later chapters how to compute the complete set of controllers achieving prescribed design specifications. By *design* we mean that several performance objectives can be intersected to simultaneously satisfy multiple performance objectives. In this framework, the performance measures that can be handled analytically include guaranteed gain and phase margins as well as H_∞ norm specifications. The computations involved in most of the cases are linear programming or solutions of linear equations with a sweeping parameter. Simultaneous satisfaction of multiple performance criteria amounts to the simultaneous solution of more extensive sets of linear inequalities. The results provide an alternative to model-based control while at the same time overcoming the limitations of classical control theory. In this sense, they offer a combination of the best of the classical and modern approaches.

5.2 Mathematical Preliminaries

In this section, we develop some notation and technical results which will be used later. First consider a real rational function

$$R(s) = \frac{A(s)}{B(s)}, \tag{5.1}$$

where $A(s)$ and $B(s)$ are polynomials with real coefficients and of degrees m and n, respectively. We assume that $A(s)$ and $B(s)$ have no zeros on the $j\omega$-axis. Let z_R^+, p_R^+ (z_L^-, p_L^-) denote the numbers of open right half-plane (RHP) and open left half-plane (LHP) zeros and poles of $R(s)$. Also let $\Delta_0^\infty \angle R(j\omega)$ denote the net change in phase of $R(j\omega)$ as ω runs from 0 to $+\infty$. Then, we have

$$\Delta_0^\infty \angle R(j\omega) = \frac{\pi}{2}\left[z_L^- - z_R^+ - \left(p_L^- - p_R^+\right)\right]. \tag{5.2}$$

This formula follows from the fact that each LHP zero and each RHP pole contribute $+\frac{\pi}{2}$ to the net phase change whereas each RHP zero and LHP pole contribute $-\frac{\pi}{2}$ to the net phase change.

For convenience, we define the (Hurwitz) signature of $R(s)$ to be:

$$\sigma(R) := z_L^- - z_R^+ - \left(p_L^- - p_R^+\right). \tag{5.3}$$

Write

$$R(j\omega) = R_r(\omega) + jR_i(\omega), \tag{5.4}$$

where $R_r(\omega)$ and $R_i(\omega)$ are rational functions in ω with real coefficients. It is easy to see that $R_r(\omega)$ and $R_i(\omega)$ have no real poles for $\omega \in (-\infty, +\infty)$ since $R(s)$ has no imaginary axis poles. To compute the net change in phase, that is, the left-hand side of (5.2), it is convenient to develop formulas in terms of $R_r(\omega)$ and $R_i(\omega)$. Note that $\omega_0 = 0$ is always a zero of $R_i(\omega)$ since $R(s)$ is real. Let

$$0 = \omega_0 < \omega_1 < \omega_2 < \cdots < \omega_{l-1} \tag{5.5}$$

denote the real, finite nonnegative zeros of $R_i(\omega) = 0$ of odd multiplicities, let

$$\operatorname{sgn}[x] = \begin{cases} +1 & \text{if } x > 0 \\ 0 & \text{if } x = 0 \\ -1 & \text{if } x < 0 \end{cases} \tag{5.6}$$

and define $\omega_l = \infty^-$. Define, for a real function $f(t)$,

$$f\left(t_0^-\right) := \lim_{t \to t_0, t < t_0} f(t), \qquad f\left(t_0^+\right) := \lim_{t \to t_0, t > t_0} f(t). \tag{5.7}$$

Lemma 5.1 (Real Hurwitz Signature Lemma) *For $n - m$ even,*

$$\sigma(R) = \left(\operatorname{sgn}\left[R_r(\omega_0)\right] + 2\sum_{j=1}^{l-1}(-1)^j \operatorname{sgn}\left[R_r(\omega_j)\right] \right.$$
$$\left. +(-1)^l \operatorname{sgn}\left[R_r(\omega_l)\right] \right)(-1)^{l-1}\operatorname{sgn}\left[R_i(\infty^-)\right].$$

For $n - m$ odd,

$$\sigma(R) = \left(\operatorname{sgn}\left[R_r(\omega_0)\right] + 2\sum_{j=1}^{l-1}(-1)^j \operatorname{sgn}\left[R_r(\omega_j)\right] \right) \cdot (-1)^{l-1}\operatorname{sgn}\left[R_i(\infty^-)\right].$$

Proof The calculation of the signature is based on the total phase change of a frequency-dependent function as the frequency ranges from 0 to ∞. The latter in turn can be broken up into a summation of phase changes over a disjoint partition of the frequency axis.

Note first that

$$\Delta_0^\infty \angle R(j\omega) = \frac{\pi}{2}\sigma(R) \tag{5.8}$$

and

$$\Delta_0^\infty \angle R(j\omega) = \Delta_{\omega_0=0}^{\omega_1} \angle R(j\omega) + \cdots + \Delta_{\omega_{l-1}}^{\omega_l=\infty^-} \angle R(j\omega). \tag{5.9}$$

For the case when $n - m$ is even, the plot of $R(j\omega)$ approaches the negative or positive real axis as ω approaches ∞. Thus, we have

$$\Delta_{\omega_k}^{\omega_{k+1}} \angle R(j\omega) = \frac{\pi}{2}\left(\operatorname{sgn}[R_r(\omega_k)] - \operatorname{sgn}[R_r(\omega_{k+1}] \right) \cdot \operatorname{sgn}[R_i(\omega_{k+1}^-)], \tag{5.10}$$

for $k = 0, \ldots, l-2$ and

$$\Delta_{\omega_{l-1}}^{\omega_l=\infty^-} \angle R(j\omega) = \frac{\pi}{2}\left(\operatorname{sgn}[R_r(\omega_{l-1})] - \operatorname{sgn}[R_r(\omega_l)] \right) \cdot \operatorname{sgn}[R_i(\infty^-)]. \tag{5.11}$$

From (5.8) and (5.9),

$$\begin{aligned}\sigma(R) = &\Big(\text{sgn}[R_r(\omega_0)] - \text{sgn}[R_r(\omega_1)]\Big)\text{sgn}[R_i(\omega_1^-)]\\ &+\Big(\text{sgn}[R_r(\omega_1)] - \text{sgn}[R_r(\omega_2)]\Big)\text{sgn}[R_i(\omega_2^-)] + \cdots\\ &+\Big(\text{sgn}[R_r(\omega_{l-2})] - \text{sgn}[R_r(\omega_{l-1})]\Big)\text{sgn}[R_i(\omega_{l-1}^-)]\\ &+\Big(\text{sgn}[R_r(\omega_{l-1})] - \text{sgn}[R_r(\omega_l)]\Big)\text{sgn}[R_i(\infty^-)].\end{aligned}$$

Since

$$\text{sgn}[R_i(\omega_i^-)] = -\text{sgn}[R_i(\omega_{i+1}^-)]$$

and

$$\text{sgn}[R_i(\omega_{l-1}^-)] = -\text{sgn}[R_i(\infty^-)],$$

we have

$$\begin{aligned}\text{sgn}[R_i(\omega_{l-2}^-)] &= (-1)^2 R_i(\infty^-),\\ \text{sgn}[R_i(\omega_{l-3}^-)] &= (-1)^3 R_i(\infty^-)],\\ &\vdots\\ \text{sgn}[R_i(\omega_2^-)] &= (-1)^{l-2} R_i(\infty^-),\\ \text{sgn}[R_i(\omega_1^-)] &= (-1)^{l-1} R_i(\infty^-),\end{aligned} \tag{5.12}$$

so that

$$\begin{aligned}\sigma(R) &= \Big(\text{sgn}[R_r(\omega_0)] - 2\text{sgn}[R_r(\omega_1)] + 2\text{sgn}[R_r(\omega_2)]\\ &\quad + \cdots + (-1)^{l-1}\text{sgn}[R_r(\omega_{l-1})] + (-1)^l\text{sgn}[R_r(\omega_l)]\Big)\cdot(-1)^{l-1}\text{sgn}[R_i(\infty^-)]\\ &= \Big(\text{sgn}[R_r(\omega_0)] + 2\sum_{j=1}^{l-1}(-1)^j\,\text{sgn}\left[R_r(\omega_j)\right]\\ &\quad +(-1)^l\text{sgn}\left[R_r(\omega_l)\right]\Big)(-1)^{l-1}\text{sgn}[R_i(\infty^-)].\end{aligned}$$

For the case when $n - m$ is odd, the plot of $R(j\omega)$ approaches the negative or positive imaginary axis as ω approaches ∞. Thus, we have

$$\Delta_{\omega_k}^{\omega_{k+1}} \angle R(j\omega) = \frac{\pi}{2}\Big(\text{sgn}[R_r(\omega_k)] - \text{sgn}[R_r(\omega_{k+1})]\Big) \cdot \text{sgn}[R_i(\omega_{k+1}^-)],$$
$$\text{for } k = 0, \ldots, l-2$$
$$\Delta_{\omega_{l-1}}^{\omega_l=\infty^-} \angle R(j\omega) = \frac{\pi}{2}\text{sgn}[R_r(\omega_{l-1})]\text{sgn}[R_i(\infty^-)].$$

Then, using (5.12), we have

$$\begin{aligned}
\sigma(R) &= \Big(\text{sgn}[R_r(\omega_0)] - \text{sgn}[R_r(\omega_1)]\Big)\text{sgn}[R_i(\omega_1^-)] \\
&\quad + \Big(\text{sgn}[R_r(\omega_1)] - \text{sgn}[R_r(\omega_2)]\Big)\text{sgn}[R_i(\omega_2^-)] \\
&\quad + \cdots + \Big(\text{sgn}[R_r(\omega_{l-2})] - \text{sgn}[R_r(\omega_{l-1})]\Big) \cdot \text{sgn}[R_i(\omega_{l-1}^-)] \\
&\quad + (-1)^{l-1}\text{sgn}[R_r(\omega_{l-1})]\text{sgn}[R_r(\infty^-)] \\
&= \Big(\text{sgn}[R_r(\omega_0)] - 2\text{sgn}[R_r(\omega_1)] + \cdots \\
&\quad + (-1)^{l-1}2\text{sgn}[R_r(\omega_{l-1})]\Big)(-1)^{l-1}\text{sgn}[R_i(\infty^-)] \\
&= \left(\text{sgn}[R_r(\omega_0)] + 2\sum_{j=1}^{l-1}(-1)^j\text{sgn}[R_r(\omega_j)]\right) \\
&\quad \cdot(-1)^{l-1}\text{sgn}[R_i(\infty^-)].
\end{aligned}$$

□

Now consider a complex rational function

$$Q(s) = \frac{D(s)}{E(s)}, \tag{5.13}$$

where $D(s)$ and $E(s)$ are polynomials with complex coefficients of degrees n and m, respectively. As before, we assume that $D(s)$ and $E(s)$ do not have zeros on the $j\omega$-axis. Let $\Delta_{-\infty}^{+\infty}\angle Q(j\omega)$ denote the net change in phase of $Q(j\omega)$ as ω runs from $-\infty$ to $+\infty$. Also let z_Q^+, p_Q^+ (z_Q^-, p_Q^-) denote the numbers of RHP (LHP) zeros and poles of $Q(s)$. Then we have

$$\Delta_{-\infty}^{+\infty}\angle Q(j\omega) = \pi\left[z_Q^- - z_Q^+ - \left(p_Q^- - p_Q^+\right)\right]. \tag{5.14}$$

This easily follows from the fact that each LHP zero (RHP zero) and each RHP pole (LHP pole) contribute $+\pi$ $(-\pi)$ to the net phase change. Summing over all poles and zeros, we obtain the formula given. Analogous to the real case, we define the signature of the complex rational function $Q(s)$:

$$\sigma(Q) := z_Q^- - z_Q^+ - \left(p_Q^- - p_Q^+\right). \tag{5.15}$$

Write

$$Q(j\omega) = Q_r(\omega) + jQ_i(\omega), \tag{5.16}$$

where $Q_r(\omega)$ and $Q_i(\omega)$ are rational functions with real coefficients. Moreover, $Q_r(\omega)$ and $Q_i(\omega)$ have no real poles for $\omega \in (-\infty, +\infty)$. It is easy to show that the numerators of $Q_r(\omega)$ and $Q_i(\omega)$ are generically of the same degree. Indeed, if this is not so, multiplying $Q(s)$ by an arbitrary complex number $\alpha + j\beta$ will render this condition to be true without changing poles, zeros or signature of $Q(s)$. Therefore, we henceforth assume this to be true. Let $\omega_0, \ldots, \omega_{l-1}$ ordered as

$$-\infty < \omega_0 < \omega_1 < \cdots < \omega_{l-1} < +\infty := \omega_l$$

denote the real, distinct, finite zeros of $Q_i(\omega) = 0$ with odd multiplicities.

Lemma 5.2 (Complex Hurwitz Signature Lemma)

$$\sigma(Q) = \left(\sum_{j=1}^{l-1}(-1)^{l-1-j}\operatorname{sgn}\left[Q_r(\omega_j)\right]\right)\operatorname{sgn}[Q_i(\infty^-)].$$

Proof As before we compute the total phase change as a sum over a disjoint partition of the frequency axis.

Let us begin by noting that

$$\pi\sigma(Q) = \Delta_{-\infty}^{+\infty}\angle Q(j\omega). \tag{5.17}$$

To evaluate the right-hand side of (5.17), write

$$\Delta_{-\infty}^{+\infty}\angle Q(j\omega) = \Delta_{-\infty}^{\omega_1}\angle Q(j\omega) + \sum_{i=1}^{l-2}\Delta_{\omega_i}^{\omega_{i+1}}\angle Q(j\omega) + \Delta_{\omega_{l-1}}^{+\infty}\angle Q(j\omega). \tag{5.18}$$

Let

$$Q(j\omega) = Q_r(\omega) + jQ_i(\omega) := \frac{A_r(\omega)}{B(\omega)} + j\frac{A_i(\omega)}{B(\omega)},$$

where $A_r(\omega)$, $A_i(\omega)$, $B(\omega)$ are polynomials with real coefficients and

$$B(\omega) \neq 0, \quad \omega \in (-\infty, \infty). \tag{5.19}$$

We assume that $A_r(\omega)$ and $A_i(\omega)$ are of equal degree. If this fails, it can be restored by multiplying $Q(s)$ by almost any complex number without changing its signature.

Introduce

$$A(j\omega) := A_r(\omega) + jA_i(\omega) \tag{5.20}$$

and note that in view of (5.19),

$$\Delta_{-\infty}^{+\infty}\angle Q(j\omega) = \Delta_{-\infty}^{\omega_1}\angle A(j\omega) + \sum_{k=1}^{l-2}\Delta_{\omega_k}^{\omega_{k+1}}\angle A(j\omega) + \Delta_{\omega_{l-1}}^{+\infty}\angle A(j\omega). \tag{5.21}$$

The finite zeros of $Q_i(\omega)$ are identical to zeros of $A_i(\omega)$ and therefore

$$\Delta_{\omega_k}^{\omega_{k+1}}\angle Q(j\omega) = \frac{\pi}{2}\text{sgn}\left[\dot{A}_i(\omega_k)\right]\cdot\left(\text{sgn}\left[A_r(\omega_k)\right] - \text{sgn}\left[A_r(\omega_{k+1})\right]\right), \tag{5.22}$$

for $k = 1, 2, \ldots, l-2$. Also, we have

$$\begin{aligned}&\Delta_{-\infty}^{\omega_1}\angle A(j\omega) + \Delta_{\omega_{l-1}}^{+\infty}\angle A(j\omega)\\ &= \frac{\pi}{2}\text{sgn}\left[\dot{A}_i(\omega_1)\right]\text{sgn}\left[A_r(\omega_1)\right] + \frac{\pi}{2}\text{sgn}\left[\dot{A}_i(\omega_{l-1})\right]\text{sgn}\left[A_r(\omega_{l-1})\right].\end{aligned} \tag{5.23}$$

The proof of the lemma is completed by noting that

$$\begin{aligned}\text{sgn}\left[\dot{Q}_i(\omega_{l-1})\right] &= \text{sgn}\left[Q_i(\infty^-)\right]\\ \text{sgn}\left[A_r(\omega_k)\right] &= \text{sgn}\left[Q_r(\omega_k)\right]\\ \text{sgn}\left[\dot{Q}_i(\omega_k)\right] &= \text{sgn}\left[\dot{A}_i(\omega_k)\right]\\ \text{sgn}\left[\dot{Q}_i(\omega_k)\right] &= (-1)^{l-1-k}\text{sgn}\left[Q_i(\infty^-)\right]\end{aligned} \tag{5.24}$$

for $k = 1, 2, \ldots, l-1$ and substituting (5.21)–(5.24) in (5.18). □

5.3 Phase, Signature, Poles, Zeros, and Bode Plots

Let P denote an LTI plant and $P(s)$ its rational transfer function with z^+, p^+ (z^-, p^-) denoting the numbers of RHP (LHP) zeros and poles, and $n(m)$ the denominator (numerator) degrees. Let the relative degree be denoted r_P:

$$r_P := n - m$$

As defined earlier the signature of P is

$$\sigma(P) = \left(z^- - z^+\right) - \left(p^- - p^+\right). \tag{5.25}$$

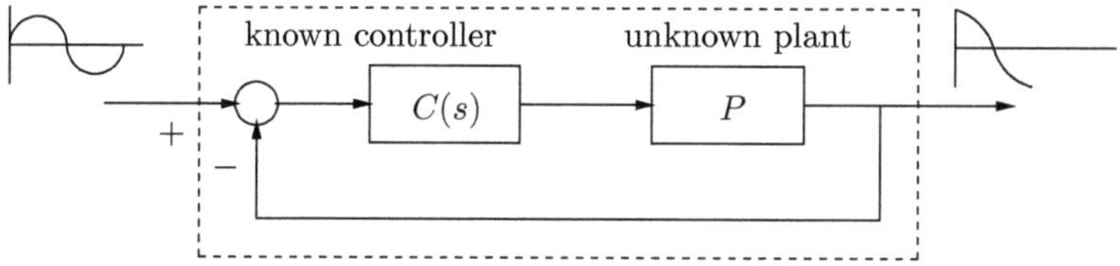

Fig. 5.1 Frequency response measurement on an unstable plant. © Taylor & Francis LLC Books. Reproduced from [1] with permission

Lemma 5.3

$$r_P = -\frac{1}{20}\cdot\left.\frac{dP_{\mathrm{db}}(\omega)}{d(\log_{10}\omega)}\right|_{\omega\to\infty} \tag{5.26}$$

$$\sigma(P) = \frac{2}{\pi}\Delta_0^{\infty}\angle P(j\omega), \tag{5.27}$$

where

$$P_{\mathrm{dB}}(\omega) := 20\log_{10}|P(j\omega)|.$$

Proof Equation (5.26) states that the relative degree is the high frequency slope of the Bode magnitude plot and (5.27) states that the signature can be found from the net change in phase from the phase plot. □

Assuming that $P(s)$ has no $j\omega$-axis poles and zeros, we can also write

$$\sigma(P) = -(n-m) - 2\left(z^+ - p^+\right) \tag{5.28}$$

or

$$\sigma(P) = -r_P - 2\left(z^+ - p^+\right). \tag{5.29}$$

Therefore, using Lemma 5.3 and Eq. (5.29) $z^+ - p^+$ can be determined from the Bode plot of P. In particular if $P(s)$ is stable the Bode plot can often be obtained experimentally by measuring the frequency response of the system. Then, the above relations with $p^+ = 0$ determine z^+ from the Bode plot data.

Unstable Plant

Now suppose that P is an *unstable* LTI plant with a rational transfer function *unknown* to us and assume that P does not have imaginary axis poles and zeros. We assume, however, that a known feedback controller $C(s)$ stabilizes P and the closed-loop frequency response can be *measured* and is denoted by $G(j\omega)$ for $\omega \in [0,\infty)$ (Fig. 5.1).

Then

$$P(j\omega) = \frac{G(j\omega)}{C(j\omega)(1-G(j\omega))} \tag{5.30}$$

is the *computed* frequency response of the unstable plant. The next result shows that knowledge of $C(s)$ and $G(j\omega)$ is sufficient to determine the numbers z^+ and p^+, that is the numbers of RHP zeros and poles of the plant. Let z_c^+ denote the number of RHP zeros of $C(s)$.

Theorem 5.1

$$z^+ = \frac{1}{2}\left[-r_P - r_C - 2z_c^+ - \sigma(G)\right] \tag{5.31}$$

$$p^+ = \frac{1}{2}\left[\sigma(P) - \sigma(G) - r_C\right] - 2z_c^+. \tag{5.32}$$

Proof We have

$$G(s) = \frac{P(s)C(s)}{1 + P(s)C(s)}$$

and, since $G(s)$ is stable,

$$\begin{aligned}\sigma(G) &= \left(z^- + z_c^-\right) - \left(z^+ + z_c^+\right) - (n + n_c)\\ &= -r_P - r_C - 2z_c^+ - 2z^+,\end{aligned}$$

which implies (5.31). From (5.25) applied to $P(s)$, we have

$$p^+ = z^+ + \frac{\sigma(P)}{2} + \frac{r_P}{2}. \tag{5.33}$$

Substituting (5.31) in (5.33), we have (5.32). □

Remark 5.1 In the above theorem, we assume that $C(s)$, a stabilizing controller is known, and the corresponding closed-loop frequency response $G(j\omega)$ for $\omega \in [0, \infty)$ can be *measured*. Thus, $P(j\omega)$ can be computed from (5.30). Now r_P and $\sigma(G)$ can be found by applying the results of Lemma 5.3 to $P(j\omega)$ and $G(j\omega)$, respectively. r_C and z_c^+ are known as $C(s)$ is known. Therefore, z^+ and p^+ can be found.

Remark 5.2 In the above analysis, we have assumed for simplicity that the plant is devoid of imaginary axis poles and zeros. When such poles and zeros are present their numbers may be known from physical considerations or their numbers and locations may be ascertained from the experimentally determined or computed Bode plot. Once identified we can lump these poles and zeros with the controller and proceed with the design procedure. At an imaginary axis zero (pole) of multiplicity k, away from the origin, the magnitude plot is zero (infinity) and the phase plot undergoes an instantaneous change of phase $k\pi$. If such poles or zeros occur at the origin there is a corresponding phase shift of $\frac{k\pi}{2}$ at zero frequency. It is straightforward to establish that, in this case, the relations (5.31) and (5.32) need to be modified to the following:

$$z^+ = \frac{1}{2}\left[-r_P - r_C - 2z_c^+ - z_c^i - \sigma(G)\right] \tag{5.34}$$

$$p^+ = \frac{1}{2}\left[\sigma(P) - \sigma(G) - r_C\right] - 2z_c^+ - z_c^i + z^i - p^i, \tag{5.35}$$

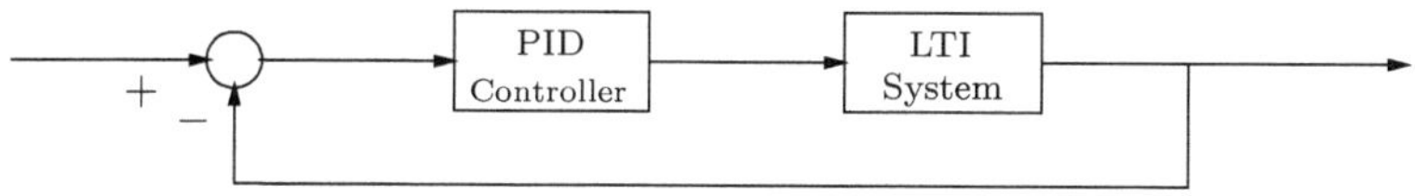

Fig. 5.2 A unity feedback system with a PID controller. © Taylor & Francis LLC Books. Reproduced from [1] with permission

where z^i, z_c^i, p^i, p_c^i denote the numbers of imaginary axis zeros and poles of plant and controller.

5.4 PID Synthesis for Delay-Free Continuous-Time Systems

In this section, we consider the synthesis and design of PID controllers for a continuous-time LTI plant, with underlying transfer function $P(s)$ with $n(m)$ poles (zeros). (See Fig. 5.2).

We assume that the *only* information available to the designer are the following:
1. Knowledge of the frequency response magnitude and phase, equivalently, $P(j\omega)$, $\omega \in [0, \infty)$ if the plant is stable.
2. Knowledge of a known stabilizing controller and the corresponding closed-loop frequency response $G(j\omega)$, if the plant is unstable.

Such assumptions are reasonable for most systems. We also make the technical assumption that the plant has no $j\omega$ poles or zeros so that the magnitude, its inverse and phase are well-defined for all $\omega \geq 0$. As we have seen from the discussion in the last section, the numbers and locations of RHP poles and zeros can be found from the above data for any LTI plant and either "divided out" or lumped with the controller.

Write

$$P(j\omega) = |P(j\omega)|e^{j\phi(\omega)} = P_r(\omega) + jP_i(\omega), \tag{5.36}$$

where $|P(j\omega)|$ denotes the *magnitude* and $\phi(\omega)$ the *phase* of the plant, at the frequency ω.

Let the PID controller be of the form

$$C(s) = \frac{k_i + k_p s + k_d s^2}{s(1+sT)}, \quad \text{for } T > 0, \tag{5.37}$$

where T is assumed to be fixed and small. We now present results for determining the stabilizing set.

Lemma 5.4 *Let*

$$F(s) := s(1+sT) + \left(k_i + k_p s + k_d s^2\right) P(s).$$

and

$$\bar{F}(s) := F(s)P(-s).$$

Then, the closed-loop stability is equivalent to

$$\sigma\left(\bar{F}(s)\right) = n - m + 2z^+ + 2. \tag{5.38}$$

Proof Closed-loop stability is equivalent to the condition that all zeros of $F(s)$ lie in the LHP. This, in turn, is equivalent to the condition

$$\sigma\left(F(s)\right) = n + 2 - \left(p^- - p^+\right).$$

Now, consider the rational function

$$\bar{F}(s) = F(s)P(-s).$$

Note that

$$\sigma\left(\bar{F}(s)\right) = \sigma(F(s)) + \sigma(P(-s)).$$

Therefore, the stability condition becomes

$$\begin{aligned}\sigma\left(\bar{F}(s)\right) &= n + 2 - \left(p^- - p^+\right) + \left(z^+ - z^-\right) - \left(p^+ - p^-\right)\\ &= n + 2 + z^+ - z^- = n - m + 2z^+ + 2.\end{aligned}$$

□

Write

$$\begin{aligned}\bar{F}(j\omega) &= j\omega(1 + j\omega T)P(-j\omega)\\ &\qquad + \left(k_i + j\omega k_p - \omega^2 k_d\right) P(j\omega)P(-j\omega)\\ &= \underbrace{\left(k_i - k_d\omega^2\right)|P(j\omega)|^2 - \omega^2 T P_r(\omega) + \omega P_i(\omega)}_{\bar{F}_r(\omega, k_i, k_d)}\\ &\qquad + j\omega \underbrace{\left(k_p|P(j\omega)|^2 + P_r(\omega) + \omega T P_i(\omega)\right)}_{\bar{F}_i(\omega, k_p)}\\ &= \bar{F}_r\left(\omega, k_i, k_d\right) + j\omega \bar{F}_i\left(\omega, k_p\right).\end{aligned}$$

Theorem 5.2 *Fix $k_p = k_p^*$ and let $\omega_1 < \omega_2 < \cdots < \omega_{l-1}$ denote the distinct frequencies of odd multiplicities which are solutions of*

$$\bar{F}_i(\omega, k_p^*) = 0, \tag{5.39}$$

or, equivalently of,

$$\begin{aligned} k_p^* &= -\frac{P_r(\omega) + \omega T P_i(\omega)}{|P(j\omega)|^2} \\ &= -\frac{\cos\phi(\omega) + \omega T \sin\phi(\omega)}{|P(j\omega)|} \\ &=: g(\omega) \end{aligned} \tag{5.40}$$

for fixed $k_p = k_p^*$. *Let* $\omega_0 = 0$ *and* $\omega_l = \infty$, *and* $j := \mathrm{sgn}\left[\bar{F}_i(\infty^-, k_p^*)\right]$. *Determine strings of integers*

$$I = [i_0, i_1, i_2, \ldots, i_l],$$

with $i_t \in \{+1, -1\}$ *such that*
For $n - m$ *even:*

$$\begin{aligned} &\left[i_0 - 2i_1 + 2i_2 + \cdots + (-1)^{l-1} 2i_{l-1} + (-1)^l i_l\right](-1)^{l-1} j \\ &\qquad = n - m + 2z^+ + 2. \end{aligned} \tag{5.41}$$

For $n - m$ *odd*

$$\begin{aligned} &\left[i_0 - 2i_1 + 2i_2 + \cdots + (-1)^{l-1} 2i_{l-1}\right](-1)^{l-1} j \\ &\qquad = n - m + 2z^+ + 2. \end{aligned} \tag{5.42}$$

Then for the fixed $k_p = k_p^*$, *the* (k_i, k_d) *values corresponding to closed-loop stability are given by*

$$\bar{F}_r(\omega_t, k_i, k_d) i_t > 0, \quad \text{for } t = 0, \cdots, l, \tag{5.43}$$

where the i_t*'s are taken from strings satisfying (5.41) and (5.42), and* ω_t*'s are taken from the solutions of (5.39).*

Proof By Lemma 5.4, the stability condition has been reduced to the signature condition in (5.38). The theorem follows from applying Lemma 5.1 to compute the signature of $\bar{F}(s)$. □

The following result clarifies what range the parameters k_p must be swept over.

Theorem 5.3 *For the given function* $g(\omega)$ *in (5.40) determined completely by the plant data* $P(j\omega)$ *and* T*:*

(1) A necessary condition for PID stabilization is that there exists k_p *such that the function*

$$k_p = g(\omega) \tag{5.44}$$

has at least k distinct roots of odd multiplicities, where

$$\begin{aligned} k &\geq \frac{n - m + 2z^+ + 2}{2} - 1 \quad \text{if } n - m \text{ even} \\ k &\geq \frac{n - m + 2z^+ + 3}{2} - 1 \quad \text{if } n - m \text{ odd.} \end{aligned}$$

(2) There exists a unique range of ω, $\underline{\omega} = (\omega_{\min}, \omega_{\max})$ *over which the condition (1) occurs and this determines the range of* ω *to be swept.*

(3) Every k_p *belonging to the stabilizing set of PID parameters lies in the range*

$$k_p \in \left(k_p^{\min}, k_p^{\max}\right),$$

where

$$k_p^{\min} := \min_{\omega \in \underline{\omega}} g(\omega), \qquad k_p^{\max} := \max_{\omega \in \underline{\omega}} g(\omega),$$

where $\underline{\omega} = (\omega_{\min}, \omega_{\max})$.

Remark 5.3 The function $g(\omega)$ is well-defined due to the assumption that the plant either has no $j\omega$ zeros or these have been lumped with the controller. The frequency $\omega_{\max}$ can be selected as any frequency after which $g(\omega)$ continues to grow monotonically. This determines the range of frequencies over which the $P(j\omega)$ data of the plant should be known accurately. Note that the ranges $(\omega_{\min}, \omega_{\max})$ and $(k_p^{\min}, k_p^{\max})$ can consist of multiple segments.

The computation of the Stabilizing Set implied by the above results is summarized as the following procedure.

5.5 Computation of PID Stabilizing Sets from Frequency Response Data

The complete set of stabilizing PID gains can be computed by the following procedure:

For stable systems: The available data is the frequency response of the plant $P(j\omega)$.

0.1 Determine the relative degree of the plant $r_P = n - m$ from the high frequency slope of the Bode magnitude plot of $P(j\omega)$.

0.2 Let $\Delta_0^\infty[\phi(\omega)]$ denote the net change of phase of $P(j\omega)$ for $\omega \in [0, \infty)$. Determine z^+ from

$$\Delta_0^\infty[\phi(\omega)] = -\left[(n-m) + 2z^+\right]\frac{\pi}{2},$$

which follows from (5.28) with $p^+ = 0$.

For unstable systems: The available data are a stabilizing controller transfer function $C(s)$ and the frequency response of the corresponding stable closed-loop system $G(j\omega)$.

0.1 Compute the frequency response $P(j\omega)$ from (5.30).

0.2 Determine relative degree of the plant r_P from the high-frequency slope of the Bode magnitude plot of $P(j\omega)$.

0.3 Determine z_c^+ and r_C from $C(s)$.

0.4 Compute $\sigma(G)$ from (5.27) applied to $G(j\omega)$.
0.5 Compute z^+ using (5.31) in Theorem 5.1.
0.6 Compute $g(\omega)$ using (5.40) and the frequency response measured data.

1. Fix $k_p = k_p^*$, solve (5.40) and let $\omega_1 < \omega_2 < \cdots < \omega_{l-1}$ denote the distinct frequencies of odd multiplicities which are solutions of (5.40).
2. Set $\omega_0 = 0$, $\omega_l = \infty$ and $j = \mathrm{sgn} \bar{F}_i(-\infty^-, k_p^*)$. Determine all strings of integers $i_t \in \{+1, -1\}$ such that:

For $n - m$ even:

$$\left[i_0 - 2i_1 + 2i_2 + \cdots + (-1)^{i-1} 2i_{l-1} + (-1)^l i_l\right](-1)^{l-1} j \\ = n - m + 2z^+ + 2. \tag{5.45}$$

For $n - m$ odd:

$$\left[i_0 - 2i_1 + 2i_2 + \cdots + (-1)^{i-1} 2i_{l-1}\right](-1)^{l-1} j \\ = n - m + 2z^+ + 2. \tag{5.46}$$

3. For the fixed $k_p = k_p^*$ chosen in Step 1, solve for the stabilizing (k_i, k_d) values from

$$\left[k_i - k_d\omega_t^2 + \frac{\omega_t \sin\phi(\omega_t) - \omega_t^2 T \cos\phi(\omega_t)}{|P(j\omega_t)|}\right] i_t > 0, \tag{5.47}$$

for $t = 0, 1, \ldots, l$.
4. Repeat the previous three steps by updating k_p over prescribed ranges. The ranges over which k_p must be swept is determined from the requirements that (5.45) or (5.46) is satisfied for at least one string of integers as in Theorem 5.3.

We emphasize that all computations are based on the data $P(j\omega)$ and knowledge of the transfer function $P(s)$ is not required. For well-posedness of the loop, it is necessary that

$$k_d \neq -\frac{T}{P(\infty)}.$$

For strictly proper plants, $P(\infty) = 0$ and this constraint vanishes.

5.6 PID Synthesis for Systems with Delay

In this section, we show how the previous results can be extended to systems with delay. Consider the finite dimensional LTI plant P_L with a cascaded delay in Fig. 5.3.

Here, P_0 represents an LTI delay-free system with a proper transfer function. The transfer functions of P_0 and P_L are denoted $P_0(s)$ and $P_L(s)$, respectively. We assume that frequency response measurements can be made at terminals "a" and "b," that is

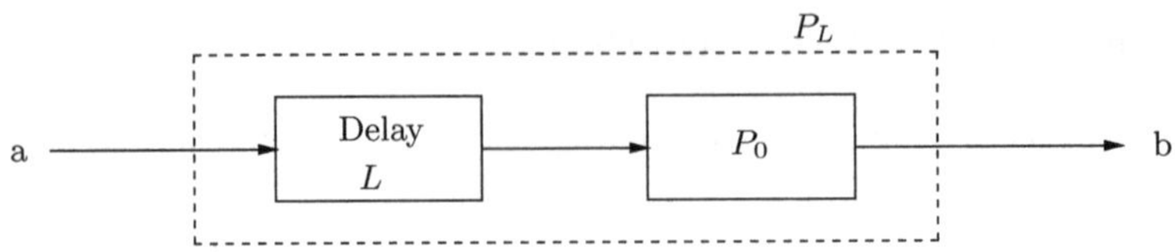

Fig. 5.3 A plant with cascaded delay. © Taylor & Francis LLC Books. Reproduced from [1] with permission

on the delay system P_L. Thus, the data we have is:

$$P_L(j\omega) = e^{-j\omega L} P_0(j\omega) = m_L(\omega)e^{j\phi_L(\omega)}, \quad 0 \leq \omega < \infty.$$

Write

$$P_0(j\omega) = m_0(\omega)e^{j\phi_0(\omega)}.$$

Therefore,

$$m_0(\omega) = m_L(\omega), \tag{5.48}$$

and

$$\phi_L(\omega) = \phi_0(\omega) - \omega L, \quad 0 \leq \omega < \infty. \tag{5.49}$$

It is clear that $m_0(\omega)$ and $\phi_0(\omega)$ can be determined from the frequency response measured data $m_L(\omega)$ and $\phi_L(\omega)$ on the system with embedded delay L when L is known. If L is unknown it can be determined from the high-frequency slope of the phase.

Now let $C(s, \mathbf{k})$ denote the PID controller

$$C(s, \mathbf{k}) = \frac{k_i + k_p s + k_d s^2}{s(1 + sT)}, \qquad \mathbf{k} = \left[k_i,\ k_p,\ k_d\right].$$

Let $\mathcal{S}_0$ denote the set of stabilizing PID controllers for the delay-free plant

$$\mathcal{S}_0 = \{\mathbf{k} \ : \ C(s, \mathbf{k}) \text{ stabilizes } P_0\}.$$

We have seen how $\mathcal{S}_0$ can be found in the previous section when $P_0(j\omega)$ is known. It follows from (5.49) that $\mathcal{S}_0$ can be determined from $P_L(j\omega)$ the data for the embedded delay system. We denote by $\mathcal{S}_L$ the set of PID controllers stabilizing the plant P_0 with cascaded delay ranging from 0 to L seconds.

Introduce the sets

$$\mathcal{S}_\infty = \{\mathbf{k} \ : \ |C(s, \mathbf{k})P_0(s)|_{s=\infty} \geq 1\},$$
$$\mathcal{S}_\mathcal{B} = \left\{\mathbf{k} \ : \ C(j\omega, \mathbf{k}) = \frac{-e^{j\omega \underline{L}}}{P_0(j\omega)}, \quad \begin{array}{l} \omega \in [0, \infty), \\ 0 \leq \underline{L} \leq L \end{array}\right\}.$$

The following theorem is obtained from the results.

Theorem 5.4 *The set $\mathcal{S}_L$ can be found from*

$$\mathcal{S}_L = \mathcal{S}_0 \backslash \left(\mathcal{S}_\infty \cup \mathcal{S}_\mathcal{B}\right). \tag{5.50}$$

The set $\mathcal{S}_\infty$ consists of those PID gains for which the Nyquist plot approaches points outside the unit circle as $s \to \infty$. The set $\mathcal{S}_\mathcal{B}$ consists of those PID gains for which an imaginary axis characteristic root occurs for delays less than L. The relationship (5.50) states that the exclusion of $\mathcal{S}_\infty$ and $\mathcal{S}_\mathcal{B}$ from $\mathcal{S}_0$ determines the stabilizing set $\mathcal{S}_L$ for the system with cascaded delay up to L seconds.

The computation of $\mathcal{S}_0$ has been described in the previous section. The set $\mathcal{S}_\infty$ is easy to calculate. In fact,

$$\mathcal{S}_\infty = \left\{\mathbf{k} \, : \, |k_d| \geq \frac{T}{|P_0(\infty)|}\right\}.$$

To determine $\mathcal{S}_\mathcal{B}$, let

$$\theta(\omega, \mathbf{k}) := \angle\left(\frac{k_i - k_d\omega^2 + j\omega k_p}{j\omega(1 + j\omega T)}\right).$$

Then the conditions defining $\mathcal{S}_\mathcal{B}$ can be written as the magnitude and phase conditions

$$k_i - k_d\omega^2 = \pm\sqrt{\frac{\omega^2(1+T^2)}{m_0^2(\omega)} - \omega^2 k_p^2}, \tag{5.51}$$

$$\theta(\omega, \mathbf{k}) \geq \pi - \phi_0(\omega) - \omega L, \quad \omega \in [0, \infty), \tag{5.52}$$

so that

$$\mathcal{S}_\mathcal{B} = \{\mathbf{k} \, : \, \mathbf{k} \text{ satisfies (5.51) and (5.52) for some } \omega \in [0, \infty)\}.$$

Note that (5.51) represents a straight line in (k_i, k_d) space for each fixed k_p and ω. The calculation of the set $\mathcal{S}_\mathcal{B}$ is tedious but straightforward if one sweeps over the frequency variable.

Example 5.1 (*An Illustrative Example of PID Synthesis*) To illustrate the previous results, we take a set of frequency response data points for a *stable* delay-free plant:

$$\mathbf{P}(j\omega) := \{P(j\omega), \quad \omega \in (0, 10) \text{ sampled every } 0.01\}.$$

The Nyquist and Bode plots are shown in Figs. 5.4 and 5.5.

The high-frequency slope of the Bode magnitude plot is $-40db/decade$ and thus $n - m = 2$. The total change of phase is -540 degrees and so

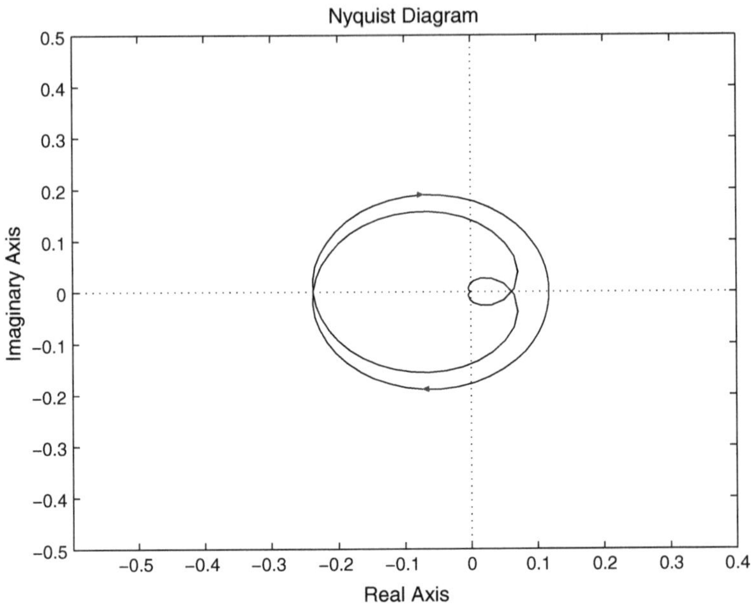

Fig. 5.4 Nyquist plot of the plant. © Taylor & Francis LLC Books. Reproduced from [1] with permission

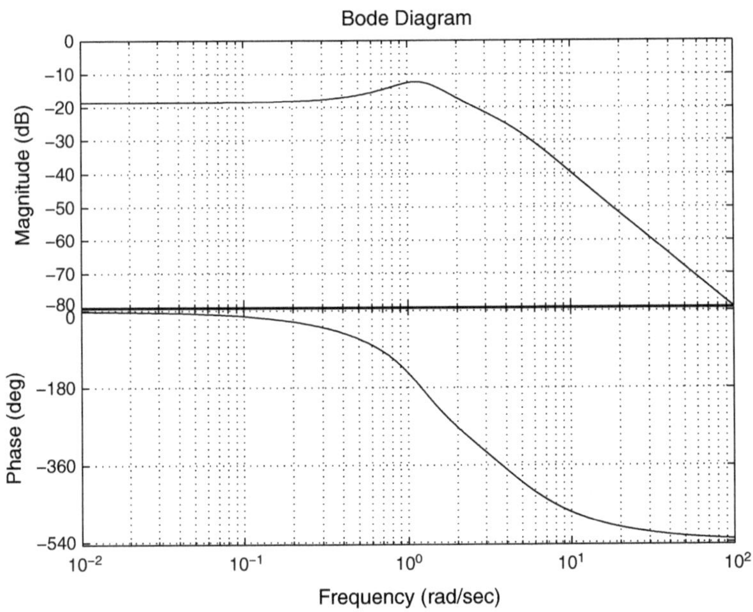

Fig. 5.5 Bode plot of the plant. © Taylor & Francis LLC Books. Reproduced from [1] with permission

$$-6\frac{\pi}{2} = -\left((n-m) - 2(p^+ - z^+)\right)\frac{\pi}{2},$$

and since the plant is stable, $p^+ = 0$, giving $z^+ = 2$ The required signature for stability can now be determined and is

$$\begin{aligned}\sigma(\bar{F}(s)) &= (n-m) + 2z^+ + 2\\ &= (2) + 2(2) + 2\\ &= 8.\end{aligned}$$

Since $n - m$ is even, we must have

$$\left[i_0 - 2i_1 + 2i_2 - 2i_3 + 2i_4 - \cdots + (-1)^l i_l\right](-1)^{l-1} j = 8,$$

where

$$j = \text{sgn}\left[\bar{F}_i(\infty^-, k_p)\right] = -\text{sgn}\left[\lim_{\omega\to\infty} g(\omega)\right] = -1$$

and it is clear that at least five terms are required to satisfy the above. In other words $l \geq 4$. From Fig. 5.6 it is easy to see that (5.40) has at most three positive frequencies as solutions and therefore we have $i_0 - 2i_1 + 2i_2 - 2i_3 + i_4 = 8$. Also $i_4 = \text{sgn}\left[\bar{F}_r(\infty^-, k_i, k_d)\right] = 1$ independent of k_i and k_d. This means that k_p must be chosen so that $\bar{F}_i(\omega, k_p^*) = 0$ has exactly three positive real zeros. This gives the feasible range of k_p values as shown in Fig. 5.6 which depicts the function:

$$g(\omega) = -\frac{\cos\phi(\omega) + \omega T \sin\phi(\omega)}{|P(j\omega)|}.. \tag{5.53}$$

The feasible range of k_p is such that k_p intersects the graph of $g(\omega)$ three times. This feasible range is shown in Fig. 5.6. In Fig. 5.6, we also observe that the *frequency range over which plant data must accurately be known for PID control* is [0, 8.2]. We now fix $k_p^* = 1$ and compute the set of ω's that satisfies

$$-\frac{\cos\phi(\omega) + \omega T \sin\phi(\omega)}{|P(j\omega)|} = 1.$$

To find the set of ω's satisfying the above, we plot the function $g(\omega)$ as shown in Fig. 5.7.

The three frequencies ω_1, ω_2, and ω_3 are required to compute the stability set in (k_i, k_d) space.

From this, we found the solutions $\{\omega_1, \omega_2, \omega_3\} = \{0,\ 0.742,\ 1.865,\ 7.854\}$. This leads to the requirement $i_0 - 2i_1 + 2i_2 - 2i_3 = 7$ giving the only feasible string

$$\mathcal{F} = \{i_0, i_1, i_2, i_3\} = \{1,\ -1,\ 1,\ -1\}.$$

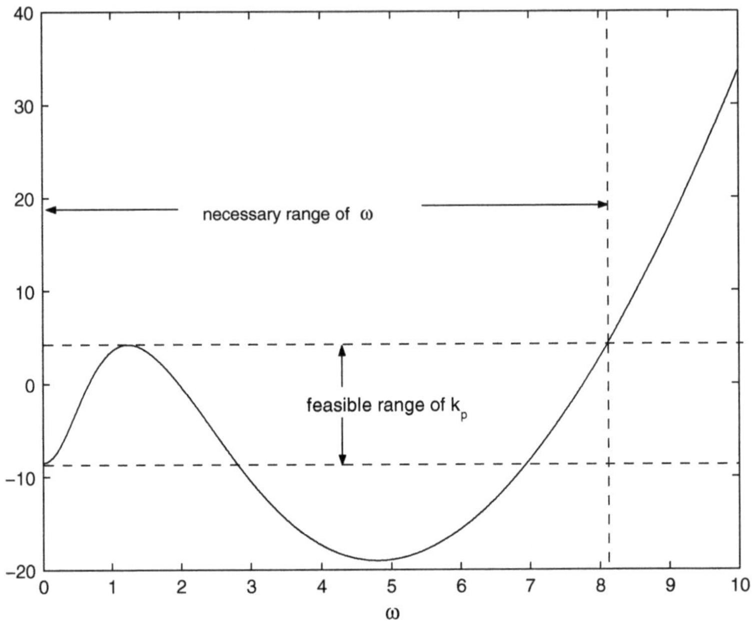

Fig. 5.6 Graph of the function $g(\omega)$ in (5.53). © Taylor & Francis LLC Books. Reproduced from [1] with permission

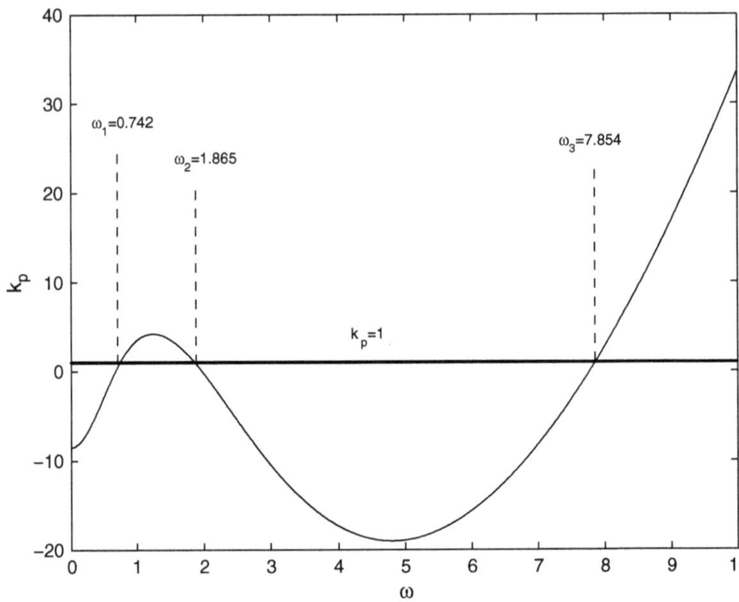

Fig. 5.7 Finding the set of ω's satisfying (5.40) with $k_p = 1$. © Taylor & Francis LLC Books. Reproduced from [1] with permission

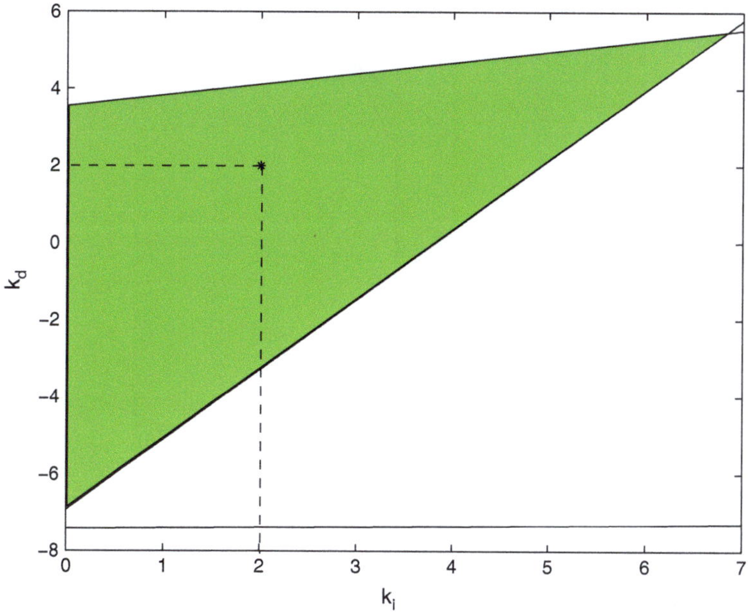

Fig. 5.8 The complete set of stabilizing PID gains in (k_i, k_d) space when $k_p = 1$. © Taylor & Francis LLC Books. Reproduced from [1] with permission

Thus, we have the following set of linear inequalities for stability:

$$\begin{aligned} k_i &> 0 \\ -3.8114 + k_i - 0.5506k_d &< 0, \\ 12.2106 + k_i - 3.4782k_d &> 0, \\ -457.0235 + k_i - 61.6853k_d &< 0. \end{aligned}$$

The complete set of stabilizing PID gains for $k_p^* = 1$ is given in Fig. 5.8. This set is determined by finding the string of integers $\{i_0, i_1, i_2, i_3\}$ satisfying the stability conditions (5.45) and (5.46). The corresponding linear inequalities (5.47) determine the stabilizing set shown in the (k_i, k_d) space. The point marked with "*" was used as a test point to verify stability.

By sweeping k_p we have the entire set of stabilizing PID gains as shown in Fig. 5.9. This set is determined by sweeping over the feasible range of k_p determined in Fig. 5.7, and solving the corresponding linear inequalities given by (5.47) in (k_i, k_d) space for a fixed k_p^* in this range.

Note that the range of k_p over which the search needs to be carried out is also obvious from Fig. 5.6 as discussed and it is $k_p \in [-8.5, 4.2]$.

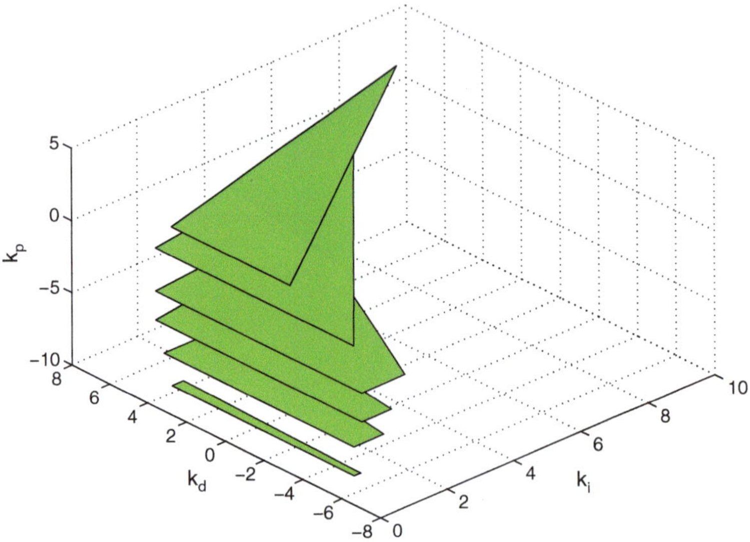

Fig. 5.9 Entire set of stabilizing PID gains. © Taylor & Francis LLC Books. Reproduced from [1] with permission

5.7 Model-Free Synthesis of First-Order Controllers

Consider the feedback configuration with an LTI plant with a first-order controller as shown in Fig. 5.1.

We consider the synthesis and design of a first-order controller of the form:

$$C(s) = \frac{x_1 s + x_2}{s + x_3}, \tag{5.54}$$

for an LTI plant for which we only know the frequency response data $P(j\omega)$, $\omega \in [0, \infty)$ and number, p^+, of the RHP poles of the plant. We also assume that the plant has no $j\omega$-axis poles or zeros. Suppose that the plant transfer function is $P(s)$ and

$$P(j\omega) = P_r(\omega) + jP_i(\omega) := |P(j\omega)|(\cos\phi(\omega) + j\sin\phi(\omega)),$$

where $\phi(\omega) = \angle P(j\omega)$. Consider the real rational function

$$F(s) := (s + x_3) + (sx_1 + x_2)P(s). \tag{5.55}$$

For closed-loop stability of the plant with a first-order controller, it is necessary and sufficient that

$$\sigma(F(s)) = n + 1 - (p^- - p^+). \tag{5.56}$$

Let

$$\bar{F}(s) := F(s)P(-s). \tag{5.57}$$

The stability condition can be restated as

$$\sigma(\bar{F}(s)) = n + 1 + (z^+ - z^-) = n - m + 2z^+ + 1. \tag{5.58}$$

Now

$$\bar{F}(s) = (s + x_3)P(-s) + (sx_1 + x_2)P(s)P(-s), \tag{5.59}$$

so that

$$\begin{aligned}\bar{F}(j\omega, x_1, x_2, x_3) = &\underbrace{x_2|P(j\omega)|^2 + \omega P_i(\omega) + x_3 P_r(\omega)}_{\bar{F}_r(\omega, x_2, x_3)} \\ &+ j\omega \underbrace{\left(x_1|P(j\omega)|^2 - x_3 P_i(\omega) + P_r(\omega)\right)}_{\bar{F}_i(\omega, x_1, x_3)}.\end{aligned}$$

It is easy to show that the curves $\bar{F}_r(\omega, x_2, x_3) = 0$ and $\bar{F}_i(\omega, x_1, x_3) = 0$, $0 \le \omega < \infty$ along with the $\bar{F}(0, x_1, x_2, x_3) = 0$ and $\bar{F}(\infty, x_1, x_2, x_3) = 0$ partition the (x_1, x_2, x_3) parameter space into signature invariant regions. By plotting these curves and selecting a test point from each of these regions, we can determine the stability regions corresponding to those with signature equal to $n - m + 2z^+ + 1$. To summarize, we have the following procedure.

Computation of First-Order Stabilizing Set for Continuous-Time Systems

1. Determine the relative degree $n - m$ from the high frequency slope of the Bode magnitude plot.
2. Let $\Delta_0^\infty[\phi(\omega)]$ denote the net change of phase $P(j\omega)$ for $\omega \in [0, \infty)$. Determine z^+ from knowledge of p^+ and

$$\Delta_0^\infty[\phi(\omega)] = -\left[(n - m) - 2(p^+ - z^+)\right]\frac{\pi}{2}. \tag{5.60}$$

3. Plot the curves below in the (x_1, x_2) plane for a fixed x_3.

$$x_3 + x_2 P(0) = 0, \tag{5.61}$$

$$\begin{cases} x_1(\omega) = \frac{1}{|P(j\omega)|}\left(\frac{\sin\phi(\omega)}{\omega}x_3 - \cos\phi(\omega)\right), \text{ for } 0 < \omega < \infty, \\ x_2(\omega) = -\frac{1}{|P(j\omega)|}\left(\cos\phi(\omega)x_3 + \omega\sin\phi(\omega)\right), \text{ for } 0 < \omega < \infty, \end{cases} \tag{5.62}$$

$$1 + P(\infty)x_2 = 0. \tag{5.63}$$

4. The curves $x_1(\omega)$ and $x_2(\omega)$ partition the (x_1, x_2) plane into disjoint signature-invariant regions. The stabilizing regions correspond to those for which $\bar{F}(s)$ has a signature of $n - m + 2z^+ + 1$.

The procedure follows from the preceding discussion. Equations in (5.62) are just

$$\bar{F}_r(\omega, x_1, x_2, x_3) = 0 \quad \text{and} \quad \bar{F}_i(\omega, x_1, x_2, x_3) = 0. \tag{5.64}$$

The signature associated with a region can be found by picking an arbitrary point in the region and using the formulas in Lemma 5.1. An alternative way to check the signature is to pick any one test point from each region and draw the Nyquist plot of $C(j\omega)P(j\omega)$. Stabilizing regions correspond to those points which give $(p^+ + 1)$ counterclockwise encirclements of the $-1 + j0$ point if $x_3 > 0(<0)$.

Example 5.2 For illustration, we have collected the frequency domain (Nyquist–Bode) data of a stable plant:

$$\mathbf{P}(j\omega) = \{P(j\omega) : \omega \in (0, 10) \text{ sampled every } 0.01\}.$$

The Nyquist plot of the plant obtained is shown in Figs. 5.10 and 5.11.

From the data $\mathbf{P}(j\omega)$, we have $P(0) = 13.333$ and $P(\infty) = 0$. Then it is easy to see that the straight line (5.63) is not applicable. After fixing $x_3 = 0.2$, the data points representing the straight line in (5.61) and the curve in (5.62) are depicted in Fig. 5.12. By testing a point for each root-invariant region, we obtained the stabilizing regions shown in Fig. 5.12.

5.8 Data-Based Design Versus Model-Based Design

In this section, we discuss some differences between model-based design and the data-based designs described here. In model-based design, mathematical models are obtained from the laws of physics that describe the dynamic behavior of a system to be controlled. On the other hand, the most common way of obtaining mathematical models in engineering systems is through a system identification process. Let us assume that the frequency domain data is obtained by exciting a linear time-invariant plant by sinusoidal signals. In theory, a system identification procedure should exactly be able to determine the unknown rational transfer function. In this ideal situation there should be no distinction between model-based and data-based synthesis methods. However, typical system identification procedures can fail to find an exact rational function even if exact (or perfect) data is available. This is especially true when the order of the plant is high. The following example illustrates that this in turn can lead to drastic differences in control, synthesis, and design.

Example 5.3 Let us assume that the frequency domain data $P(j\omega)$ shown below is obtained from a 20th order plant. Note that the plant is unstable with two RHP

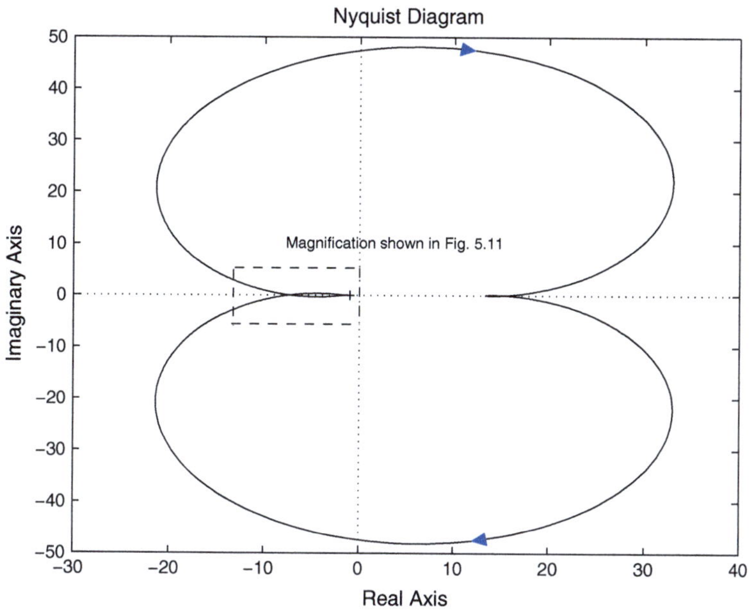

Fig. 5.10 Nyquist plot of $P(j\omega)$. © Taylor & Francis LLC Books. Reproduced from [1] with permission

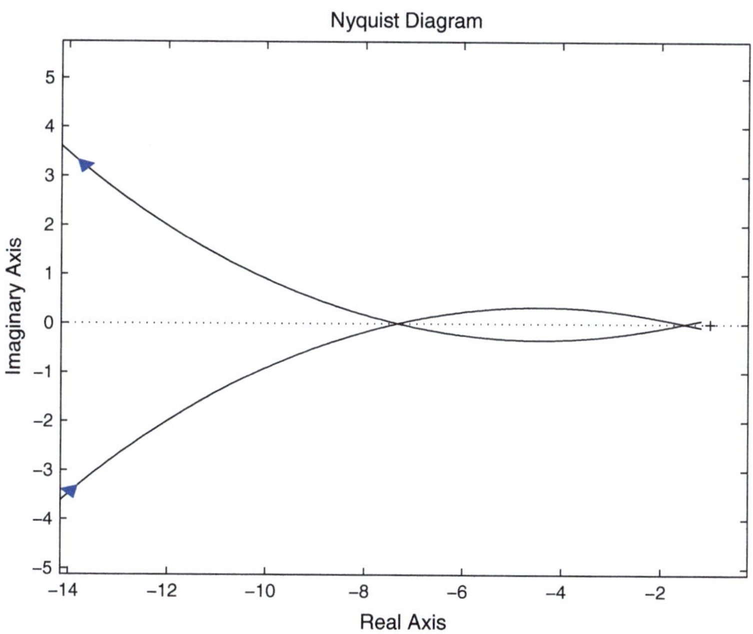

Fig. 5.11 Nyquist plot of $P(j\omega)$ (Area magnified). © Taylor & Francis LLC Books. Reproduced from [1] with permission

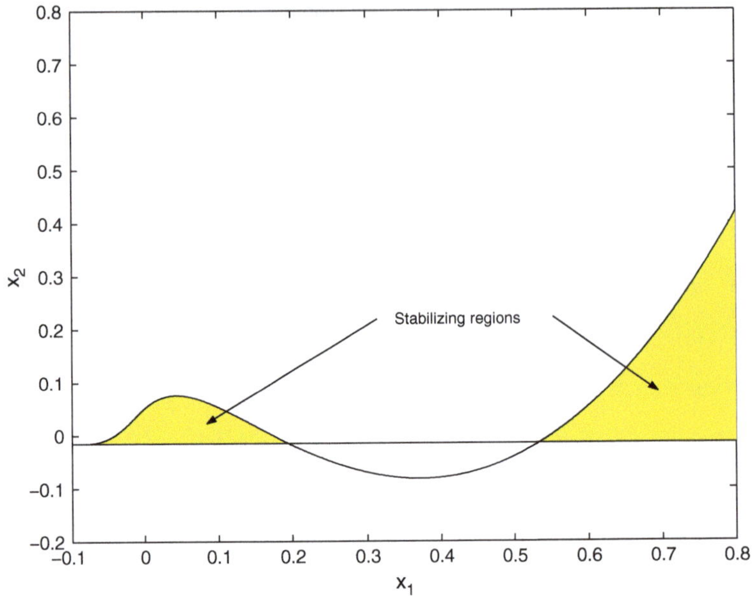

Fig. 5.12 Stabilizing regions for $x_3 = 0.2$. © Taylor & Francis LLC Books. Reproduced from [1] with permission

poles. Then mathematical models of 20th, 10th, 7th, and 4th orders were obtained by a system identification process applied to $P(j\omega)$. Figure 5.13 shows that the Bode plots of the four identified models along with the frequency domain data collected from the 20th order plant considered here. It is observed that the Bode plots of these are almost identical except the fact that the fourth-order identified model is relatively crude. We now compute the stabilizing PID parameter regions of each of these systems. Figure 5.14 shows the stabilizing regions in the PID controller parameter space.

We can make the following observations. For convenience, let us denote by $G_{20}(s)$, $G_{10}(s)$, $G_7(s)$ and $G_4(s)$ the 20th, 10th, 7th, and 4th order models identified, respectively.

1. The models $G_{20}(s)$ and $G_4(s)$ are found to be not PID stabilizable for the chosen $k_p = 5$. In other words, the stabilizing region in PID parameter space for the given plant is empty with $k_p = 5$. This is consistent with the data-driven case as is evident from Fig. 5.15. The signature condition requires that the line representing $k_p = 5$ must intersect the $g(\omega)$ graph a minimum of three times for stability. Figure 5.15 shows that such a necessary condition cannot be satisfied with $k_p > 4.5$.
   This is not unexpected for the model $G_4(s)$ since there is some difference between the identified and actual Bode (frequency response) plots.

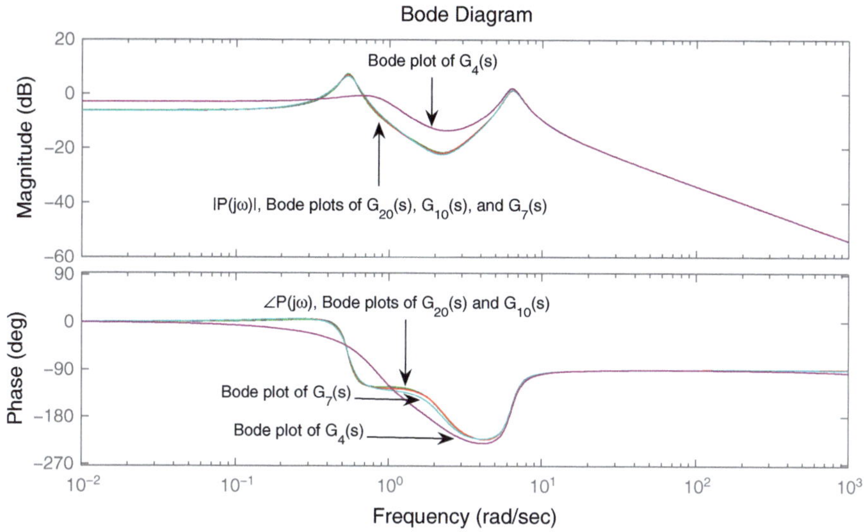

Fig. 5.13 Frequency domain data and the Bode plots of 20th, 10th, 7th, 4th order identified models. © Taylor & Francis LLC Books. Reproduced from [1] with permission

2. We have verified that $G_{20}(s)$ is also not PID stabilizable for any value of k_p. This may seem surprising since the Bode magnitude and phase plot of $G_{20}(s)$ is indistinguishably close to the data $P(j\omega)$. In fact, $G_{20}(s)$ has additional RHP poles and zeros over those in the original plant model which makes stabilization difficult.
3. The stabilizing regions are found for $G_{10}(s)$ and $G_7(s)$. These regions overlap, but are not the same. It suggests that selection of controllers might be done inside the intersection of the stabilizing regions for $G_{10}(s)$ and $G_7(s)$.
4. The stabilizing region obtained from the data-based method given here differs from those for $G_{10}(s)$ and $G_7(s)$. Thus, a reasonable selection of controller may be done inside the intersection of the stabilizing regions for $G_{10}(s)$, $G_7(s)$, and the region obtained from the data based method.
5. In general, the accuracy of system identification depends on accuracy of the data considered. On the other hand, the data based approach will work effectively as long as the roots of $g(\omega) = k_p^*$ are found reliably.

In practice, experimentally obtained frequency domain data always contains noise and measurement errors. As discussed above, the stability regions determined by the model-based design and the data-based design will generally be different. Although the accuracy of the regions depends on each particular case, the data-based design gives useful alternatives to model-based design methods and in general the two complement each other. In particular, it gives new guidelines for identification when it is to be used for controller design.

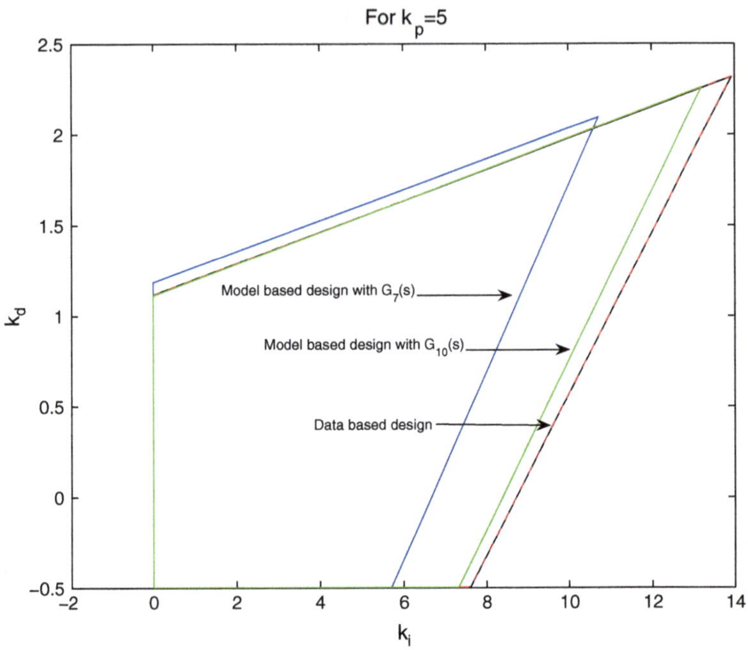

Fig. 5.14 Stabilizing regions. © Taylor & Francis LLC Books. Reproduced from [1] with permission

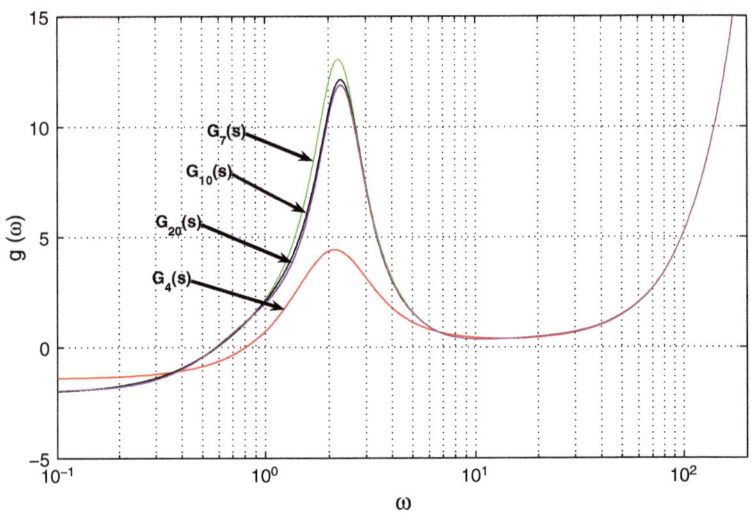

Fig. 5.15 $g(w)$ versus ω plot for the identified systems. © Taylor & Francis LLC Books. Reproduced from [1] with permission

5.9 Notes and References

The main results are taken from [1] and closely follow [2–5].

References

1. Bhattacharyya, S.P., Datta, A., Keel, L.H.: Linear Control Theory Structure, Robustness, and Optimization. CRC Press Taylor & Francis Group, Boca Raton (2009)
2. Keel, L.H., Bhattacharyya, S.P.: Direct synthesis of first order controllers from frequency response measurements. In: Proceedings of the American Control Control, Portland, Oregon, 8–10 June 2005
3. Keel, L.H., Bhattacharyya, S.P.: PID controller synthesis free of analytical models. In: Proceedings of the 16th IFAC World Congress, Prague, Czech Republic, 4–8 July 2005
4. Keel, L.H., Bhattacharyya, S.P.: Controller synthesis free of analytical models: three term controllers. IEEE Trans. Autom. Control **53**(6), 1353–1369 (2008)
5. Keel, L.H., Mitra, S., Bhattacharyya, S.P.: Data driven synthesis of three term digital controllers. SICE J. Control Meas. Syst. Integr. **1**(2), 102–110 (2008)

Part II
Robust Design Based on Gain and Phase Margins

Chapter 6
Gain and Phase Margin-Based Design for Continuous-Time Plants

Abstract In this chapter, we introduce an approach to the robust design of PID controllers for continuous-time plants based on gain and phase margin specifications. This design is based on a simple parametrization of constant magnitude and constant phase loci of the PID controller. This parametrization produces ellipses and straight lines for first-order and PI controllers and cylinder and plane for PID controllers. These geometric figures are computed by considering a prescribed but arbitrary gain crossover frequency and prescribed but arbitrary phase margin for the closed-loop system with the given plant. These graphical representations enable the retrieval of PI, PID, and first-order controller designs with simultaneous specifications on gain and phase margins.

6.1 Introduction

In classical control theory, for linear time-invariant systems, it is customary to apply a sinusoidal input signal to a physical system with the objective of obtaining the plant frequency response. This measurement is used subsequently for controller design. A comparison of the input signal and the output signal is made to obtain the ratio and difference, respectively, in magnitude and phase, and this information, ranging over all frequencies, is the frequency response of the system. Bode plots, Nyquist plots, and Nichols charts are examples of tools to represent the frequency response of a linear time-invariant system with the objective of reshaping the closed-loop frequency response by controller design.

In this chapter, we introduce a graphical representation of the magnitude and phase of the controller in the space of design parameters or controller gains when the gain crossover frequency and the desired phase margin are prescribed. The graphical representation of a first-order or PI controller is an ellipse for constant magnitude of the controller and a straight line for constant phase of the controller at a prescribed frequency. For the PID controller, the graphical representation is given by a cylinder and a plane for constant magnitude and constant phase, respectively.

The chapter is organized as follows. In Sect. 6.2, we develop constant magnitude and constant phase loci for continuous-time three term controllers. In Sect. 6.3, we

I. D. Díaz-Rodríguez et al., *Analytical Design of PID Controllers*,
https://doi.org/10.1007/978-3-030-18228-1_6

present the procedure to compute the gain–phase margin design curves. These curves represent the achievable performance of the system using a PI, PID, or first-order controller. Section 6.4 presents a brief introduction to time-delay tolerance as additional information that can be represented graphically. Section 6.5 presents a procedure to simultaneously specify desired performance specifications and the retrieval of the controller gains from the achievable performance set. In Sect. 6.6, we present a controller design procedure for delay-free systems. Also, we give examples for PI, PID, and first-order controller designs. In Sect. 6.8, we extend the description of constant magnitude and constant phase loci for PI and PID controllers considering systems with time-delay.

6.2 Magnitude and Phase Loci

Consider the general feedback system shown in Fig. 6.1.

Here $P(s)$ is the transfer function of the plant to be controlled and $C(s)$ is the transfer function of the controller to be designed. The plant $P(s)$ will be assumed to be rational and strictly proper. Therefore

$$P(s) = \frac{N(s)}{D(s)}, \tag{6.1}$$

where $N(s)$, $D(s)$ are polynomials in the Laplace variable s with degree $N(s) <$ degree $D(s)$. Let us assume that the rational proper controller $C(s)$ stabilizes $P(s)$. We denote by $P(j\omega)$ and $C(j\omega)$ the frequency responses of the plant and controller, respectively, where ω, the frequency in radians, runs from 0 to ∞. The Nyquist and Bode plots of

$$G(j\omega) := C(j\omega)\,P(j\omega) \tag{6.2}$$

are shown in Fig. 6.2 and Fig. 6.3, respectively.

Let ω_g denote the gain crossover frequency, ω_p the phase crossover frequency, GM the gain margin, and PM the phase margin of the closed-loop system. As explained in the previous chapter, gain and phase margins determine how stable or robust the control system is and they are the basis of classical control designs. From the frequency responses shown in Figs. 6.2 and 6.3, we have

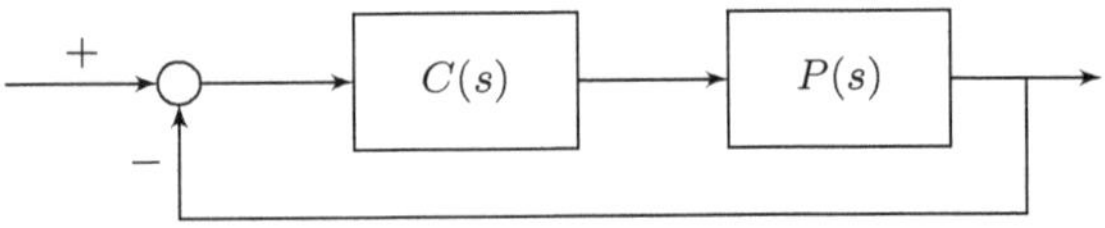

Fig. 6.1 Feedback control system

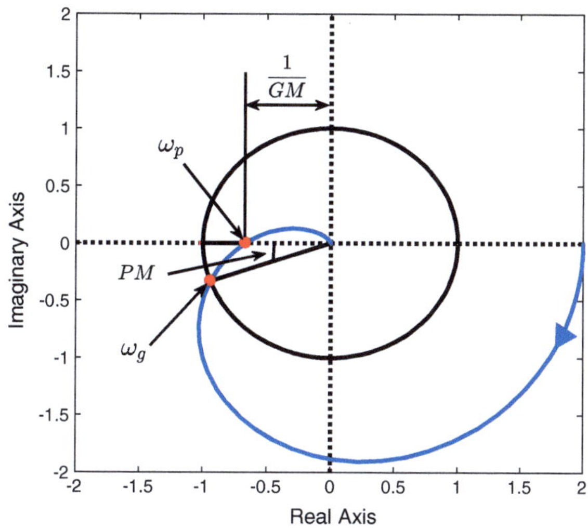

Fig. 6.2 Nyquist plot of $G(s)$ in (6.2)

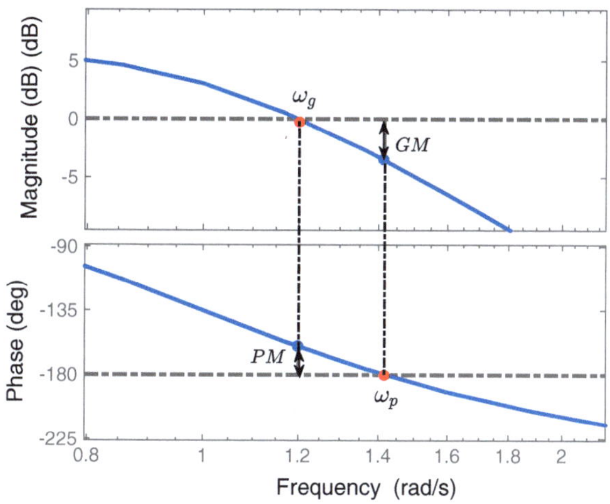

Fig. 6.3 Bode plot of $G(s)$ in (6.2)

$$|C(j\omega_g)P(j\omega_g)| = 1 \tag{6.3}$$

$$\angle(C(j\omega_p)P(j\omega_p)) = -n\pi,\ n = 1,\ 3,\ 5,\ ... \tag{6.4}$$

$$GM := \frac{1}{|C(j\omega_p)P(j\omega_p)|} \tag{6.5}$$

$$PM := \angle(C(j\omega_g)P(j\omega_g)) - \pi. \tag{6.6}$$

For a prescribed value of ω_g, Eqs. (6.3) and (6.6) represent the requirements on the magnitude and phase of the controller:

$$|C(j\omega_g)| = \frac{1}{|P(j\omega_g)|} \tag{6.7}$$

$$\angle(C(j\omega_g)) = n\pi + PM - \angle P(j\omega_g)),\ n \text{ odd}. \tag{6.8}$$

Equations (6.7) and (6.8), will lead to a graphical parametrization in the space of controller parameters represented by an ellipse and straight line for PI and first-order controllers and a cylinder and a plane for continuous-time PID controllers. In the following sections, we describe this parametrization.

We remark that ω_g is sometimes considered the bandwidth (BW) of the closed-loop system and can also represent a design specification.

6.2.1 PI Controllers

For a PI controller

$$C(s) = \frac{k_p s + k_i}{s}, \tag{6.9}$$

where k_p k_i are design parameters. Then, setting $s = j\omega$

$$C(j\omega) = \frac{k_p(j\omega) + k_i}{j\omega} \tag{6.10}$$

so that

$$|C(j\omega)|^2 = k_p^2 + \frac{k_i^2}{\omega^2} =: M^2 \tag{6.11}$$

and

$$\angle C(j\omega) = \arctan\left(\frac{-k_i}{\omega k_p}\right) := \Phi. \tag{6.12}$$

Equations (6.11) and (6.12) can be rewritten as

$$\frac{(k_p)^2}{a^2} + \frac{(k_i)^2}{b^2} = 1 \tag{6.13}$$

$$k_i = c k_p, \tag{6.14}$$

where

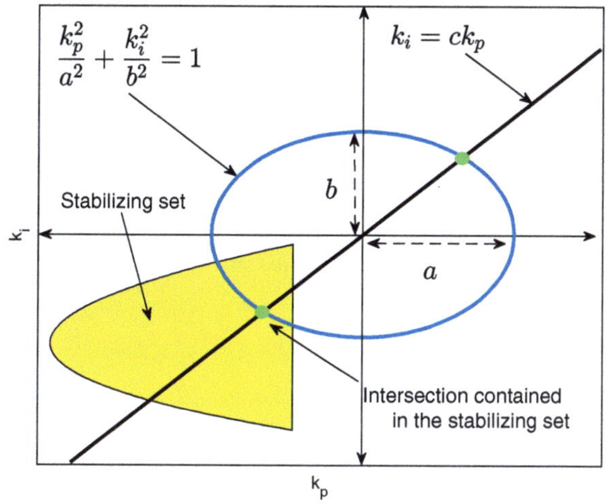

Fig. 6.4 Ellipse and straight line intersecting with a stabilizing set

$$a^2 = M^2 \tag{6.15}$$

$$b^2 = M^2\omega^2 \tag{6.16}$$

$$c = \omega \tan \Phi. \tag{6.17}$$

Thus, for given ω, constant M loci are ellipses and constant phase loci are straight lines in k_p, k_i space. The major and minor axes of the ellipse are given by a and b (see Fig. 6.4). The slope of the line is represented by c (see Fig. 6.4). Suppose ω_g is the prescribed closed-loop gain crossover frequency. Then

$$M = M_g := \frac{1}{|P(j\omega_g)|} \tag{6.18}$$

and if ϕ_g^* is the desired closed-loop phase margin in radians,

$$\Phi = \Phi_g := \pi + \phi_g^* - \angle P(j\omega_g). \tag{6.19}$$

From (6.11) and (6.12), we obtain the ellipse and straight line corresponding to $M = M_g$ and $\Phi = \Phi_g$, giving the design point (k_p^*, k_i^*). If this intersection point lies in the stabilizing set S, the design is feasible; otherwise, the specifications are unattainable and have to be revised.

6.2.2 PID Controllers

Let $P(s)$ and $C(s)$ denote the plant and controller transfer functions. The frequency response of the plant and controller are $P(j\omega)$, $C(j\omega)$, respectively, where $\omega \in$

$[0, \infty]$. For a PID controller

$$C(s) = \frac{k_d s^2 + k_p s + k_i}{s}, \tag{6.20}$$

where k_p, k_i, and k_d are the design parameters. Then setting $s = j\omega$

$$C(j\omega) = \frac{k_d (j\omega)^2 + k_p (j\omega) + k_i}{j\omega}. \tag{6.21}$$

From (6.21), we have

$$|C(j\omega)|^2 = k_p^2 + \left(k_d\omega - \frac{k_i}{\omega}\right)^2 := M^2 \tag{6.22}$$

and

$$\angle C(j\omega) = \arctan\left(\frac{k_d\omega - \frac{k_i}{\omega}}{k_p}\right) := \Phi. \tag{6.23}$$

Equation (6.22) represents an ellipse in the (k_p, k_d) space with center at $(0, \frac{k_i}{\omega^2})$ and in (k_p, k_i, k_d) space is an elliptical cylinder. From (6.22) and (6.23), we have

$$k_p^2 = M^2 - \left(k_d\omega - \frac{k_i}{\omega}\right)^2 = \frac{\left(k_d\omega - \frac{k_i}{\omega}\right)^2}{\tan^2\Phi}, \tag{6.24}$$

which leads to the following expressions:

$$k_i = k_d\omega^2 \pm k_p\omega \tan\Phi \tag{6.25}$$

$$k_p = \pm\sqrt{\frac{M^2}{1 + \tan^2\Phi}}. \tag{6.26}$$

Suppose ω_g is the prescribed closed-loop gain crossover frequency. Then

$$M = \frac{1}{|P(j\omega_g)|}. \tag{6.27}$$

If ϕ_g^* is the desired phase margin in radians,

$$\Phi = \pi + \phi_g^* - \angle P(j\omega_g). \tag{6.28}$$

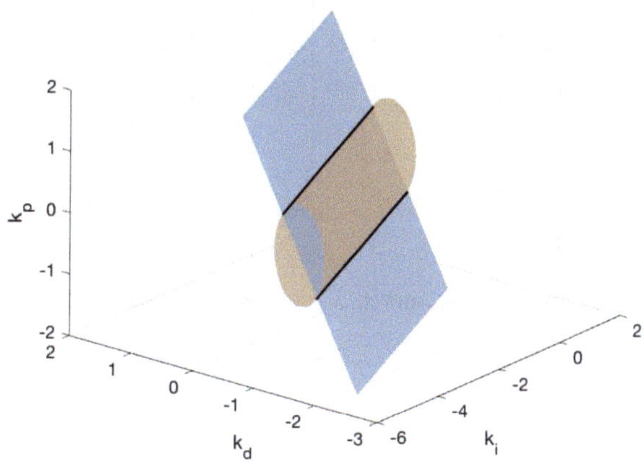

Fig. 6.5 Cylinder and plane intersecting in the (k_p, k_i, k_d) space

Thus, from Eqs. (6.22) and (6.23) with $M = M_g$, $\Phi = \Phi_g$, we have a cylinder and a plane (see Fig. 6.5) in (k_p, k_i, k_d) space. The plane is represented by Eq. (6.25) and the cylinder by (6.22).

6.2.3 First-Order Controllers

Let $P(s)$ and $C(s)$ denote the plant and controller transfer functions. The frequency responses of the plant and controller are denoted $P(j\omega)$, $C(j\omega)$, respectively. For the first-order controller

$$C(s) = \frac{x_1 s + x_2}{s + x_3}, \tag{6.29}$$

where x_1, x_2, and x_3 are real design parameters. Then setting $s = j\omega$

$$C(j\omega) = \left(\frac{x_1(j\omega) + x_2}{j\omega + x_3}\right)\left(\frac{x_3 - j\omega}{x_3 - j\omega}\right) = \frac{(x_1\omega^2 + x_2 x_3) + j\omega(x_1 x_3 - x_2)}{x_3^2 + \omega^2}. \tag{6.30}$$

Let

$$L_0 := x_1\omega^2 + x_2 x_3, \quad L_1 := x_1 x_3 - x_2. \tag{6.31}$$

From (6.30) and (6.31)

$$|C(j\omega)|^2 = \frac{L_0^2}{(x_3^2+\omega^2)^2} + \frac{L_1^2}{\left(\frac{x_3^2+\omega^2}{\omega}\right)^2} =: M^2 \tag{6.32}$$

and

$$\angle C(j\omega) = \arctan\left(\frac{\omega L_1}{L_0}\right) =: \Phi. \tag{6.33}$$

Equations (6.32) and (6.33) can be rewritten as

$$\frac{L_0^2}{a^2} + \frac{L_1^2}{b^2} = 1, \tag{6.34}$$
$$L_1 = cL_0, \tag{6.35}$$

where

$$a^2 = M^2(x_3^2+\omega^2)^2, \tag{6.36}$$
$$b^2 = \frac{M^2}{\omega}(x_3^2+\omega^2)^2, \tag{6.37}$$
$$c = \frac{\tan\Phi}{\omega}. \tag{6.38}$$

Thus, for a given ω and x_3, constant M loci are ellipses and constant phase loci are straight lines in L_0, L_1 space. The major and minor axes of the ellipse are given by a and b. The slope of the line is represented by c. The mapping from x_1, x_2 to L_0, L_1 and vice versa for prescribed x_3 is given by (6.31) and

$$x_1 = \frac{L_0 + L_1 x_3}{x_3^2+\omega^2}, \quad x_2 = \frac{L_0 x_3 - \omega^2 L_1}{x_3^2+\omega^2}. \tag{6.39}$$

Suppose ω_g is the prescribed closed-loop gain crossover frequency. This corresponds roughly to another design specification, namely, bandwidth (BW). Then

$$M := \frac{1}{|P(j\omega_g)|} \tag{6.40}$$

and if ϕ_g^* is the desired phase margin in radians,

$$\Phi := \pi + \phi_g^* - \angle P(j\omega_g). \tag{6.41}$$

From (6.34) and (6.35), we obtain the ellipse and straight lines corresponding to $M = M_g$ and $\Phi = \Phi_g$, giving the design point (x_1^*, x_2^*, x_3^*). If this intersection point lies in the stabilizing set S for first-order controllers for the given plant, the design is feasible; otherwise, the specifications have to be altered.

6.3 Achievable Gain–Phase Margin Design Curves

The gain–phase margin design curves represent the set of achievable gain margin GM), phase margin PM), and gain crossover frequencies (ω_g) for our plant with a PI, or PID controller. The procedure for constructing these design curves (see Fig. 6.11, for example) is as follows:

1. Set a range of PM $\phi_g^* \in [\phi_g^-, \phi_g^+]$ and gain crossover frequency $\omega_g \in [\omega_g^-, \omega_g^+]$.
2. For prescribed values of ϕ_g^* and ω_g, plot the corresponding ellipse and straight line.
3. If the intersection point of the ellipse and straight line lies outside of the stabilizing set, then this point is rejected and we go to step 2.
4. If the intersection of the ellipse and straight line is contained in the stabilizing set, it represents a feasible design point with the PI, or PID controller gains (k_p^*, k_i^*) or (k_p^*, k_i^*, k_d^*) that satisfies the prescribed ϕ_g^* and ω_g.
5. Given the selected PI or PID controller gains (k_p^*, k_i^*) or (k_p^*, k_i^*, k_d^*), the upper and lower GM of the system are given by

$$GM_{upper} = \frac{k_p^{ub}}{k_p^*} \quad \text{and} \quad GM_{lower} = \frac{k_p^{lb}}{k_p^*}, \tag{6.42}$$

   where k_p^{ub} and k_p^{lb} are the controller gains at the further and close boundary, respectively, of the stabilizing set following the straight line intersecting the ellipse.
6. Go to step 2 and repeat for all values of ϕ_g^* and ω_g in the ranges.

6.4 Time-Delay Tolerance

The time-delay tolerance design curves represent the actual time-delay tolerance achievable with a given PI or PID controller. This set of design curves is an extension of the previous gain–phase design curves because we can use the information calculated before to create this new time-delay tolerance design set. The time-delay tolerance for a chosen design point can be calculated by

$$\tau := \frac{PM}{\omega_g}, \tag{6.43}$$

where PM is the phase margin in radians and ω_g is the gain crossover frequency in radians per second. Then, taking all the points calculated from the gain–phase margin design set, we can find the values of time-delay tolerance and express the new plot with the x-axis as phase margin and y-axis as time-delay tolerance. Similarly to gain–phase margin design curves, these time-delay tolerance curves are indexed by a prescribed value of gain crossover frequency.

6.5 Simultaneous Performance Specifications and Retrieval of Controller Gains

The designer can select a desired design point from the achievable gain–phase margin curves and retrieve the controller gains corresponding to this simultaneous specification of desired GM, PM, and ω_g. The controller gain retrieval process is as follows:

1. Select desired GM, PM, and ω_g from the achievable gain–phase margin curves.
2. For the specified point, construct the ellipse and straight line for a PI controller and a cylinder and plane for PID controller by using the selected PM and ω_g from the constant gain and constant phase loci.
3. Select the intersection of the ellipse and straight line or cylinder and plane contained in the stabilizing set. This will provide the gains (k_p^*, k_i^*) or (k_p^*, k_i^*, k_d^*).
4. The controller that satisfies the prescribed gain and phase margins and ω_g specifications is $C(s) = \frac{k_p^* s + k_i^*}{s}$ or $C(s) = \frac{k_d^* s^2 + k_p^* s + k_i^*}{s}$.

The same procedure is used for retrieving the controller gains using the time-delay tolerance design curves. We can select a point from the time-delay tolerance design curves and use the value of phase margin and gain crossover frequency to compute the ellipse and straight line and find the intersection contained in the stabilizing set, if it exists.

6.6 Gain–Phase Margin-Based Controller Design for Delay-Free Systems

The PI, PID, or first-order controller design approach to satisfy simultaneous specifications of gain margin, phase margin, and gain crossover frequency for a linear time-invariant single-input single-output plant can be summarized as follows:

1. Compute the PI, first-order, or PID Stabilizing Set.
2. Parametrize constant gain and phase loci.
3. Construct the gain–phase margin design curves.
4. Select simultaneous gain margin, phase margin, and gain crossover frequency design specifications from the achievable gain–phase margin design curves.
5. Retrieve the PI or PID controller gains satisfying the design specifications from the intersection of gain and phase loci points, lying in the stabilizing set.

6.7 Examples

Example 6.1 (Continuous-Time PI Controller Design) Consider Fig. 6.1 with an unstable plant with transfer function:

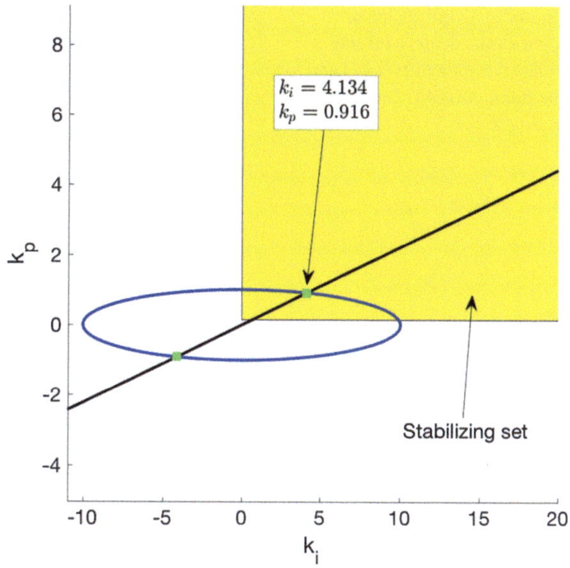

Fig. 6.6 Intersection of ellipse and straight line superimposed on the stabilizing set for Example 6.1

$$P(s) = \frac{10}{s-1} \tag{6.44}$$

and the PI controller in Eq. (6.9). In this example, we show how to find the controller gains that satisfy the desired performance by computing an ellipse and a straight line and superimposing on the stabilizing set. For this example, the stabilizing set can be found from the characteristic equation

$$\delta(s) = s^2 + (10k_p - 1)s + 10k_i. \tag{6.45}$$

We find the stabilizing region is given by

$$k_p > \frac{1}{10} \qquad \text{and} \qquad k_i > 0. \tag{6.46}$$

This means the stabilizing region is unbounded and contained in the first quadrant in the (k_p, k_i) plane. Now, consider a design value for a specific gain crossover frequency and phase margin. For illustration purposes, consider $\omega_g^* = 10$ rad/s and $\phi_g^* = 60°$. Using Eqs. (6.13) and (6.14), representing the constant magnitude and constant phase of the controller, we can compute the corresponding ellipse and straight line for the prescribed values of ω_g^* and ϕ_g^*. In Fig. 6.6, we can see the intersection of the ellipse and straight line superimposed on the stabilizing set.

The intersection of the ellipse and straight line contained in the stabilizing set provides the desired values of ω_g^* and ϕ_g^*. The corresponding PI controller gains are

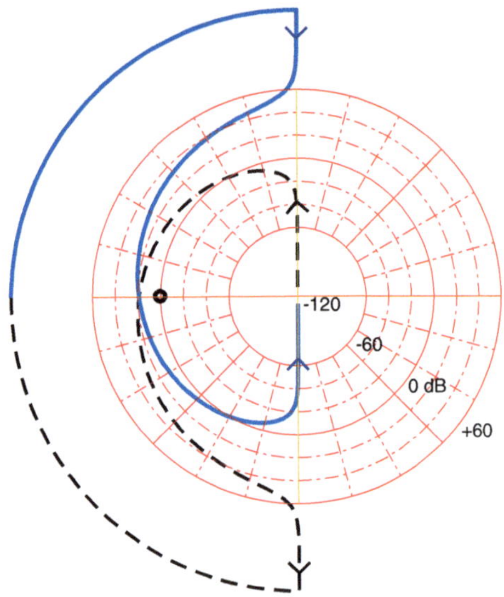

Fig. 6.7 Nyquist plot in logarithmic scale with the controller gains in (6.47) and the plant in (6.44) for Example 6.1

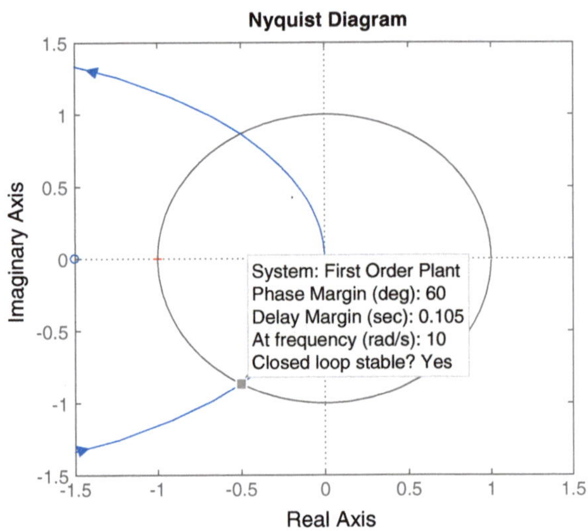

Fig. 6.8 Nyquist plot with the controller gains in (6.47) and the plant in (6.44) for Example 6.1

$$k_p = 0.916 \quad \text{and} \quad k_i = 4.134. \tag{6.47}$$

Considering the controller gains in (6.47) and the plant in (6.44), we can inspect the Nyquist plot and verify stability, phase margin, and gain crossover frequency.

In Fig. 6.7, we can see the Nyquist plot using a logarithmic scale. In this figure, we notice that we have one counterclockwise encirclement of $-1 + j0$. Therefore, the closed-loop system is asymptotically stable. In Fig. 6.8, we can verify that the phase margin of the system is 60° and the gain crossover frequency is 10 rad/s.

Example 6.2 (Continuous-Time PI controller Design) Let us consider the continuous-time system represented in Fig. 6.1 using the plant

$$P(s) = \frac{s-5}{s^2+1.6s+0.2} \tag{6.48}$$

and the controller

$$C(s) = \frac{k_p s + k_i}{s}. \tag{6.49}$$

A. Computation of the Stabilizing Set

The first step in the controller design procedure is to obtain the stabilizing set of PI controllers for the given plant. The closed-loop characteristic polynomial is

$$\delta(s, k_p, k_i) = s^3 + (k_p + 1.6)s^2 + (k_i - 5k_p + 0.2)s - 5k_i. \tag{6.50}$$

Here $n = 2$, $m = 1$, and $N(-s) = -5 - s$. Therefore, we obtain

$$\begin{aligned} \nu(s) &= \delta(s, k_p, k_i)N(-s) \\ &= -s^4 - (6.6 + k_p)s^3 - (8.2 + k_i)s^2 + (25k_p - 1)s + 25k_i \end{aligned} \tag{6.51}$$

so that

$$\begin{aligned} \nu(j\omega, k_p, k_i) &= (-\omega^4 + (k_i + 8.2)\omega^2 + 25k_i) + j[(k_p + 6.6)\omega^3 + (25k_p - 1)\omega] \\ &= p(\omega) + jq(\omega). \end{aligned} \tag{6.52}$$

We find that $z^+ = 1$ so that the signature requirement on $\nu(s)$ for stability is

$$n - m + 1 + 2z^+ = 4. \tag{6.53}$$

Since the degree of $\nu(s)$ is even, we see from the signature formulas that $q(\omega)$ must have at least one positive real root of odd multiplicity. The range of k_p such that $q(\omega, k_p)$ has at least one real, positive, distinct, finite zero with odd multiplicity was determined to be $k_p \in (-1.6, 0.04)$ which is the allowable range for k_p. By sweeping over different k_p values within the interval $(-1.6, 0.04)$, we can generate the set of stabilizing (k_p, k_i) values. This set is shown in Fig. 6.9.

B. Construction of Achievable Gain–Phase Margin Design Curves

For a range of phase margins and gain crossover frequencies, we superimpose ellipses and straight lines on the stabilizing set (see Fig. 6.10).

We can see in Fig. 6.10 the intersection points of ellipses and straight lines for different values of phase margin and gain crossover frequency. We notice how for different values of phase margin, the gain crossover frequency limit is different. In

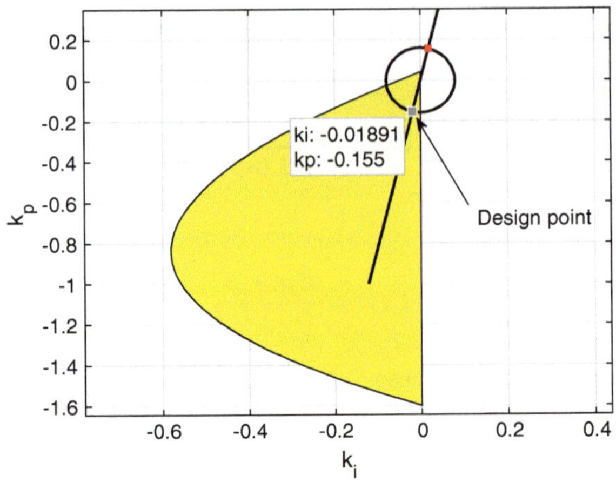

Fig. 6.9 PI stabilizing set for Example 6.2 and intersection of ellipse and straight line for the final design point. © 2016 IEEE. Reproduced from [4] with permission

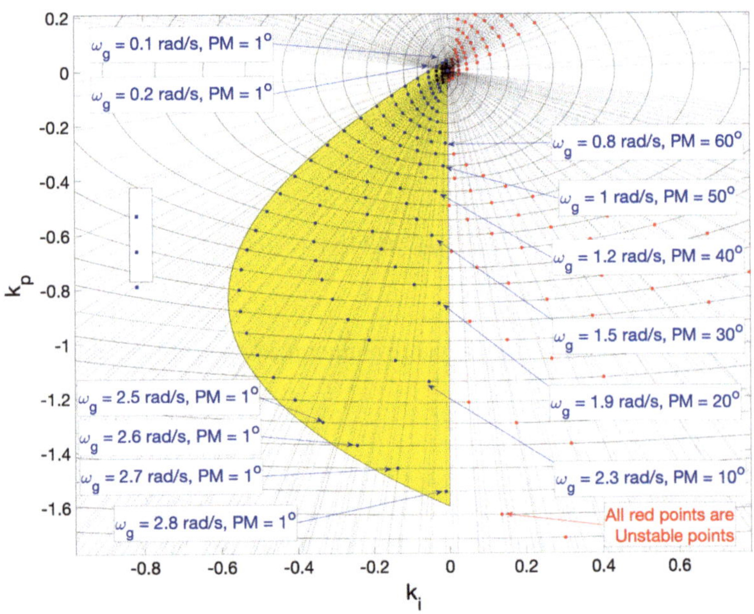

Fig. 6.10 Construction of the gain–phase margin design curves for PI controller design in Example 6.2 by intersection points of ellipses and straight lines. © 2016 IEEE. Reproduced from [4] with permission

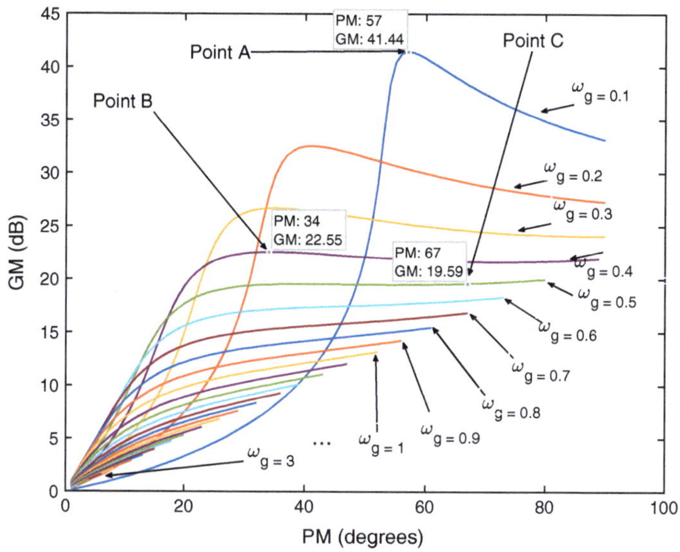

Fig. 6.11 Achievable gain–phase margin design curves in the gain–phase plane for PI controller design in Example 6.2. © 2016 IEEE. Reproduced from [4] with permission

this way, we obtain the maximum achievable values for the gain crossover frequency. For example, the maximum value of gain crossover frequency is 2.3 rad/s when considering a $PM = 10°$ and the maximum gain crossover frequency is 0.8 rad/s when considering $PM = 60°$. All the intersection points, contained in the stabilizing set, determine the PI controller gains and are used to construct the gain–phase design curves shown in Fig. 6.11. For this example, the evaluated range for phase margin is from 1° to 90° and the range for the gain crossover frequency is from 0.1 to 3 rad/s.

C. Simultaneous Performance Specifications and Retrieval of Controller Gains

As shown in Fig. 6.11, we can clearly see the achievable performance for Example 6.2. In this case, the maximum gain margin that we can get is 41.44 dB. The phase margin, corresponding to this maximum gain margin, is 57° with a gain crossover frequency of 0.1 rad/s. We represent this point as Point A in Fig. 6.11. We can see how increasing the gain crossover frequency leads to a decrease in our achievable values for gain and phase margins. For example, for a gain crossover frequency of 0.4 rad/s, the corresponding maximum gain margin we can get is 22.55 dB, and the corresponding phase margin is 34°. We call this Point B in Fig. 6.11. In Fig. 6.11, we have chosen a candidate design (Point C). The controller corresponding to these design specifications can be recovered by constructing the straight line and ellipse corresponding to these specifications (see Fig. 6.9). The PI controller gains for these specifications are

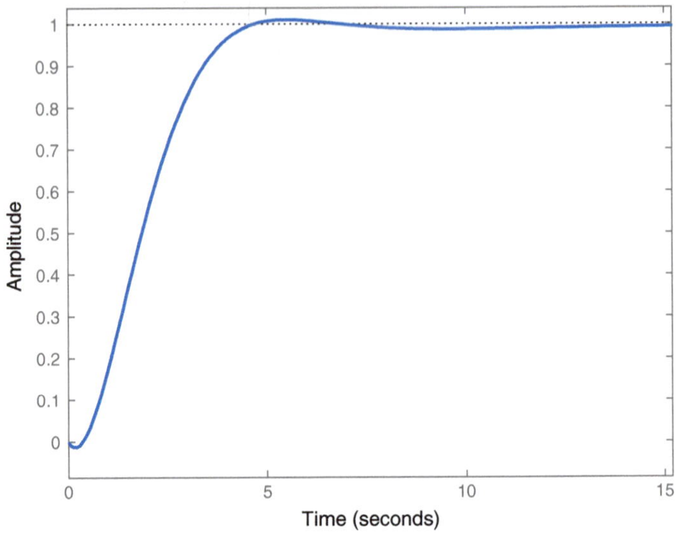

Fig. 6.12 Step response for the system in Example 6.2 using the PI controller design $C(s)^* = \frac{k_p^* s + k_i^*}{s}$. © 2016 IEEE. Reproduced from [4] with permission

$$k_p^* = -0.1556 \tag{6.54}$$
$$k_i^* = -0.0189. \tag{6.55}$$

The step response for this controller is given in Fig. 6.12. These controller gains correspond to the point of $\omega_g = 0.5$, $PM = 67°$, and $GM = 19.6$ dB in the gain–phase margin design plane (see Point C in Fig. 6.11).

In Fig 6.13, we can see the Nyquist plot for the controller gains selected. Here, we can see that those controller gains satisfy the desired performance specifications, $PM = 67°, GM = 19.6\,\text{dB}$.

We can also compute the time-delay tolerance design curves. Following Eq. (6.43) and taking the values from Fig. 6.11, we get Fig. 6.14.

In Fig. 6.14, we can see the achievable time-delay tolerances for the system using the proposed controller. We can select any point from the curves and retrieve the controller gains following the same procedure as selecting a point from the gain–phase margin design curves. In this case, we selected the same $PM = 67°$ and $\omega_g = 0.5$ rad/s. The time-delay tolerance is

$$\tau = 2.339\,\text{s}. \tag{6.56}$$

Example 6.3 (*Continuous-Time PI Controller Design*) Let us consider the continuous-time LTI plant

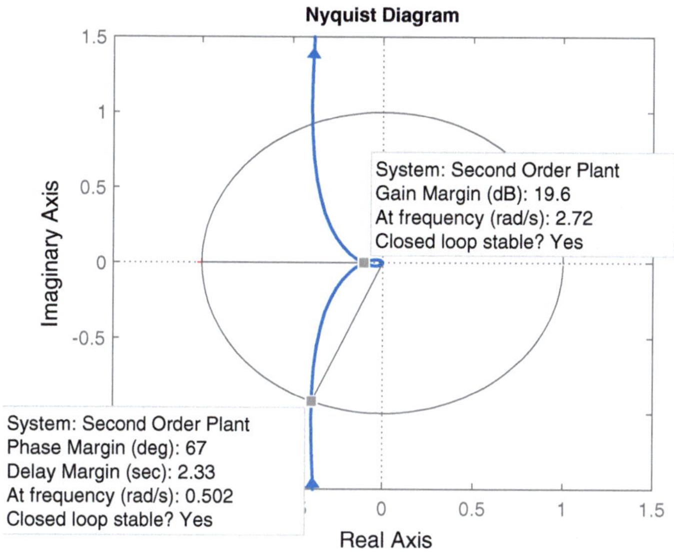

Fig. 6.13 Nyquist plot for $k_p = -0.1556$, $k_i = -0.0189$ in the PI controller design in Example 6.2

$$P(s) = \frac{s^3 - 4s^2 + s + 2}{s^5 + 8s^4 + 32s^3 + 46s^2 + 46s + 17} \tag{6.57}$$

and the controller

$$C(s) = \frac{k_p s + k_i}{s}. \tag{6.58}$$

A. Computation of the Stabilizing Set

The closed-loop characteristic polynomial is

$$\begin{aligned}\delta(s, k_p, k_i) &= s^6 + 8s^5 + (k_p + 32)s^4 + (k_i - 4k_p + 46)s^3 \\ &\quad + (k_p - 4k_i + 46)s^2 + (k_i + 2k_p + 17)s + 2k_i.\end{aligned} \tag{6.59}$$

Here $n = 5$, $m = 3$, and $N(-s) = -s^3 - 4s^2 - s + 2$. Therefore, we obtain

$$\begin{aligned}\nu(s) &= \delta(s, k_p, k_i)N(-s) \\ &= -s^9 - 12s^8 + (-k_p - 65)s^7 + (-k_i - 180)s^6 \\ &\quad + (14k_p - 246)s^5 + (14k_i - 183)s^4 + (-17k_p - 22)s^3 \\ &\quad + (75 - 17k_i)s^2 + (4k_p + 34)s + 4k_i\end{aligned} \tag{6.60}$$

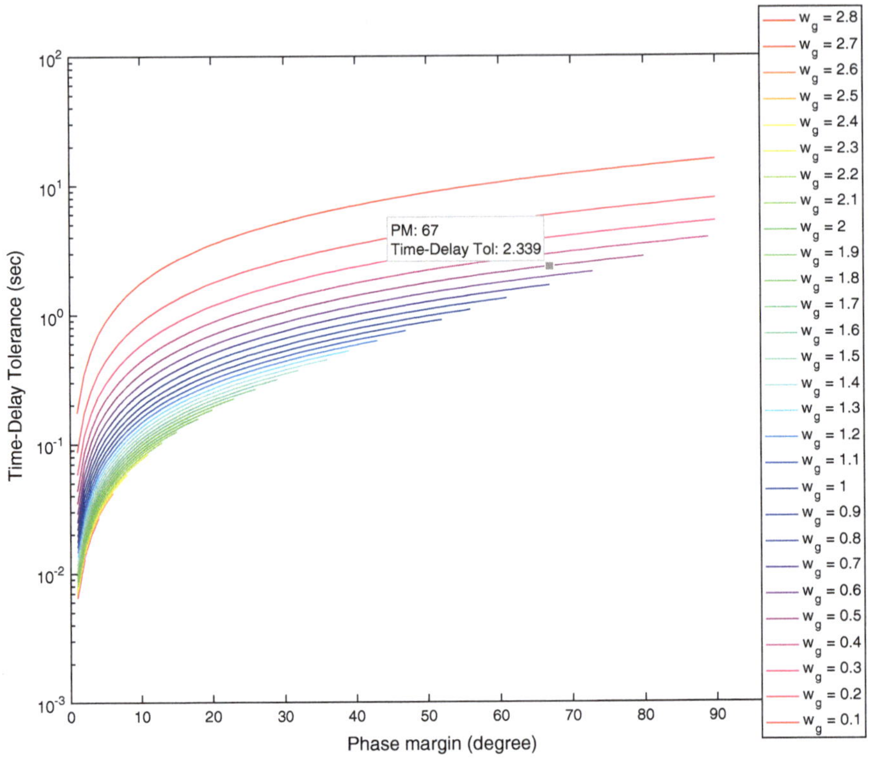

Fig. 6.14 Time-delay tolerance design curves for Example 6.2

so that

$$\begin{aligned}\nu(j\omega, k_p, k_i) =& -12\omega^8 + (k_i + 180)\omega^6 + (14k_i - 183)\omega^4 + (17k_i - 75)\omega^2 \\ &+ 4k_i + j[-\omega^9 + (k_p + 65)\omega^7 + (14k_p - 246)\omega^5 \\ &+ (17k_p + 22)\omega^3 + (4k_p + 34)\omega] \\ =& \; p(\omega) + jq(\omega). \end{aligned} \tag{6.61}$$

We find that $z^+ = 2$ so that the signature requirement on $\nu(s)$ for stability is

$$n - m + 1 + 2z^+ = 7. \tag{6.62}$$

Since the degree of $\nu(s)$ is odd, we see from the signature formulas that $q(\omega)$ must have at least three positive real zeros of odd multiplicity. The range of k_p such that $q(\omega, k_p)$ has at least one real, positive zero with odd multiplicity was determined to be $k_p \in (-8.5, 4.2)$ which is the allowable range for k_p. By sweeping over different

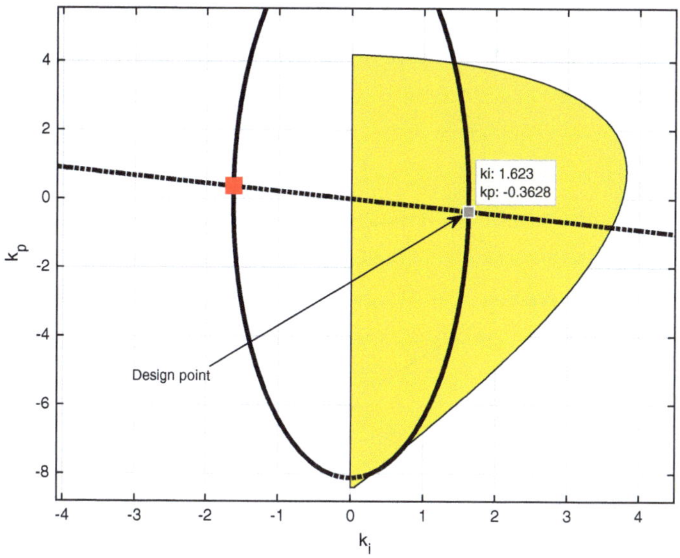

Fig. 6.15 PI stabilizing set for Example 6.3 and intersection of ellipse and straight line for the final design point. © 2016 IEEE. Reproduced from [4] with permission

k_p values within the interval $(-8.5, 4.2)$ we can generate the set of stabilizing (k_p, k_i) values. This set is shown in Fig. 6.15.

B. Construction of Achievable Gain–Phase Margin Design Curves

As in Example 6.2, for a prescribed range of phase margins and gain crossover frequencies, we superimpose ellipses and straight lines on the stabilizing set. The intersection points, contained in the stabilizing set, determine the PI controller gains. For this example, the evaluated range for achievable phase margin is from 1° to 90° and the range for achievable gain crossover frequency is from 0.1 to 1 rad/s.

C. Simultaneous Specifications and Retrieval of Controller Gains

Figure 6.16 displays the achievable performance for Example 6.3. In this case, the maximum gain margin that we can get is 13.14 dB. The phase margin, corresponding to this maximum gain margin, is 79° with a gain crossover frequency of 0.1 rad/s as shown in Fig. 6.16 marked as Point A. We can see how when increasing the gain crossover frequency, our values for achievable gain and phase margins decrease. For example, for a gain crossover frequency of 0.4 rad/s, the maximum gain margin we can get is 2.522 dB, and the corresponding phase margin is 44°, marked as Point B. In Fig. 6.16, we have chosen a candidate design labeled as Point C. The controller corresponding to this design specification can be recovered by constructing the straight line and ellipse corresponding to these specifications (see Fig. 6.15). The PI controller gains for these specifications are

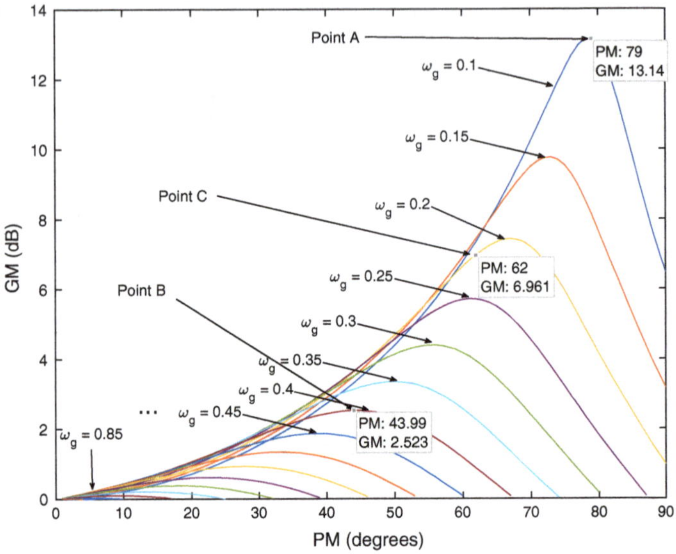

Fig. 6.16 Achievable gain–phase margin design curves in the gain–phase plane for Example 6.3.

$$k_p^* = -0.36283 \tag{6.63}$$

$$k_i^* = 1.6228. \tag{6.64}$$

The step response for this controller is given in Fig. 6.17. These controller gains correspond to the point $\omega_g = 0.2$, $PM = 62°$, and $GM = 6.96$ dB in the gain–phase margin design plane (see Point C in Fig. 6.16).

In Fig. 6.18, we display the Nyquist plot for the controller gains selected. Here, we can see that those controller gains satisfy the desired performance specifications, $PM = 62°$, $GM = 6.96$ dB.

We can also compute the time-delay tolerance design curves. Following Eq. (6.43) and taking the values from Fig. 6.16, we get Fig. 6.19 where we can see the achievable time-delay tolerance for the system using the proposed controller. We can select any point from the curves and retrieve the controller gains following the same procedure as taking a point from the gain–phase margin design curves. In this case, we selected $PM = 62°$ and $\omega_g = 0.2$ rad/s. The time-delay tolerance is

$$\tau = 5.411 \text{ s}. \tag{6.65}$$

Example 6.4 (*Continuous-Time PID Controller Design*) Consider the continuous-time LTI system represented in Fig. 6.1 with the plant

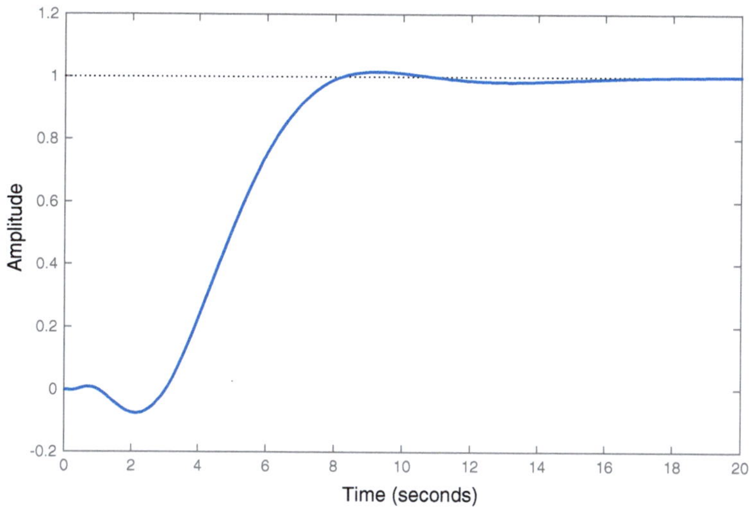

Fig. 6.17 Step response for the system in Example 6.3 using $C(s)^* = \frac{k_p^* s + k_i^*}{s}$. © 2016 IEEE. Reproduced from [4] with permission

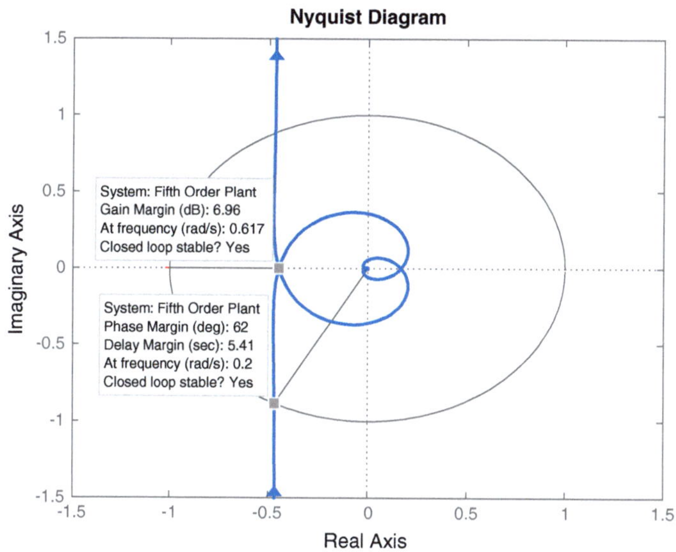

Fig. 6.18 Nyquist plot for $k_p = -0.36283$, $k_i = 1.6228$ in the PI controller design in Example 6.3

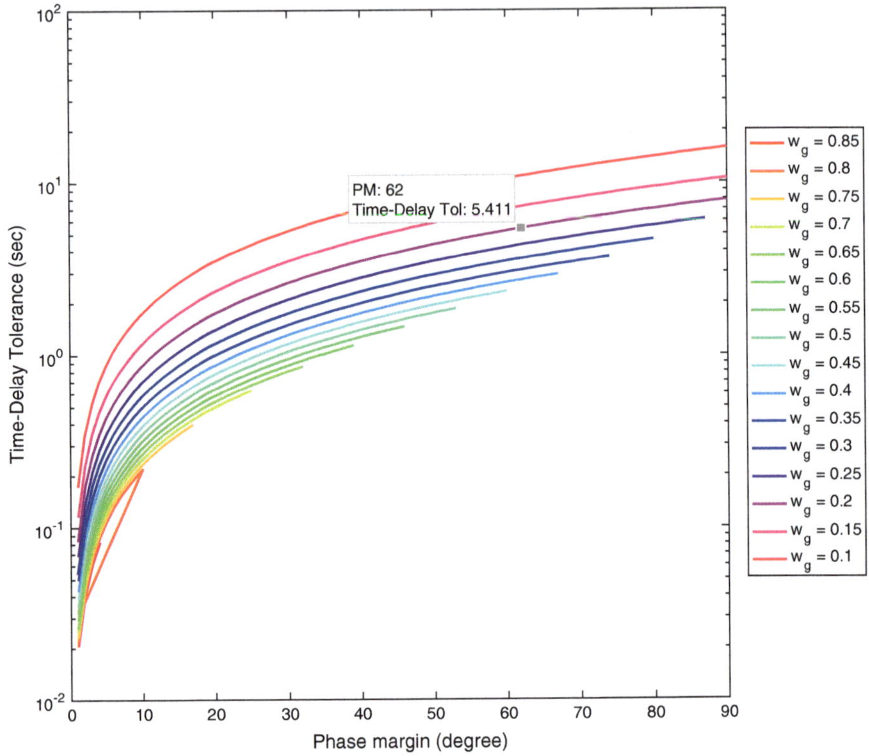

Fig. 6.19 Time-delay tolerance design curves for Example 6.3

$$P(s) = \frac{s-3}{s^3 + 4s^2 + 5s + 2} \tag{6.66}$$

and the controller

$$C(s) = \frac{k_d s^2 + k_p s + k_i}{s}. \tag{6.67}$$

A. Computation of the Stabilizing Set

The closed-loop characteristic polynomial is

$$\delta(s, k_p, k_i) = s^4 + (k_d + 4)s^3 + (k_p - 3k_d + 5)s^2 + (k_i - 3k_p + 2)s - 3k_i. \tag{6.68}$$

Here $n = 3$, $m = 1$, and $N(-s) = -s - 3$. Therefore, we obtain

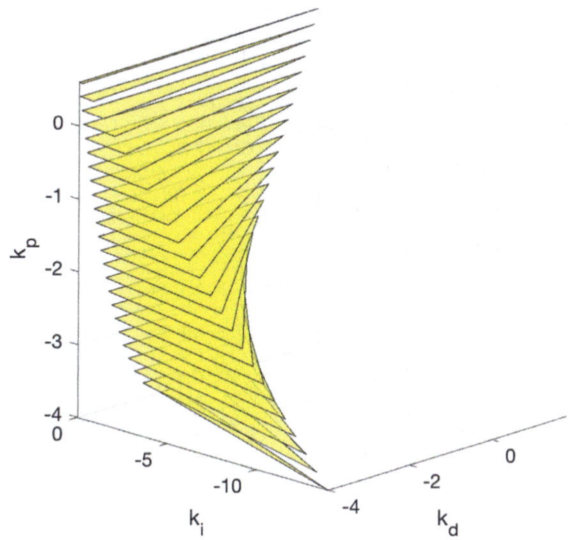

Fig. 6.20 PID stabilizing set for Example 6.4

$$
\begin{aligned}
\nu(s) &= \delta(s, k_p, k_i)N(-s) \\
&= -s^5 + (-k_d - 7)s^4 + (-k_p - 17)s^3 + (9k_d - k_i - 17)s^2 \\
&\quad + (9k_p - 6)s + 9k_i
\end{aligned} \tag{6.69}
$$

so that

$$
\begin{aligned}
\nu(j\omega, k_p, k_i) &= (-k_d - 7)\omega^4 + (k_i - 9k_d + 17)\omega^2 + 9k_i \\
&\quad + j[\omega^5 + (k_p + 17)\omega^3 + (9k_p - 6)\omega] \\
&= p(\omega) + jq(\omega).
\end{aligned} \tag{6.70}
$$

We find that $z^+ = 1$ so that the signature requirement on $\nu(s)$ for stability is

$$\sigma(\nu) = n - m + 1 + 2z^+ = 5. \tag{6.71}$$

Since the degree of $\nu(s)$ is odd, we see from the signature formulas that $q(\omega)$ must have at least two positive real roots of odd multiplicity. The range of k_p such that $q(\omega, k_p)$ has at least two real, positive, distinct, finite zero with odd multiplicity was determined to be $k_p \in (-4, 0.65)$ which is the allowable range for k_p. By sweeping over different k_p values within the interval $(-4, 0.65)$ we can generate the set of stabilizing (k_p, k_i) values. This set is shown in Fig. 6.20.

B. Construction of Achievable Gain–Phase Margin Design Curves

For the construction of the achievable gain–phase curves for the PID controller, the evaluated range of ω_g is $[0.1, 1.2]$ and the range for PM is from 1 to 100°. Using the constant gain and constant phase loci equations, (6.22) and (6.23) we

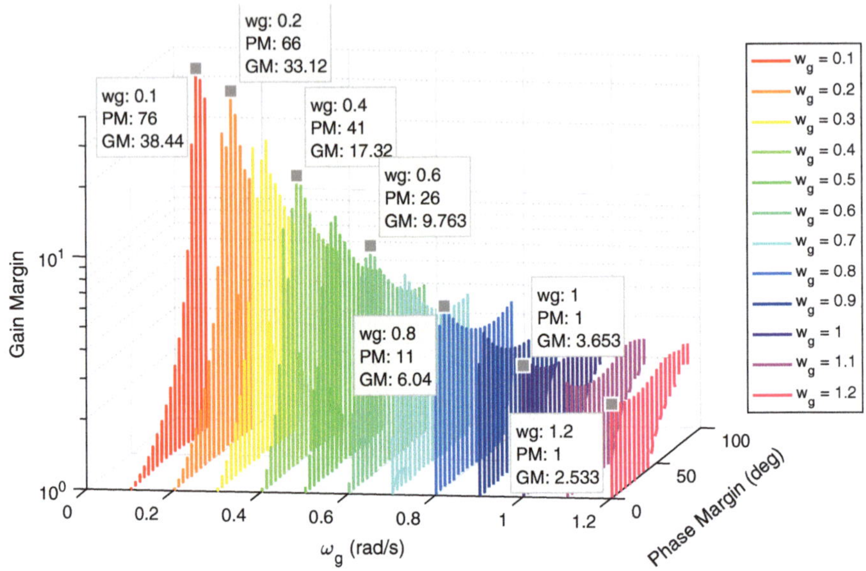

Fig. 6.21 Achievable performance in terms of GM, PM, and ω_g for PID controller design in Example 6.4

now get a cylinder and a plane in the (k_p, k_i, k_d) space. The cylinder and the plane, superimposed on the stabilizing set (see Fig. 6.37) will have two intersection line segments in the (k_i, k_d) plane. The specific value where the intersection occurs can be obtained using (6.26). Equation (6.26) will give us two values for k_p, but only one is contained in the stabilizing set. The intersection line segment in the (k_p, k_i, k_d) represents the PID controller gains that satisfy the PM and ω_g. Evaluating for a range of PM and ω_g, we can construct the achievable gain–phase margin set represented in 3D in Fig. 6.21. If we fix $\omega_g = 0.8$ rad/s, we can see the achievable performance in 2D in Fig. 6.22. Here we can see that the maximum GM we can get is 6.269 with a PM of 9° and for a PM of 60° the GM is 3.548.

C. Simultaneous Specifications and Retrieval of Controller Gains

In Fig. 6.21, we can see the achievable gain–phase margin set of curves indexed by ω_g in different colors. Notice that we can get more GM and PM for lower values of ω_g. For example, for $\omega_g = 0.1$ rad/s, the maximum GM that we can get is 38.44 with a PM of 76°. For $\omega_g = 0.2$ rad/s, the maximum GM is 33.12 with a $PM = 66°$. For a larger value of ω_g, we get lower values for GM and PM. For example, for $\omega_g = 1.2$ rad/s we get a maximum $GM = 2.533$ and $PM = 1°$. The designer has the liberty by using Fig. 6.21 to choose values for GM, PM, and ω_g that best suits his design needs.

After the selection of simultaneous GM, PM, and ω_g from the achievable gain–phase margin set, the designer can retrieve the controller gains corresponding to the point. For illustration purposes, let us say that the desired performance values

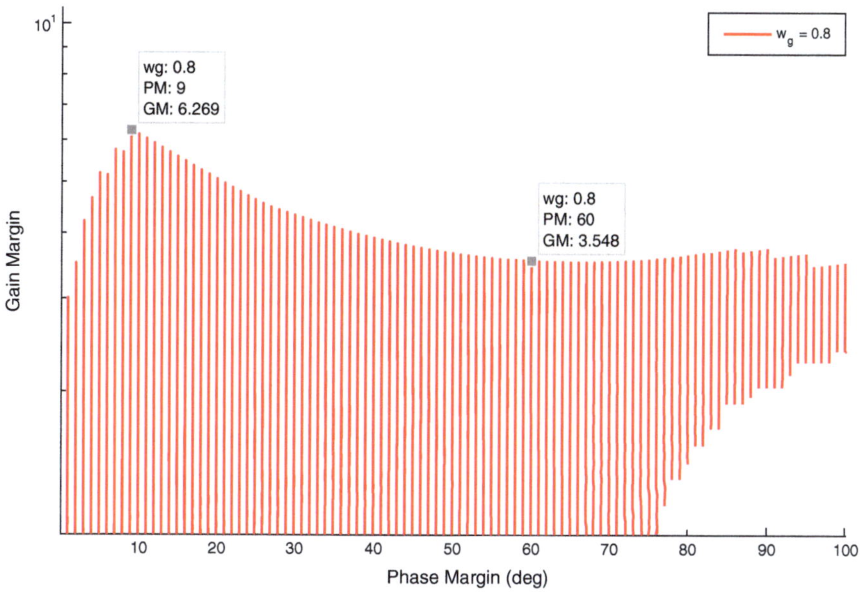

Fig. 6.22 Achievable gain–phase margin set for $\omega_g = 0.8$ rad/s for PID controller design in Example 6.4

chosen for this example are a PM of 60°, GM of 3.548, and a ω_g of 0.8 rad/s (see Fig. 6.22.) Then, using these values and the constant gain and constant phase loci for PID controllers, we can find the intersection of the cylinder and the plane in the (k_p, k_i, k_d) in 3D space shown in Fig. 6.23. The controller gains are $k_p^* = -1.1317$, $k_i^* = -0.4783$, and $k_d^* = -0.6$. In Fig. 6.24, we can see the Nyquist plot for the controller gains selected. Here, we can see that those controller gains satisfy the desired performance specifications, $PM = 60°$, $GM = 3.5482$ (11 dB).

Example 6.5 (*Continuous-Time First-Order Controller Design*) Consider the system configuration in Fig. 6.1 with an unstable, non-minimum phase plant

$$P(s) = \frac{s-2}{s^2+0.6s-0.1}, \quad C(s) = \frac{x_1 s + x_2}{s + x_3}. \tag{6.72}$$

A. Computation of the Stabilizing Set

Considering $P(s)$ in (6.72), we determine the distribution of the root invariant regions. For $\omega = 0$, we have

$$-2x_2 - 0.1x_3 = 0. \tag{6.73}$$

Then, there exists a real root boundary at

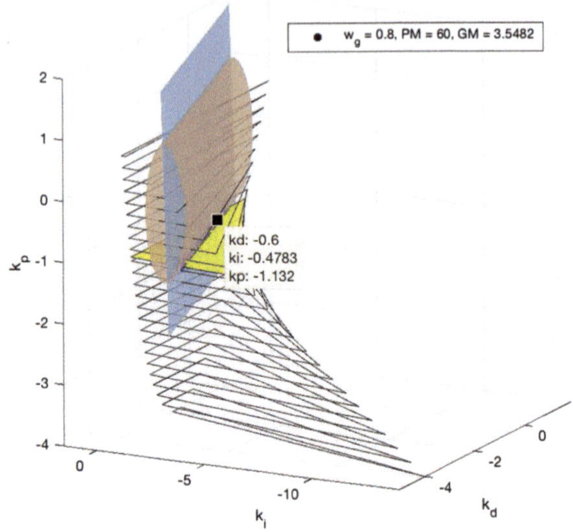

Fig. 6.23 Intersection of a cylinder and a plane superimposed on the PID stabilizing Set and PID controller design in Example 6.4

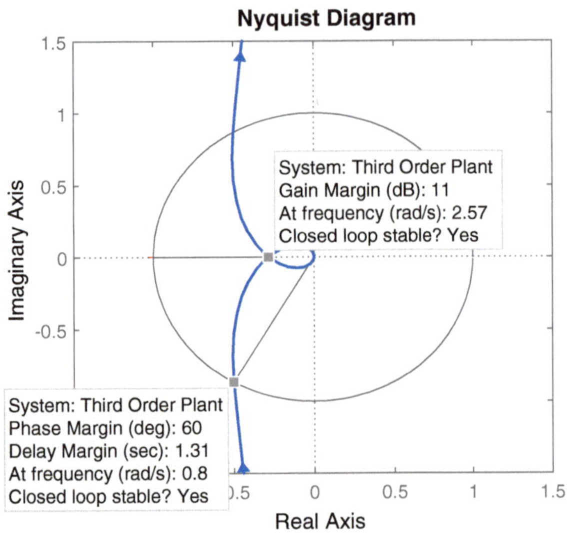

Fig. 6.24 Nyquist plot for $k_p^* = -1.1317$, $k_i^* = -0.4783$, and $k_d^* = -0.6$ in the PID controller design in Example 6.4

$$x_2 = -0.05x_3. \tag{6.74}$$

For $\omega > 0$, we have boundaries:

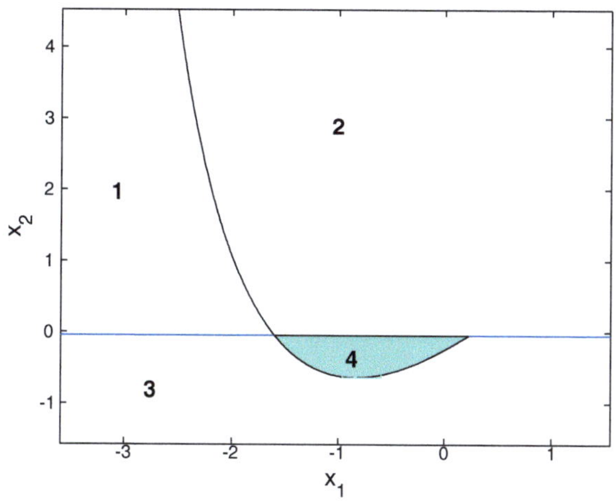

Fig. 6.25 Root invariant regions for $x_3 = 1$ in Example 6.5. ©2015 IEEE. Reproduced from [1] with permission

$$x_1(\omega) = \frac{(-\omega^2 - 0.1)x_3 - 2.6\omega^2 - 0.2}{\omega^2 + 4}$$

$$x_2(\omega) = \frac{(-2.6\omega^2 - 0.2)x_3 + \omega^4 - 1.1\omega^2}{\omega^2 + 4}. \tag{6.75}$$

The stability region for $x_3 = 1$ and the curves and lines are shown in Fig. 6.25. The regions numbered represent invariant root regions. Then, we can pick any value contained in the regions and check the roots. It was found that the region numbered 4 is the stabilizing region. By sweeping x_3 from -0.4 to 8, we obtain the following three-dimensional figure shown in Fig. 6.26, which represents the stabilizing set for our plant $P(s)$ in (6.72).

B. Construction of Achievable Gain–Phase Margin Design Curves

For the construction of the achievable gain–phase margin curves in this example, the evaluated range of gain crossover frequencies $\omega_g \in [0.1, 2]$ and the range of phase margin $PM \in [1°, 120°]$. Using the ellipse and straight line intersection points, we can construct the achievable gain–phase margin curves represented in Fig. 6.27.

C. Simultaneous Specifications and Retrieval of Controller Gains

In Fig. 6.27, we can observe the achievable gain–phase margin set of curves indexed by ω_g^* in different colors. Notice that the curves above the 10^0 GM represent the upper GM and the curves below 10^0 GM represent the lower GM. We notice that the maximum PM that we can get is 100° for a $\omega_g = [0.3, 0.4, 0.5]$ rad/s with a value of upper $GM = [2.68, 2.589, 2.446]$ and lower $GM = [0.9974, 0.9929, 0.9059]$, respectively. Another example of the values of GM and PM that we can get is the point with a PM of 40° with an upper GM of 13.3 and a $\omega_g = 0.2$ rad/s. However, for this value, we get a lower GM of 0.5484. We notice that for a larger GM from the achievable gain–phase margin set, we get a lower PM. The designer has the liberty

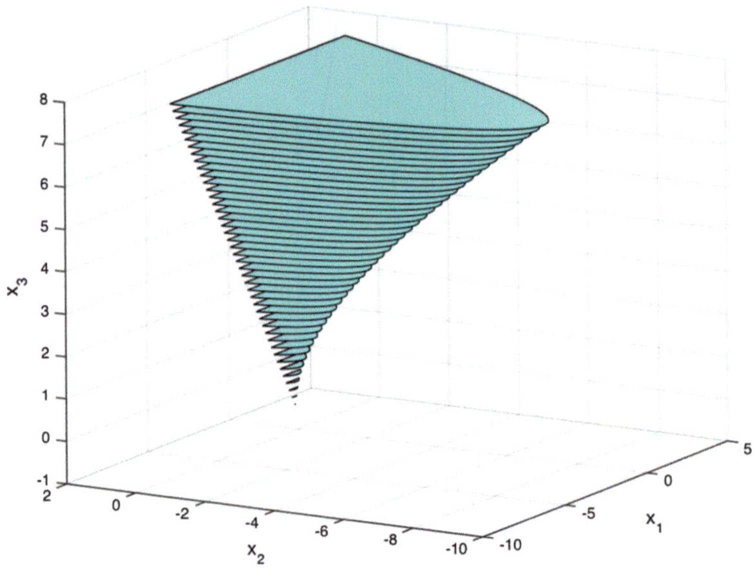

Fig. 6.26 Stability region for $-0.4 \le x_3 \le 8$ in Example 6.5. © 2015 IEEE. Reproduced from [1] with permission

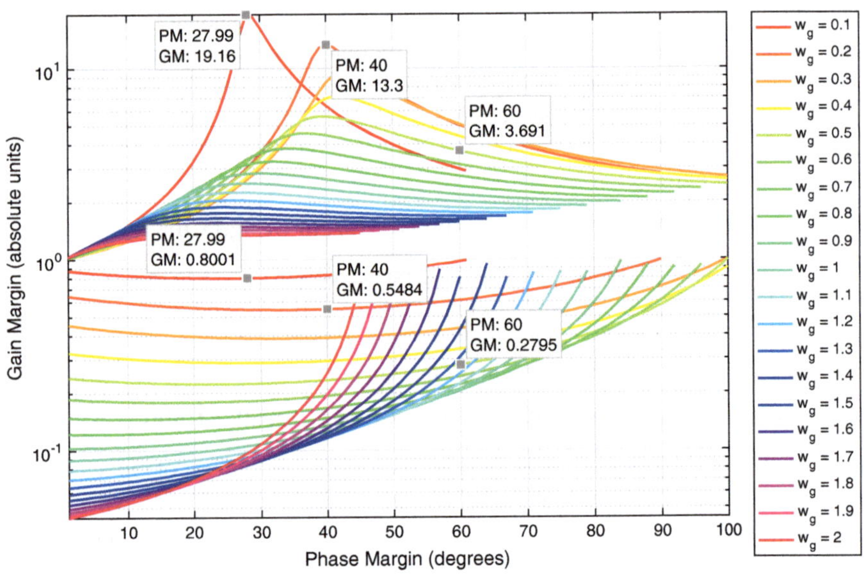

Fig. 6.27 Achievable performance in terms of GM, PM, and ω_g for first-order controller design in Example 6.5

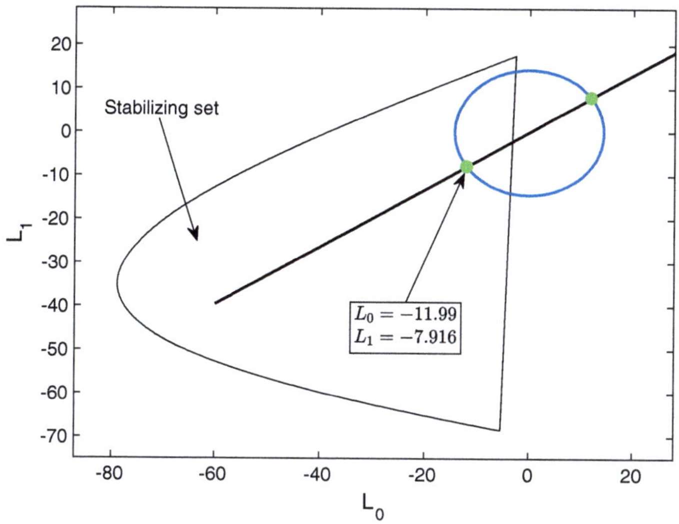

Fig. 6.28 Intersection of the ellipse and straight line superimposed on the stabilizing set corresponding to the GM, PM, and ω_g specified in Example 6.5

to choose values for GM, PM, and ω_g from Fig. 6.27 that best suits his design needs from these achievable sets.

Now, for illustration purposes, suppose that our desired phase margin specification is $PM = 60°$, gain margin of $GM = 3.691$ with a gain crossover frequency of $\omega_g^* = 0.5$ rad/s from Fig. 6.27 for a $x_3 = 8$. Then, taking these values for the constant gain and constant phase loci presented in Sect. 6.2.3, we can find the intersection of an ellipse and straight line to get the controller gains (see Fig. 6.28). Using (6.34) and (6.35), we have

$$\frac{L_0^2}{(14.3667)^2} + \frac{L_1^2}{(14.3667)^2} = 1 \tag{6.76}$$

and

$$L_1 = 0.6603 L_0. \tag{6.77}$$

Using (6.39), we can find the values $x_1 = -2.158$ and $x_2 = -1.431$ for the controller gains.

Then, our desired controller $C^*(s)$ to satisfy the specified phase margin, gain margin, and gain crossover frequency is

$$C^*(s) = \frac{-2.158s - 1.431}{s + 8}. \tag{6.78}$$

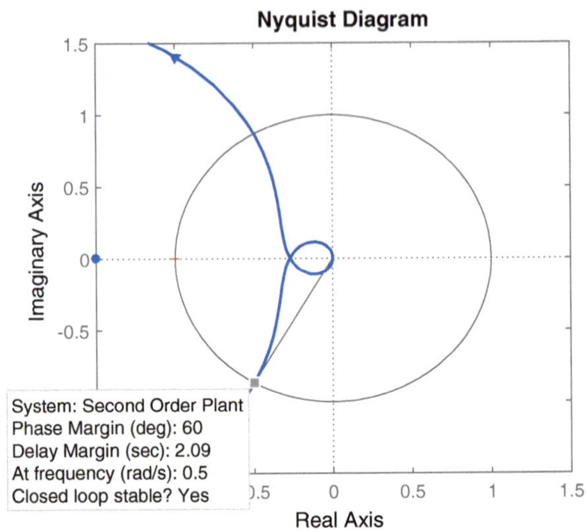

Fig. 6.29 Nyquist plot for $x_1 = -2.158$, $x_2 = -1.431$, and $x_3 = 8$ in the first-order controller design in Example 6.5

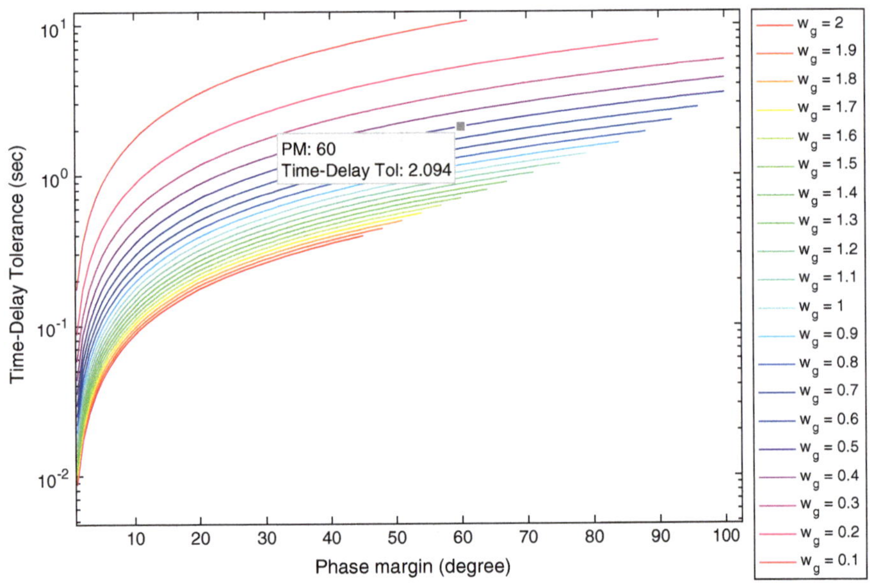

Fig. 6.30 Time-delay tolerance design curves for Example 6.5

In Fig. 6.29, we see the Nyquist plot for the controller gains selected. Here, we can see that those controller gains satisfy the desired performance specifications, $PM = 60°$, $GM = 3.691$ (11.3 dB).

We can also compute the time-delay tolerance design curves. Following Eq. (6.43) and taking the values from Fig. 6.27, we get Fig. 6.30.

In Fig. 6.30, we can see the achievable time-delay tolerance for the system using the proposed controller. We can select any point from the curves and retrieve the controller gains following the same procedure as taking a point from the gain–phase margin design curves. In this case, we selected $PM = 60^\circ$ and $\omega_g = 0.5$ rad/s. The time-delay tolerance is

$$\tau = 2.094 \ \text{s}. \tag{6.79}$$

6.8 Controller Design for Time-Delay Systems

For continuous-time systems with time-delay L secs, it is possible to describe the controller in a geometric form. For the cases of PI and PID controllers, the constant gain and phase loci result in ellipses and straight lines. The only change is that one must add a negative phase $-L\omega$ to the phase of the plant. We derive in detail for PID controllers only, as the derivation for PI controllers is similar.

Let $G(s)$ and $C(s)$ denote the plant and controller transfer functions. In this case, the plant includes a time-delay, that is $G(s) = e^{-Ls}P(s)$, where $L \geq 0$ is the time-delay. The frequency responses of the plant and controller are $e^{-Lj\omega}P(j\omega)$ and $C(j\omega)$, respectively, where $\omega \in [0, \infty]$. For a PID controller

$$C(s) = \frac{k_d s^2 + k_p s + k_i}{s}, \tag{6.80}$$

where k_p, k_i, and k_d are design parameters. Then for $s = j\omega$

$$C(j\omega) = \frac{k_d(j\omega)^2 + k_p(j\omega) + k_i}{j\omega}. \tag{6.81}$$

From (6.81), we have

$$|C(j\omega)|^2 = k_p^2 + \left(k_d\omega - \frac{k_i}{\omega}\right)^2 := M^2 \tag{6.82}$$

and

$$\angle C(j\omega) = \arctan\left(\frac{k_d\omega - \frac{k_i}{\omega}}{k_p}\right) := \Phi. \tag{6.83}$$

From (6.82) and (6.83), we have

$$k_p^2 = M^2 - \left(k_d\omega - \frac{k_i}{\omega}\right)^2 = \frac{\left(k_d\omega - \frac{k_i}{\omega}\right)^2}{\tan^2\Phi}. \tag{6.84}$$

From (6.84), we can have the following expressions:

$$k_i = k_d\omega^2 \pm k_p\omega\tan\Phi \tag{6.85}$$

$$k_p = \pm\sqrt{\frac{M^2}{1+\tan^2\Phi}}. \tag{6.86}$$

Suppose ω_g is the prescribed closed-loop gain crossover frequency. Then

$$M_g := \frac{1}{|G(e^{j\omega_g})|} = \frac{1}{|e^{-Lj\omega_g}||P(j\omega_g)|}. \tag{6.87}$$

Equation (6.87) becomes

$$M_g := \frac{1}{|P(j\omega_g)|}. \tag{6.88}$$

If ϕ_g^* is the desired phase margin in radians,

$$\Phi_g = \pi + \phi_g^* - \angle G(e^{j\omega_g}) \tag{6.89}$$

and by

$$\angle G(e^{j\omega_g}) = \angle\left[e^{-jL\omega_g}P(e^{j\omega_g})\right] = -L\omega_g + \angle P(j\omega_g) \tag{6.90}$$

(6.89) becomes

$$\Phi_g := \pi + \phi_g^* + L\omega_g - \angle P(j\omega_g). \tag{6.91}$$

With $M = M_g$, $\Phi = \Phi_g$ in (6.84) and (6.85) we again obtain an elliptical cylinder and plane as before.

Example 6.6 (Continuous-Time PI Controller Design for an Open-Loop Stable Ziegler–Nichols Plant) Let us consider an open-loop stable Ziegler–Nichols plant

$$P(s) = \frac{1}{2s+1}e^{-0.3s} \tag{6.92}$$

and the PI controller, $C(s)$. We proceed to apply the summarized procedure presented in Chap. 3.

A. Computation of the Stabilizing Set

The characteristic equation is given by

$$\delta(s) = (2s + 1)s + (k_p s + k_i)e^{-0.3s}, \tag{6.93}$$

and

$$\delta^*(s) = e^{0.3s}(2s + 1)s + (k_p s + k_i). \tag{6.94}$$

For $L = 0$, we have

$$\delta(s) = 2s^2 + (k_p + 1)s + k_i. \tag{6.95}$$

For stability, it is required $k_p > -1$, $k_i > 0$. For $L > 0$ and

$$\delta^*(j\omega) = \delta_r(\omega) + j\delta_i(j\omega), \tag{6.96}$$

where

$$\delta_r(\omega) = k_i - \omega\sin(0.3\omega) - 2\omega^2\cos(0.5\omega) \tag{6.97}$$

$$\delta_i(\omega) = \omega\left[5k_p + \cos(0.5\omega) + 12\omega\sin(0.5\omega)\right]. \tag{6.98}$$

We can calculate the range of k_p for stability

$$-1 < k_p < 6.6667\sqrt{\alpha_1^2 + 0.0225}. \tag{6.99}$$

Following all the steps presented in Chap. 3, we get the stabilizing set in Fig. 6.31.

B. Construction of Achievable Gain–Phase Margin Design Curves

For the construction of the achievable gain–phase margin set in this example, the evaluated range of ω_g is $[0.1, 2.8]$ and the range for PM is from 1° to 110°. The calculation of the GM for each case is done by (6.42). Using the ellipse and straight line intersection points, we can construct the achievable gain–phase margin curves presented in Fig. 6.32.

C. Selection of Simultaneous Desired GM, PM, and ω_g Specifications from the Achievable Gain–Phase Margin Design Curves

In Fig. 6.32, we can see the achievable gain–phase margin set of curves indexed by ω_g^* in different colors. We notice that the maximum PM that we can get is 103.8° for a $\omega_g = 0.8$ rad/s with a value of GM of 5.872. Another example of the values of GM and PM that we can get is the point with a PM of 79.72° with a GM of 149 and

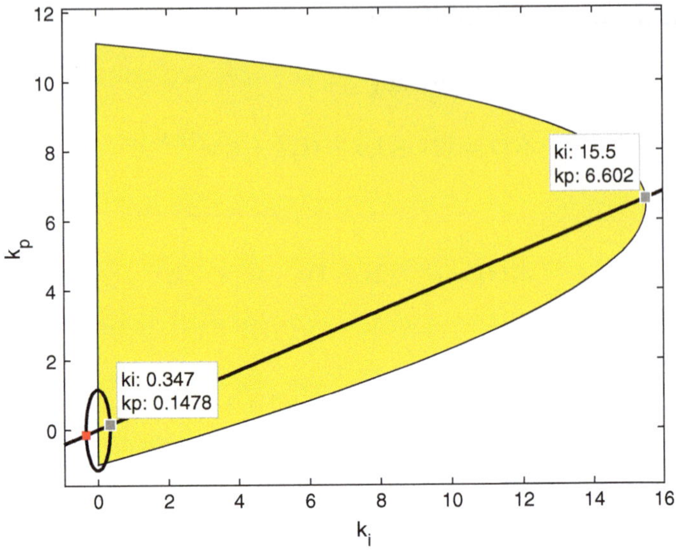

Fig. 6.31 Stabilizing set in yellow for PI controller design in Example 6.6

a $\omega_g = 0.1$ rad/s. The designer has the liberty to choose values for GM, PM, and ω_g from Fig. 6.32.

D. Simultaneous Specifications and Retrieval of Controller Gains

After the selection of simultaneous GM, PM, and ω_g from the achievable gain–phase margin curves, the designer can retrieve the controller gains corresponding to this point. For illustration purposes, let us say that the desired performance values chosen for this example are a PM of 61.16°, $GM = 44.6$ (33 dB), and a $\omega_g = 0.3$ rad/s from Fig. 6.32. Then, taking these values for the constant gain and constant phase loci, we can find the intersection of an ellipse and a straight line as shown in Fig. 6.32. The controller gains are $k_p^* = 0.1478$ and $k_i^* = 0.347$. In Fig. 6.33 we display the Nyquist plot for the controller gains selected. Here, we can see that those controller gains satisfy the desired performance specifications, $PM = 61.2°$, $GM = 44.6$ (33 dB).

Example 6.7 (*Continuous-Time PI Controller Design for an Open-Loop Unstable Ziegler–Nichols Plant*) Let us consider an open-loop unstable Ziegler–Nichols plant

$$P(s) = \frac{5}{-12s + 1} e^{-0.5s} \tag{6.100}$$

and the PI controller, $C(s)$. We proceed to apply the summarized procedure presented in Chap. 3.

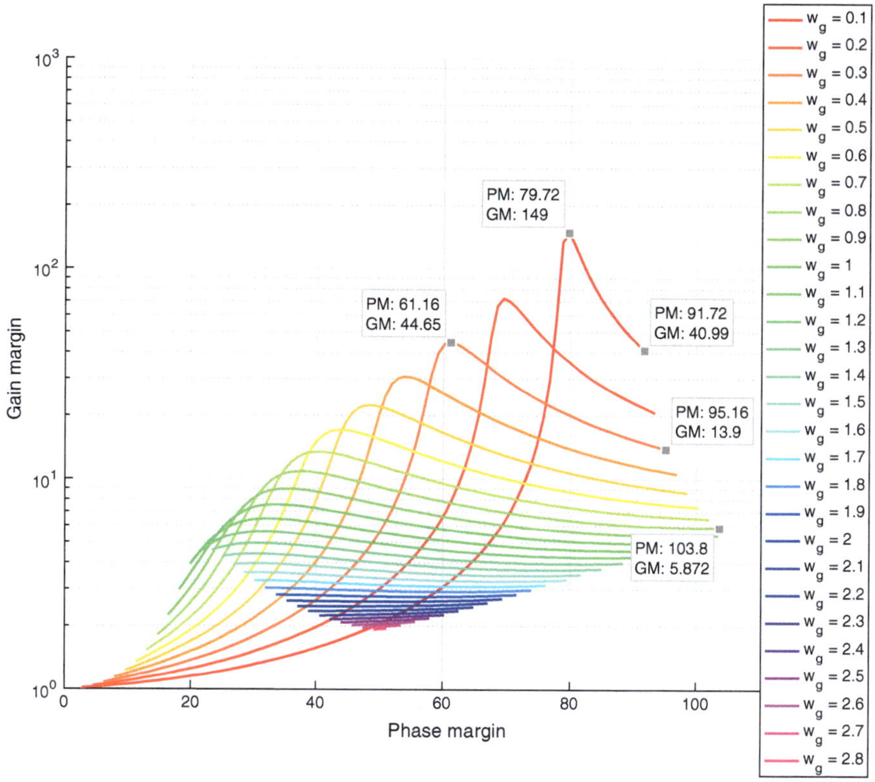

Fig. 6.32 Achievable performance in terms of GM, PM, and ω_g for PI controller design in Example 6.6, intersection of an ellipse and a straight line (dot in black), and the controller gains (k_p^{ub}, k_i^{ub}) at the upper boundary points in the stabilizing set

Fig. 6.33 Nyquist plot for $k_p^* = 0.1478$ and $k_i^* = 0.347$ in the PI controller design in Example 6.6

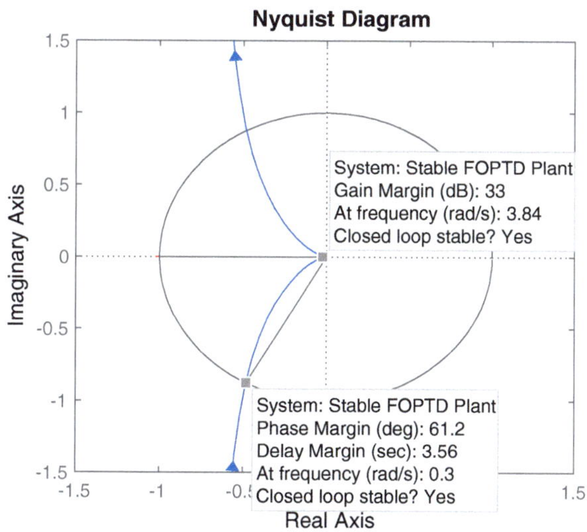

A. Computation of the Stabilizing Set

The characteristic equation is given by

$$\delta(s) = (-12s + 1)s + 5(k_p s + k_i)e^{-0.5s} \tag{6.101}$$

and

$$\delta^*(s) = e^{0.5s}(-12s + 1)s + (k_p s + k_i)5. \tag{6.102}$$

For $L = 0$, we have

$$\delta(s) = -12s^2 + (5k_p + 1)s + 5k_i. \tag{6.103}$$

For stability, it is required $k_p < -\frac{1}{5}$, $k_i < 0$. For $L > 0$,

$$\delta^*(j\omega) = \delta_r(\omega) + j\delta_i(j\omega), \tag{6.104}$$

where

$$\delta_r(\omega) = 5k_i - \omega \sin(0.5\omega) + 12\omega^2 \cos(0.5\omega) \tag{6.105}$$
$$\delta_i(\omega) = \omega \left[5k_p + \cos(0.5\omega) + 12\omega \sin(0.5\omega)\right]. \tag{6.106}$$

We can calculate the range of k_p for stability

$$-4.8\sqrt{\alpha_1^2 + 0.0017} < k_p < -\frac{1}{5}. \tag{6.107}$$

Following all the steps summarized in Chap. 3, we get the stabilizing set in Fig. 6.34.

B. Construction of Achievable Gain–Phase Margin Design Curves

For the construction of the achievable gain–phase margin set in this example, the evaluated range of ω_g is $[0.1, 3]$ and the range for PM is from 0° to 70°. The calculation of the GM for each case is done by (6.42). Using the ellipse and straight line intersection points, we can construct the achievable gain–phase margin set presented in Fig. 6.35.

C. Simultaneous Specifications and Retrieval of Controller Gains

In Fig. 6.35, we can see the achievable gain–phase margin set of curves indexed by ω_g^* in different colors. Notice that the curves above the 10^0 GM represent the upper GM and the curves below 10° GM represent the lower GM. We notice that the maximum PM that we can get is 66° for a $\omega_g = 0.4$ rad/s with a value of upper GM of 7.547 and lower GM of 0.2045. Another example of the values of GM and PM that we can get is the point with a PM of 47° with an upper GM of 23.72 and a $\omega_g = 0.1$ rad/s.

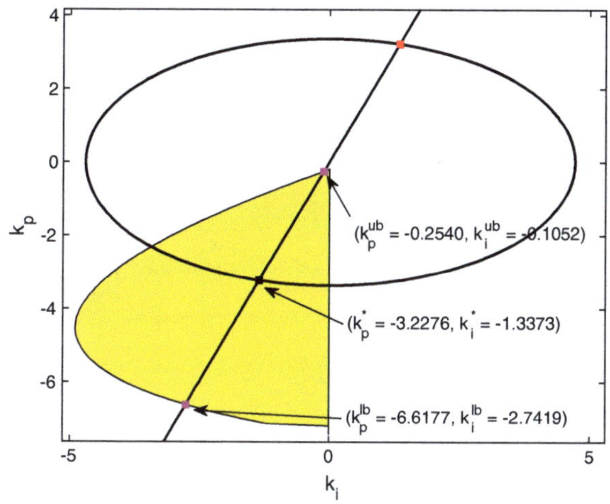

Fig. 6.34 Stabilizing set in yellow for PI controller design in Example 6.7, intersection of an ellipse and a straight line (dot in black), and the controller gains (k_p^{lb}, k_i^{lb}) and (k_p^{ub}, k_i^{ub}) at the lower and upper boundary points in the stabilizing set (dots in magenta). © 2017 IFAC. Reproduced from the original publication "Advanced Tuning for Ziegler–Nichols Plants", IFAC-PapersOnline, Volume 50, Issue 1, July 2017, Pages 1805–1810 with permission

However, for this value, we get a lower GM of 0.6404. We notice that for a larger GM from the achievable gain–phase margin set, we get a lower PM. The blue dots represent the specification points corresponding to a PM of 30°.

D. Retrieval of the PI Controller Gains Corresponding to a Selected Desired Point in the Achievable Performance Set

After the selection of simultaneous GM, PM, and ω_g from the achievable gain–phase margin set, the designer can retrieve the controller gains corresponding to these specifications. For illustration purposes, let us say that the desired performance values chosen for this example are a PM of 30°, $GM = 2.05$, and a $\omega_g = 1.4$ rad/s from Fig. 6.35. Then, taking these values for the constant gain and constant phase loci, we can find the intersection of an ellipse and a straight line as shown in Fig. 6.34. The controller gains are $k_p^* = -3.2276$ and $k_i^* = -1.3373$. In Fig. 6.36 we see the Nyquist plot for the controller gains selected. Here, we can see that those controller gains satisfy the desired performance specifications, $PM = 30°$, $GM = 2.05$ (6.23 dB).

Example 6.8 (*Continuous-Time PID Controller Design for an Open-Loop Stable Ziegler–Nichols Plant*)
Let us consider an open-loop stable Ziegler–Nichols plant

$$P(s) = \frac{1}{2s+1}e^{-2s} \tag{6.108}$$

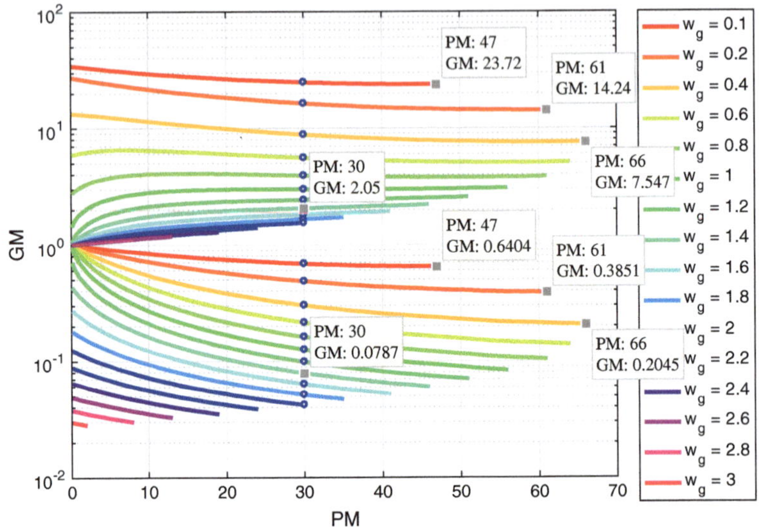

Fig. 6.35 Achievable performance in terms of GM, PM, and ω_g for PI controller design in Example 6.7. The blue dots represent the intersections of ellipses and straight lines with a PM of 30°. © 2017 IFAC. Reproduced from the original publication "Advanced Tuning for Ziegler–Nichols Plants", IFAC-PapersOnline, Volume 50, Issue 1, July 2017, Pages 1805–1810 with permission

Fig. 6.36 Nyquist plot for $k_p^* = -3.2276$ and $k_i^* = -1.3373$ in the PI controller design in Example 6.7. Reprinted from ref. © 2017 IFAC. Reproduced from the original publication "Advanced Tuning for Ziegler–Nichols Plants", IFAC-PapersOnline, Volume 50, Issue 1, July 2017, Pages 1805–1810 with permission

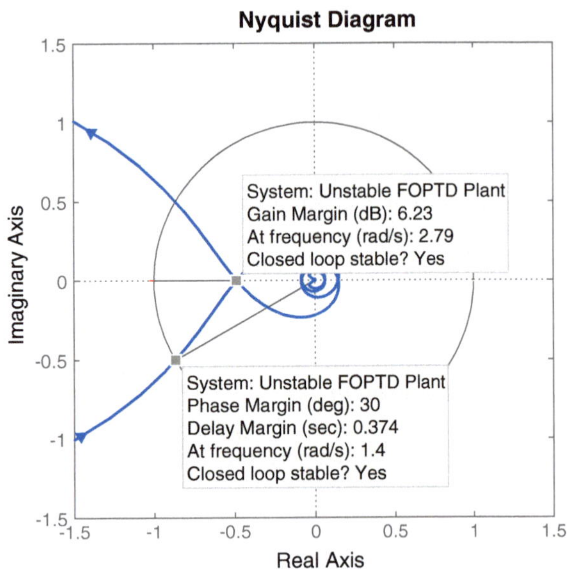

and the PID controller $C(s)$. We proceed to apply the summarized procedure presented in Chap. 3.

A. Computation of the Stabilizing Set

The characteristic equation is given by

$$\delta(s) = (2s + 1)s + (k_d s^2 + k_p s + k_i)e^{-2s} \tag{6.109}$$

and

$$\delta^*(s) = e^{2s}(2s + 1)s + (k_d s^2 + k_p s + k_i). \tag{6.110}$$

For $L = 0$, we have

$$\delta(s) = (k_d + 2)s^2 + (k_p + 1)s + k_i. \tag{6.111}$$

For stability, it is required

$$k_p > -1, \quad k_i > 0, \quad k_d > -2. \tag{6.112}$$

For $L > 0$

$$\delta^*(j\omega) = \delta_r(\omega) + j\delta_i(j\omega), \tag{6.113}$$

where

$$\delta_r(\omega) = k_i - k_d\omega^2 - \omega\sin(2\omega) - 2\omega^2\cos(2\omega) \tag{6.114}$$

$$\delta_i(\omega) = \omega\left[k_p + \cos(2\omega) - 2\omega\sin(2\omega)\right]. \tag{6.115}$$

We can calculate the range of k_p for stability

$$-1 < k_p < [\alpha_1\sin(\alpha_1) - \cos(\alpha_1)]. \tag{6.116}$$

Following the summarized procedure presented in Chap. 3, we get the stabilizing set in Fig. 6.37.

B. Construction of Achievable Gain–Phase Margin Design Curves

For the construction of the achievable gain–phase margin curves for the PID controller design case, the evaluated range of ω_g is [0.1, 1.3] and the range for PM is from 1° to 120°. For the PID case, using the constant gain and constant phase loci equations, (6.22) and (6.23) we now get a cylinder and a plane in the (k_p, k_i, k_d) 3D space, respectively. The cylinder and the plane, superimposed on the stabilizing set (see Fig. 6.37) will have two intersection line segments in the (k_i, k_d) plane. The specific value where the intersection occurs can be obtained using (6.24). Equation (6.26) will give us two values for k_p, but only one is contained in the stabilizing set.

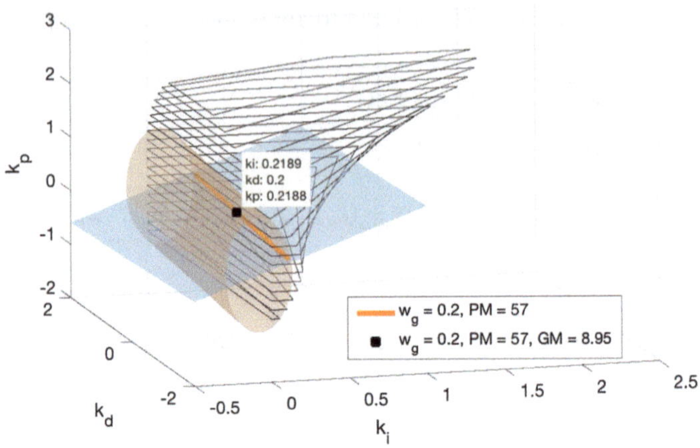

Fig. 6.37 Intersection of a cylinder and a plane superimposed on the PID stabilizing set and a PID design point for Example 6.8. © 2017 IFAC. Reproduced from the original publication "Advanced Tuning for Ziegler–Nichols Plants", IFAC-PapersOnline, Volume 50, Issue 1, July 2017, Pages 1805–1810 with permission

The intersection line segment in the (k_p, k_i, k_d) represents the PID controller gains that satisfy the PM and ω_g. Evaluating the range of PM and ω_g, we can construct the achievable gain–phase margin set represented in 3D in Fig. 6.38. If we take a value of $\omega_g = 0.2$ rad/s, we can see the achievable performance in 2D in Fig. 6.39. Here we can see that the maximum GM we can get is 8.95 with a PM of 57°.

C. Simultaneous Specifications and Retrieval of Controller Gains

In Fig. 6.38, we can see the achievable gain–phase margin set of curves indexed by ω_g in different colors. Notice that we can get more GM and PM for lower values of ω_g. For example, for $\omega_g = 0.1$ rad/s, the maximum GM that we can get is 20 with a PM of 72°. For $\omega_g = 0.2$ rad/s, the maximum GM is 8.95 with a $PM = 57°$. For a larger value of ω_g, we get lower values for GM and PM. For example, for $\omega_g = 1.3$ rad/s we get a maximum $GM = 1.012$ and $PM = 19°$. The designer has the liberty to choose values for GM, PM, and ω_g that best suits his design needs, from Fig. 6.38.

After the selection of simultaneous GM, PM, and ω_g from the achievable gain–phase margin set, the designer can retrieve the controller gains corresponding to this point. For illustration purposes, let us say that the desired performance values chosen for this example are a PM of 57°, GM of 8.95, and a ω_g of 0.2 rad/s (see Fig. 6.38.) Then, taking these values and the constant gain and constant phase loci for PID controllers, we can find the intersection of the cylinder and the plane in the (k_p, k_i, k_d) 3D space shown in Fig. 6.37. The controller gains are $k_p^* = 0.2188$, $k_i^* = 0.2189$, and $k_d^* = 0.2$. In Fig. 6.40, we can see the Nyquist plot for the controller gains selected. Here, we can see that those controller gains satisfy the desired performance specifications, $PM = 57°$, $GM = 8.95$ (19 dB).

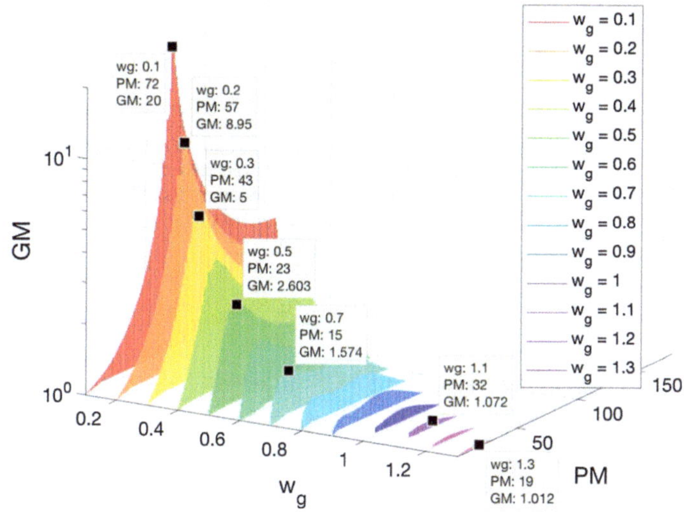

Fig. 6.38 Achievable performance in terms of GM, PM, and ω_g for PID controller design in Example 6.8. © 2017 IFAC. Reproduced from the original publication "Advanced Tuning for Ziegler–Nichols Plants", IFAC-PapersOnline, Volume 50, Issue 1, July 2017, Pages 1805–1810 with permission

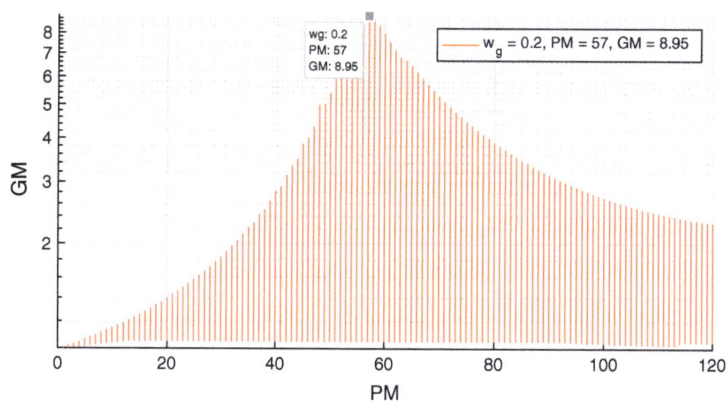

Fig. 6.39 Achievable gain–phase margin set for $\omega_g = 0.2$ rad/s for PID controller design in Example 6.8. © 2017 IFAC. Reproduced from the original publication "Advanced Tuning for Ziegler–Nichols Plants", IFAC-PapersOnline, Volume 50, Issue 1, July 2017, Pages 1805–1810 with permission

Example 6.9 (*Continuous-Time PID Controller Design for an Open-Loop Unstable Ziegler–Nichols Plant*) Let us consider an open-loop unstable Ziegler–Nichols plant

$$P(s) = \frac{2}{-3s+1}e^{-0.5s} \tag{6.117}$$

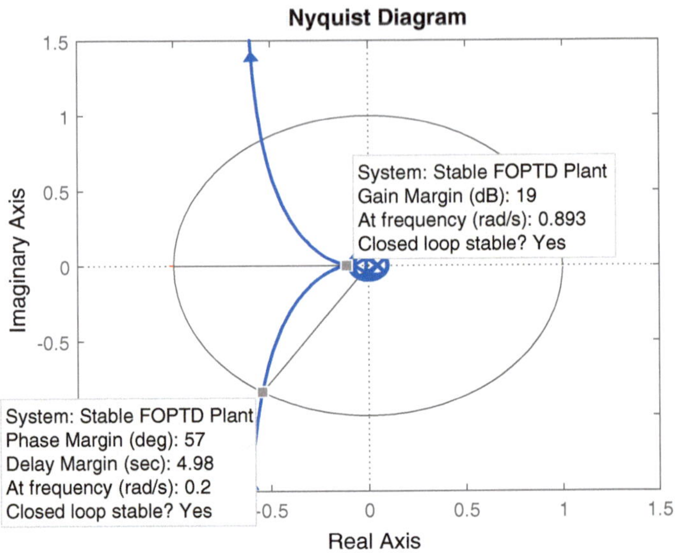

Fig. 6.40 Nyquist plot for $k_p^* = 0.2188$, $k_i^* = 0.2189$, and $k_d^* = 0.2$ in the PID controller design in Example 6.8. © 2017 IFAC. Reproduced from the original publication "Advanced Tuning for Ziegler–Nichols Plants", IFAC-PapersOnline, Volume 50, Issue 1, July 2017, Pages 1805–1810 with permission

and the PID controller $C(s)$. We proceed to apply the summarized procedure presented in Chap. 3.

A. Computation of the Stabilizing Set The characteristic equation, for $L = 0$ is given by

$$\delta(s) = (-3 + 2k_d)s^2 + (2k_p + 1)s + 2k_i. \tag{6.118}$$

For stability, it is required

$$k_p < -\frac{1}{2}, \quad k_i < 0, \quad k_d < \frac{3}{2}. \tag{6.119}$$

For $L > 0$

$$\delta^*(j\omega) = \delta_r(\omega) + j\delta_i(j\omega), \tag{6.120}$$

where

$$\delta_r(\omega) = 2k_i - 2k_d\omega^2 - \omega\sin(0.5\omega) + 3\omega^2\cos(0.5\omega) \tag{6.121}$$

$$\delta_i(\omega) = \omega\left[2k_p + \cos(0.5\omega) + 3\omega\sin(0.5\omega)\right]. \tag{6.122}$$

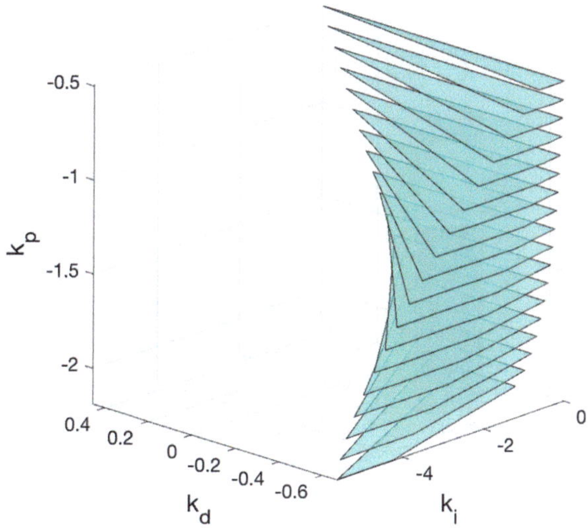

Fig. 6.41 PID stabilizing set for Example 6.9

We can calculate the range for k_p for stability

$$\frac{1}{2}\left[-\frac{3}{2}\alpha_1\sin(\alpha_1)-\cos(\alpha_1)\right] < k_p < -\frac{1}{2}. \tag{6.123}$$

Following the summarized procedure presented in Chap. 3, we get the stabilizing set in Fig. 6.41.

B. Construction of Achievable Gain–Phase Margin Design Curves

For the construction of the achievable gain–phase margin set for the PID controller design case, the evaluated range of ω_g is [0.3, 1.5] and the range for PM is from 1° to 90°. For the PID case, using the constant gain and constant phase loci equations, (6.22) and (6.23), we now get a cylinder and a plane in the (k_p, k_i, k_d) 3D space, respectively. The cylinder and the plane, superimposed on the stabilizing set (see Fig. 6.37) will have two intersection line segments in the (k_i, k_d) plane. The specific value where the intersection occurs can be obtained using (6.26). Equation (6.26) will give us two values for k_p, but only one is contained in the stabilizing set. The intersection line segment in the (k_p, k_i, k_d) represents the PID controller gains that satisfy the prescribed PM and ω_g. Evaluating the range of PM and ω_g, we can construct the achievable gain–phase margin set represented in 3D in Fig. 6.42. If we choose a value of $\omega_g = 0.7$ rad/s, we can see the achievable performance in 2D in Fig. 6.43.

C. Simultaneous Performance Specifications and Retrieval of Controller Gains

In Fig. 6.42, we can see the achievable gain–phase margin set of curves indexed by ω_g in different colors. Notice that we can get more GM and PM for lower values of ω_g.

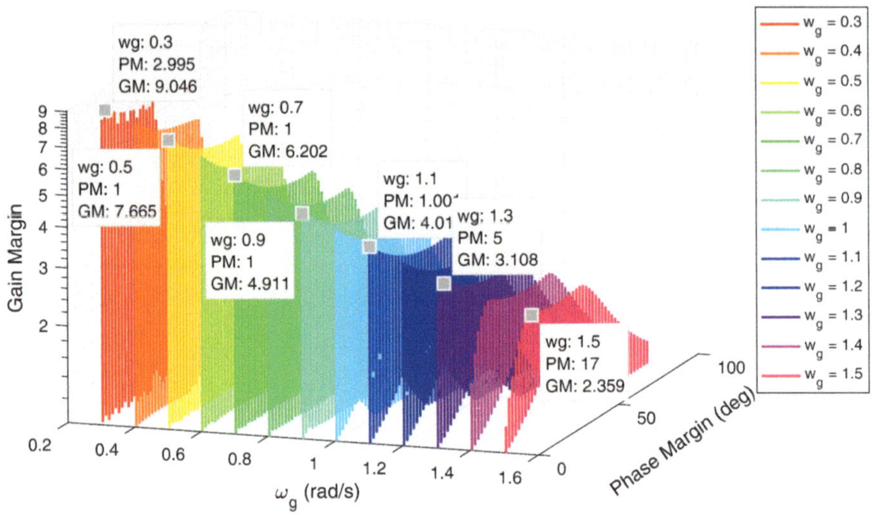

Fig. 6.42 Achievable performance in terms of GM, PM, and ω_g for PID controller design in Example 6.9

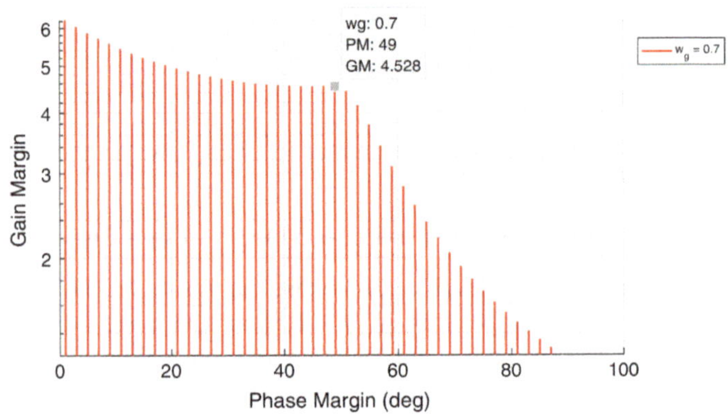

Fig. 6.43 Achievable gain–phase margin set for $\omega_g = 0.7$ rad/s for PID controller design in Example 6.9

For example, for $\omega_g = 0.1$ rad/s, the maximum GM that we can get is 20 with a PM of 72°. For $\omega_g = 0.2$ rad/s, the maximum GM is 8.95 with a $PM = 57°$. For a larger value of ω_g, we get lower values for GM and PM. For example, for $\omega_g = 1.3$ rad/s we get a maximum $GM = 1.012$ and $PM = 19°$. The designer only has the liberty to choose values for GM, PM, and ω_g that best suits his design needs from these curves.

After the selection of simultaneous GM, PM, and ω_g from the achievable gain–phase margin set, the designer can retrieve the controller gains corresponding to the point. For illustration purposes, let us say that the desired performance values chosen

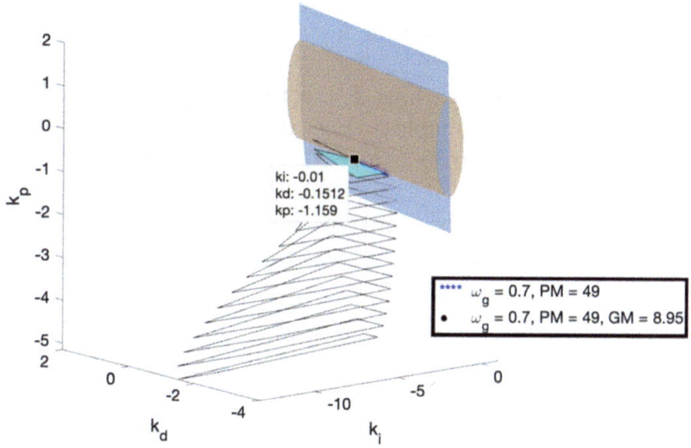

Fig. 6.44 Intersection of a cylinder and a plane superimposed on the PID stabilizing set and a PID design point in Example 6.9

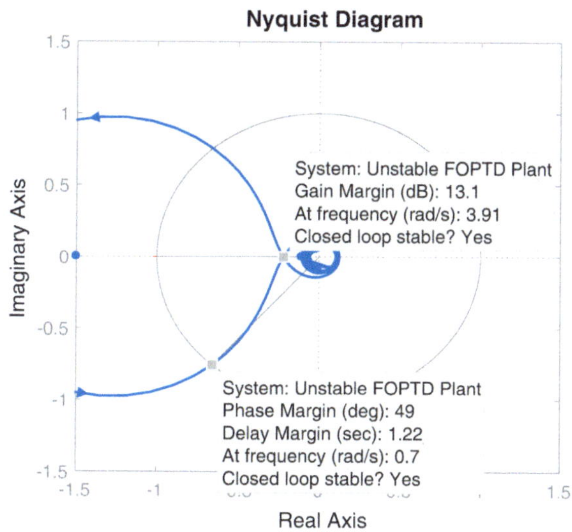

Fig. 6.45 Nyquist plot for $k_p^* = -1.1594$, $k_i^* = -0.01$, and $k_d^* = -0.1512$ in the PID controller design in Example 6.9

for this example are a PM of 49°, GM of 4.5285, and a ω_g of 0.7 rad/s (see Fig. 6.38). Then, taking these values and the constant gain and constant phase loci for PID controllers presented, we can find the intersection of the cylinder and the plane in the (k_p, k_i, k_d) 3D space shown in Fig. 6.44. The controller gains are $k_p^* = -1.1594$, $k_i^* = -0.01$, and $k_d^* = -0.1512$. In Fig. 6.45, we can see the Nyquist plot for the controller gains selected. Here, we can see that those controller gains satisfy the desired performance specifications, $PM = 49°$, $GM = 4.5285$ (13.1 dB).

6.9 Notes and References

The examples developed through Figs. 6.9, 6.10, 6.11, 6.12, 6.15, 6.16, and 6.17 were presented in [4] and the examples in Figs. 6.25 and 6.26 are taken from [2]. The examples shown in Figs. 6.34, 6.35, 6.36, 6.37, 6.38, 6.39, and 6.40 were developed in [5].

The main results of this chapter were obtained in [3].

References

1. Diaz-Rodriguez, I.D.: Modern design of classical controllers: continuous-time first order controllers. In: Proceedings of the 41st Annual Conference of the IEEE Industrial Electronics Society, Student Forum. IECON, pp. 000070–000075 (2015)
2. Díaz-Rodríguez, I.D.: Modern design of classical controllers: Continuous-time first order controllers. In: IECON 2015-41st Annual Conference of the IEEE Industrial Electronics Society, pp. 70–75. IEEE (2015)
3. Diaz Rodriguez, I.D.J.: Modern Design of Classical Controllers. Ph.D. thesis, Texas A&M University (2017)
4. Díaz-Rodríguez, I.D., Bhattacharyya, S.P.: PI controller design in the achievable gain-phase margin plane. In: IEEE, CDC 55th IEEE Conference on Decision and Control. IEEE (2016)
5. Diaz-Rodriguez, I.D., Han, S., Keel, L.H., Bhattacharyya, S.P.: Advanced tuning for Ziegler-Nichols plants. In: The 20th World Congress of the International Federation of Automatic Control, IFAC (2017)

Chapter 7
Gain–Phase Margin-Based Design of Discrete-Time Controllers

Abstract In this chapter, we develop design procedures for digital PI and PID controllers based on gain and phase margin specifications using magnitude and phase loci.

7.1 Introduction

It is a common practice to control continuous-time systems using digital controllers. In this case, the continuous-time plant is first discretized to produce a z-domain transfer function corresponding to a prescribed sampling period T. Integral control can be implemented digitally with this fixed sampling period T. Similarly, proportional and derivative actions can be implemented digitally. Integral action can guarantee zero steady-state error for constant reference and disturbance discrete-time signals if the discrete-time control system can be stabilized. In the following sections, we discuss how to design digital PI and PID controllers achieving prescribed gain and phase margins using magnitude and phase loci superimposed on the discrete-time PID stabilizing sets calculated in Chap. 4.

7.2 PI Controllers

Consider a unity feedback loop and let $P(z)$ and $C(z)$ denote the plant and controller transfer functions in the z domain. Assume that $P(1) \neq 0$ so that stabilization of the plant by a PI controller is not impossible. The frequency response of the plant and controller are $P(e^{j\omega T})$, $C(e^{j\omega T})$, respectively, where T is the sampling period and $\omega \in \left[0, \frac{2\pi}{T}\right]$. For a PI controller,

I. D. Díaz-Rodríguez et al., *Analytical Design of PID Controllers*,
https://doi.org/10.1007/978-3-030-18228-1_7

$$C(z) = \frac{K_0 + K_1 z}{z - 1}, \tag{7.1}$$

where K_0, K_1 are design parameters. Then,

$$C(e^{j\omega T}) = \frac{K_0 + K_1 e^{j\omega T}}{e^{j\omega T} - 1}. \tag{7.2}$$

With $\omega T =: \theta$

$$C(e^{j\theta}) = \frac{K_0 + K_1 e^{j\theta}}{e^{j\theta} - 1} \tag{7.3}$$

or

$$C(e^{j\theta}) = \frac{K_0 e^{-j\frac{\theta}{2}} + K_1 e^{j\frac{\theta}{2}}}{e^{j\frac{\theta}{2}} - e^{-j\frac{\theta}{2}}} = \frac{(K_1 + K_0)\cos\frac{\theta}{2} + j(K_1 - K_0)\sin\frac{\theta}{2}}{2j\sin\frac{\theta}{2}}. \tag{7.4}$$

Let

$$L_0 := K_1 + K_0 \tag{7.5}$$
$$L_1 := K_1 - K_0. \tag{7.6}$$

From (7.4)–(7.6), we have

$$|C(e^{j\theta})|^2 = \frac{L_0^2}{4\tan^2\frac{\theta}{2}} + \frac{L_1^2}{4} =: M^2, \tag{7.7}$$

$$\angle C(e^{j\theta}) = \arctan\left(\frac{-L_0}{L_1\tan\frac{\theta}{2}}\right) =: \Phi. \tag{7.8}$$

Equations (7.7) and (7.8) can be written as

$$\frac{L_0^2}{a^2} + \frac{L_1^2}{b^2} = 1, \tag{7.9}$$
$$L_1 = cL_0, \tag{7.10}$$

where

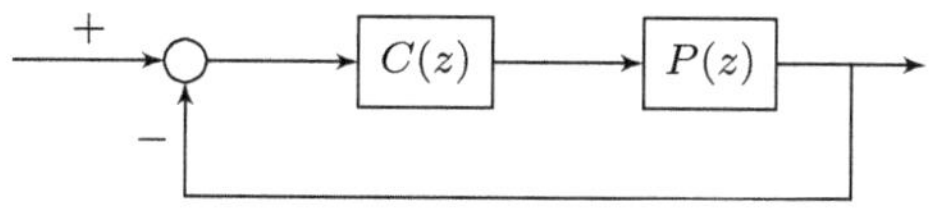

Fig. 7.1 Unity feedback block diagram

$$a^2 = 4M^2 \tan^2 \frac{\theta}{2}, \tag{7.11}$$

$$b^2 = 4M^2, \tag{7.12}$$

$$c = -\frac{1}{\tan \Phi \tan \frac{\theta}{2}}. \tag{7.13}$$

Thus, constant M loci are ellipses and constant Φ loci are straight lines in L_0, L_1 space. The major and minor axes of the ellipse are given by (7.11), (7.12) and c represents the slope of the line in (7.10). The mapping from K_0, K_1 to L_0, L_1 and viceversa is given by

$$K_0 = \frac{L_0 - L_1}{2}, \tag{7.14}$$

$$K_1 = \frac{L_0 + L_1}{2}. \tag{7.15}$$

Suppose ω_g is the prescribed closed-loop gain crossover frequency. Then $\theta_g = \omega_g T$ and

$$M_g := \frac{1}{|P(e^{j\theta_g})|} \tag{7.16}$$

and if ϕ_g^* is the desired phase margin in radians,

$$\Phi_g := \pi + \phi_g^* - \angle P(e^{j\theta_g}). \tag{7.17}$$

From (7.9) and (7.10), we obtain the ellipse and straight lines corresponding to $M = M_g$ and $\Phi = \Phi_g$, giving the design point (K_P^*, K_I^*). If this intersection point lies in the stabilizing set S, the design is feasible, otherwise, the specifications are unattainable and have to be revised.

Example 7.1 (*Discrete-Time PI Controller Design*) Consider the discrete-time system represented in Fig. 7.1, with

$$P(z) = \frac{z - 0.1}{z^3 + 0.1z - 0.25} \quad \text{and} \quad C(z) = \frac{K_0 + K_1 z}{z - 1}. \tag{7.18}$$

A. Computation of the Stabilizing Set

Using the Tchebyshev representation with $\rho = 1$, we have

$$\begin{aligned}
R_N(u) &= -u - 0.1,\\
T_N(u) &= 1,\\
R_D(u) &= -4u^3 + 2.9u - 0.25,\\
T_D(u) &= 4u^2 - 0.9,\\
P_1(u) &= 0.4u^3 + 2u^2 - 0.04u - 0.875,\\
P_2(u) &= 0.34 - 0.4u^2 - 2u,\\
P_3(u) &= 0.2u + 1.01.
\end{aligned} \tag{7.19}$$

Then, we have

$$\begin{aligned}
R(u, K_0, K_1) =& -0.8u^4 - 4.4u^3 - (1.22 + 0.2K_1)u^2\\
&+ (2.915 - 1.01K_1 + 0.2K_0)u + (0.535 + 1.01K_0),\\
T(u, K_1) =& \; 0.8u^3 + 4.4u^2 + (1.62 + 0.5K_1)u + (-1.215 + 1.01K_1).
\end{aligned} \tag{7.20}$$

Since $P(z)$ is of order 3 and $C(z)$, the PI controller, is of order 1, the number of roots of $\delta(z)$ inside the unit circle is required to be 4 for stability. Then,

$$i_1 - i_2 = \underbrace{(i_\delta + i_{N_r})}_{i_1} - \underbrace{l}_{i_2} = 3, \tag{7.21}$$

where i_δ and i_{N_r} are the number of roots of $\delta(z)$ and the reverse polynomial of $N(z)$, respectively, inside the unit circle and l is the degree of $N(z)$. Since the required i_δ is 4, $i_{N_r} = 0$, and $l = 1$, $i_1 - i_2$ is required to be 3. Therefore, we require at least two real zeros of $T(u, K_1)$ as a necessary condition to satisfy the stability condition. We can find the corresponding feasible range for K_1 to be $K_1 \in (-0.94, 1.415)$. Following the stabilizing set computation procedure for this range of K_1, we obtain the stability region shown in Fig. 7.4 in (K_0, K_1) space. To illustrate the example in detail, we fix $K_1 = 1$. Then, the real roots of $T(u, K_1)$ in $(-1, +1)$ are -0.5535 and 0.0919. Furthermore, $\text{sgn}[T(-1)] = +1$, and $i_1 - i_2 = 3$ requires that

$$\begin{aligned}
&\frac{1}{2}\text{sgn}[T(-1)](\text{sgn}[R(-1, K_0)] - 2\text{sgn}[R(-0.5535, K_0)]\\
&\qquad + 2\text{sgn}[R(0.0919, K_0)] - \text{sgn}[R(1, K_0)]) = 3.
\end{aligned} \tag{7.22}$$

Here, the only valid sequence satisfying the last equation is

$$\begin{aligned}
\text{sgn}[R(-1, K_0)] &= +1, \quad \text{sgn}[R(-0.5535, K_0)] = -1,\\
\text{sgn}[R(0.0919, K_0)] &= +1, \quad \text{sgn}[R(1, K_0)] = -1.
\end{aligned} \tag{7.23}$$

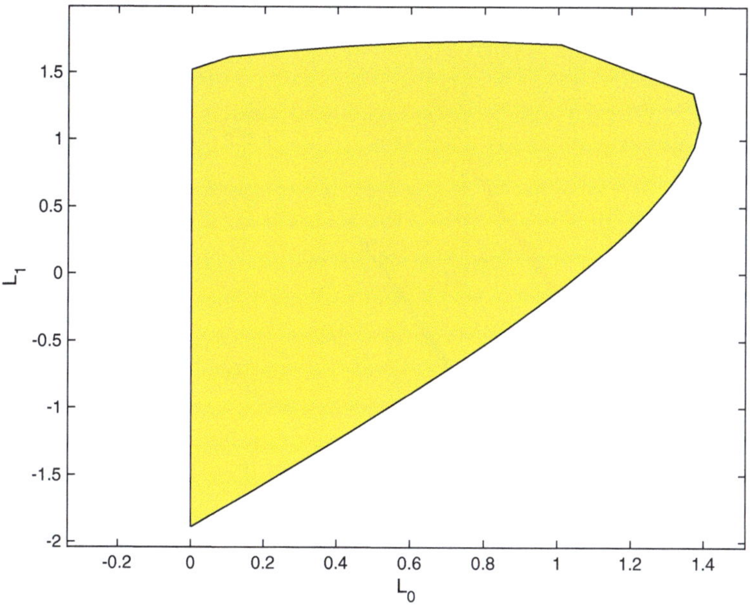

Fig. 7.2 Stabilizing set for the discrete-time PI controller in Example 7.1

Corresponding to this sequence, we have the following set of linear inequalities:

$$\begin{aligned} K_0 &> -1, & K_0 &< 0.3151, \\ K_0 &> -0.6754 & K_0 &< 3.4545. \end{aligned} \tag{7.24}$$

This set of inequalities characterizes the stability region in K_0 space for a fixed $K_1 = 1$. By repeating this procedure for the prescribed range of K_1, we obtain the stability region shown in Fig. 7.2 in (L_0, L_1) space.

B. Construction of the Achievable Gain–Phase Margin Design Curves

In Fig. 7.3, we show an example of selecting a specific phase margin ($PM = 60°$) for a range of gain crossover frequencies. In Fig. 7.4, we can see the intersection points of ellipses and straight lines superimposed on the stabilizing set. In this case, the range of gain crossover frequencies is from 0 to 12 rad/s and for phase margins PM from 1 to 60°.

C. Simultaneous Specifications and Retrieval of Controller Gains

In Fig. 7.5, we can see the achievable performance for this example. In this case, the maximum gain margin that we can get is 35 dB with a phase margin of 88° with a gain crossover frequency of 0.1 rad/s. The gain–phase margin design plane contains information about the capabilities of the system for achieving gain margin, phase

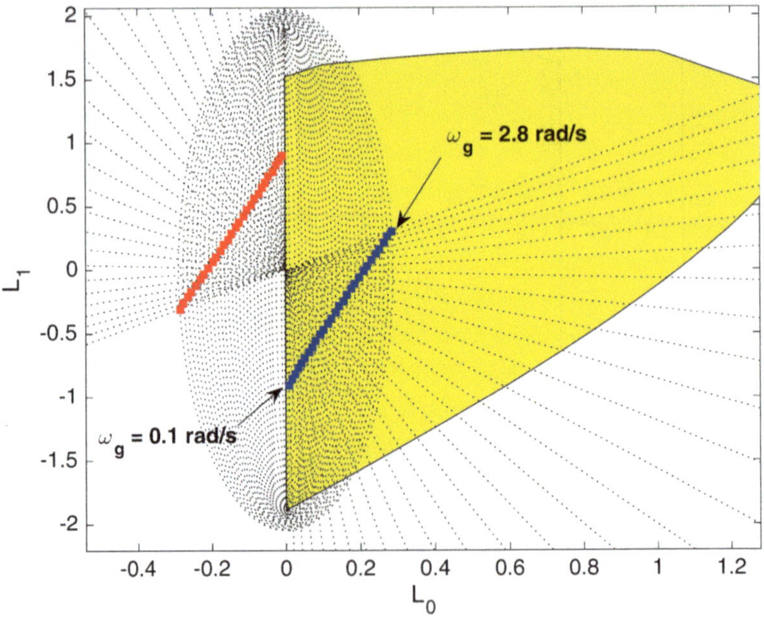

Fig. 7.3 Stabilizing set in yellow and intersection points of ellipses and straight lines in (L_0, L_1) plane for $PM = 60°$ and $\omega_g \in [0.1, 2.8]$ rad/s for the discrete-time PI controller in Example 7.1

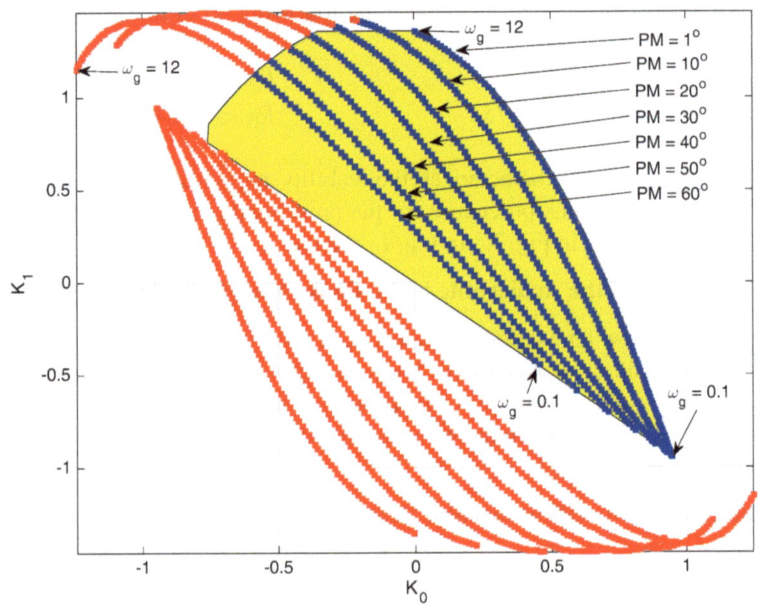

Fig. 7.4 Stabilizing set in yellow and intersection points of ellipses and straight lines in (K_0, K_1) plane for $PM \in [1, 60]$ degree and $\omega_g \in [0.1, 12]$ rad/s for the discrete-time PI controller in Example 7.1

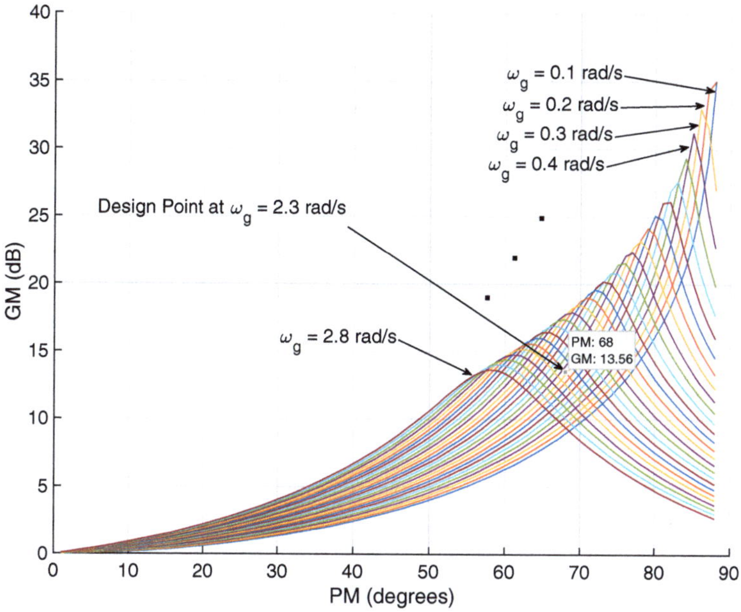

Fig. 7.5 Achievable gain–phase margin design curves in the gain–phase plane for $\omega_g \in [0.1, 2.8]$ rad/s and $PM \in [0, 90]$ degrees for the discrete-time PI controller in Example 7.1

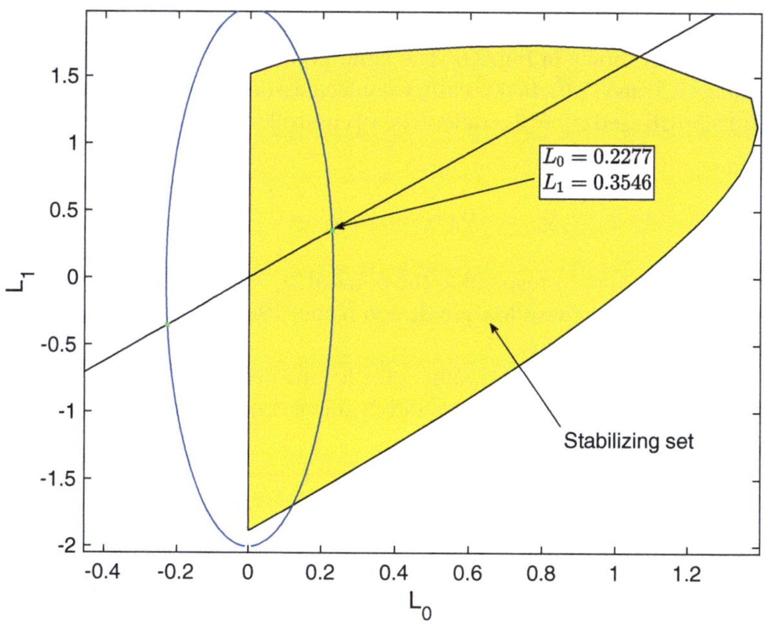

Fig. 7.6 Intersection of ellipse and straight line for the "Design" point in Example 7.1

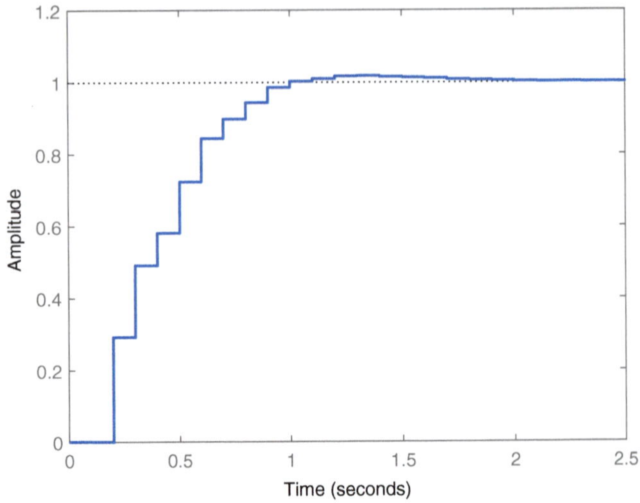

Fig. 7.7 Step response with $C^*(z) = \frac{K_1^* z + K_0^*}{z-1}$ in the Example 7.1

margins, and gain crossover frequency simultaneously. Also, it shows the limitations of the system associated with the use of a PI controller.

We can select a candidate "Design" point from Fig. 7.5. After selecting this point, this design point is mapped by the intersection of the ellipse and straight line into the (L_0, L_1) space as shown in Fig. 7.6. The latter point is then mapped to the (K_0, K_1) space using (7.5) and (7.6). In this case, we selected a design point having $PM = 68°$, $GM = 13.56$ dB, and $\omega_g = 2.3$ rad/s. The PI controller gains for these specifications are

$$K_0^* = -0.06349, \quad K_1^* = 0.2912. \tag{7.25}$$

These controller gains correspond to the point of $\omega_g = 2.3$, $PM = 68°$, and $GM =$ 13.56 dB in the Gain–Phase Margin design plane. The step response for these controller gains is shown in Fig. 7.7.

In Fig. 7.8, we can see the Nyquist plot for the controller gains selected. Here, we can see that those controller gains satisfy the desired performance specifications, $PM = 68°$, $GM = 13.56$.

7.3 PID Controllers

Let $P(z)$ and $C(z)$ denote the plant and controller transfer functions. The frequency response of the plant and controller are $P(e^{j\omega T})$, $C(e^{j\omega T})$, respectively, where T is the sampling period and $\omega \in \left[0, \frac{2\pi}{T}\right]$. For a PID controller,

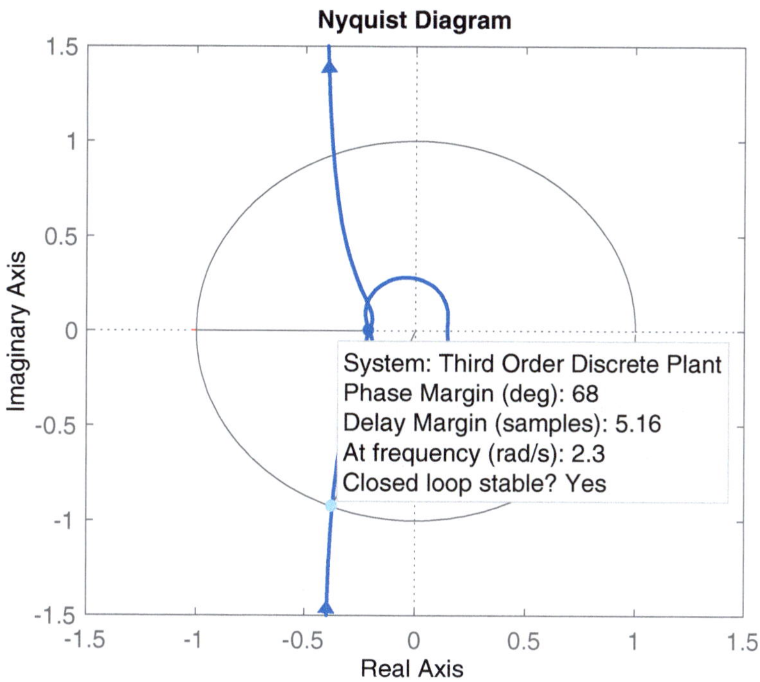

Fig. 7.8 Nyquist plot for $K_0 = -0.06349$, $K_1 = 0.2912$ in the PI controller design in Example 7.1

$$C(z) = \frac{K_0 + K_1 z + K_2 z^2}{z(z-1)}, \tag{7.26}$$

where K_0, K_1, and K_2 are design parameters. Then, letting $\omega T =: \theta$

$$C(e^{j\theta}) = \frac{K_0 e^{-j\theta} + K_1 + K_2 e^{j\theta}}{e^{j\theta} - 1}. \tag{7.27}$$

Note that,

$$C(e^{j\theta}) = \frac{(K_2 + K_0)\cos\theta + K_1 + j(K_2 - K_0)\sin\theta}{(\cos\theta - 1 + j\sin\theta)}. \tag{7.28}$$

Let

$$L_0 := K_2 + K_0, \quad L_1 := K_2 - K_0. \tag{7.29}$$

From (7.28) and (7.29), we have

$$|C(e^{j\theta})|^2 = \frac{\cos^2\theta\left(L_0 + \frac{K_1}{\cos\theta}\right)^2 + \sin^2\theta L_1^2}{(\cos\theta - 1)^2 + (\sin\theta)^2} = \frac{\left(L_0 + \frac{K_1}{\cos\theta}\right)^2}{\left(\frac{\sqrt{\mu}}{\cos\theta}\right)^2} + \frac{L_1^2}{\left(\frac{\sqrt{\mu}}{\sin\theta}\right)^2}$$
$$=: M^2, \tag{7.30}$$

where $\mu = (\cos\theta - 1)^2 + (\sin\theta)^2$ and

$$\angle C(e^{j\theta}) = \tan^{-1}\left(\frac{L_1 \sin\theta}{K_1 + L_0\cos\theta}\right) - \tan^{-1}\left(\frac{\sin\theta}{\cos\theta - 1}\right). \tag{7.31}$$

Using the relationships

$$\tan^{-1} u - \tan^{-1} v = \tan^{-1}\left(\frac{u - v}{1 + uv}\right), \tag{7.32}$$

we have

$$\angle C(e^{j\theta}) = \tan^{-1}\left(\frac{\sin\theta\,(L_1(\cos\theta - 1) - (L_0\cos\theta + K_1))}{(L_0\cos\theta + K_1)(\cos\theta - 1) + L_1\sin^2\theta}\right) =: \Phi. \tag{7.33}$$

Equations (7.30) and (7.33) can be written as

$$\frac{(L_0 + a)^2}{b^2} + \frac{L_1^2}{c^2} = 1, \quad L_1 = dL_0 + e, \tag{7.34}$$

where

$$a = \frac{K_1}{\cos\theta}, \quad b^2 = \frac{\mu M^2}{\cos^2\theta}, \quad c^2 = \frac{\mu M^2}{\sin^2\theta}, \tag{7.35}$$

$$d = \frac{\sin\theta\cos\theta + \cos\theta\tan\Phi(\cos\theta - 1)}{\sin\theta(\cos\theta - 1) - \sin^2\theta\tan\Phi}, \tag{7.36}$$

$$e = \frac{K_1(\cos\theta - 1)\tan\Phi + \sin\theta}{\sin\theta(\cos\theta - 1) - \sin^2\theta\tan\Phi}. \tag{7.37}$$

For fixed K_1, constant M loci are ellipses and constant phase loci are straight lines in L_0, L_1 space. The major and minor axes of the ellipse are given by b and c. The slope of the line is represented by d and e determines the point at which the line crosses the L_1 axis. The mapping from K_0, K_2 to L_0, L_1 for fixed K_1 and viceversa is given by

$$K_0 = \frac{L_0 - L_1}{2}, \quad K_2 = \frac{L_0 + L_1}{2}. \tag{7.38}$$

Suppose ω_g^* is the prescribed closed-loop gain crossover frequency. Then

$$M_g := \frac{1}{|P(e^{j\theta_g})|} \tag{7.39}$$

and if ϕ_g^* is the desired phase margin in radians,

$$\Phi_g := \pi + \phi_g^* - \angle P(e^{j\theta_g}). \tag{7.40}$$

From (7.34), we obtain the ellipse and straight lines corresponding to $M = M_g$ and $\Phi = \Phi_g$, giving the design point (k_p^*, k_i^*, k_d^*) which is feasible if it lies in the stabilizing set.

Example 7.2 (*Discrete-Time PID Controller Design*) Consider the unity feedback discrete-time system in Fig. 7.1 with

$$P(z) = \frac{1}{z^2 - 0.25} \quad \text{and} \quad C(z) = \frac{K_0 + K_1 z + K_2 z^2}{z(z-1)} \tag{7.41}$$

A. Computation of the Stabilizing Set

Using the Tchebyshev representation with $\rho = 1$, we have

$$\begin{aligned} R_N(u) &= 1, \\ T_N(u) &= 0, \\ R_D(u) &= 2u^2 - 1.25, \\ T_D(u) &= -2u, \\ P_1(u) &= 2u^2 - 1.25, \\ P_2(u) &= -2u, \\ P_3(u) &= 1. \end{aligned} \tag{7.42}$$

Then, we have

$$\begin{aligned} R(u, L_0, K_1) &= -4u^3 - 2u^2 + (3.25 - L_0)u + K_1 + 1.25, \\ T(u, L_1) &= 4u^2 + 2u + L_1 - 1.25. \end{aligned} \tag{7.43}$$

Since $P(z)$ is of order 2 and $C(z)$, the PID controller, is of order 2, the number of roots of $\delta(z)$ inside the unit circle is required to be 4 for stability. Then,

$$i_1 - i_2 = \underbrace{(i_\delta + i_{N_r})}_{i_1} - \underbrace{(l+1)}_{i_2}, \tag{7.44}$$

where i_δ and i_{N_r} are the number of roots of $\delta(z)$ and the reverse polynomial of $N(z)$, respectively, inside the unit circle and l is the order of $N(z)$. Since the required i_δ

is 4, $i_{N_r} = 0$, and $l = 0$, $i_1 - i_2$ is required to be 3. Therefore, we require two real roots from $T(u, L_1)$ to satisfy the stability condition. We can find the feasible range for L_1, for these two required roots. For this example, the range is $L_1 \in [-1, 1.4]$. To illustrate the example in detail, we fix $L_1 = 1$. Then, the real roots of $T(u, L_1)$ in $(-1, +1)$ are -0.6036 and 0.1036. Furthermore, $\text{sgn}[T(-1)] = +1$, and $i_1 - i_2 = 3$ requires that

$$\begin{aligned}&\frac{1}{2}\text{sgn}[T(-1)](\text{sgn}[R(-1, L_0, K_1)] - 2\text{sgn}[R(-0.6036, L_0, K_1)]\\&\qquad + 2\text{sgn}[R(0.1036, L_0, K_1)] - \text{sgn}[R(1, L_0, K_1)]) = 3.\end{aligned} \tag{7.45}$$

Here, the only valid sequence satisfying the last equations is

$$\begin{aligned}\text{sgn}[R(-1, L_0, K_1)] &= +1, \quad \text{sgn}[R(-0.6036, L_0, K_1)] = -1,\\ \text{sgn}[R(0.1036, L_0, K_1)] &= +1, \qquad\quad \text{sgn}[R(1, L_0, K_1)] = -1.\end{aligned} \tag{7.46}$$

Corresponding to this sequence, we have the following set of linear inequalities:

$$\begin{aligned}&-0.5607 + 0.6036L_0 + K_1 < 0, \qquad\quad L_0 + K_1 > 0,\\&+1.5607 + 0.1036L_0 + K_1 > 0, \quad -1.5 - L_0 + K_1 < 0.\end{aligned} \tag{7.47}$$

This set of inequalities characterize the stability region in (L_0, K_1) space for a fixed $L_1 = 1$. By repeating this procedure for the considered range of L_1, we obtain the stability region shown in Fig. 7.11.

B. Construction of the Achievable Gain–Phase Margin Design Curves

Suppose our desired phase margin specification to be $\phi_g^* = 60°$ with a gain crossover frequency of $\omega_g^* = 2.23$ rad/s and a fixed value $K_1 = 0.1$. Then, using (7.30) and (7.33), we have that

$$|C(e^{j\theta_g})|^2 = \frac{\left(L_0 + \frac{K_1}{\cos\theta_g}\right)^2}{\left(\frac{\sqrt{\mu}}{\cos\theta_g}\right)^2} + \frac{L_1^2}{\left(\frac{\sqrt{\mu}}{\sin\theta_g}\right)^2} = \frac{1}{|P(e^{j\theta_g})|^2}, \tag{7.48}$$

which implies that

$$\frac{(L_0 + K_1 1.0254)^2}{(0.1784)^2} + \frac{L_1^2}{(0.7868)^2} = 1, \tag{7.49}$$

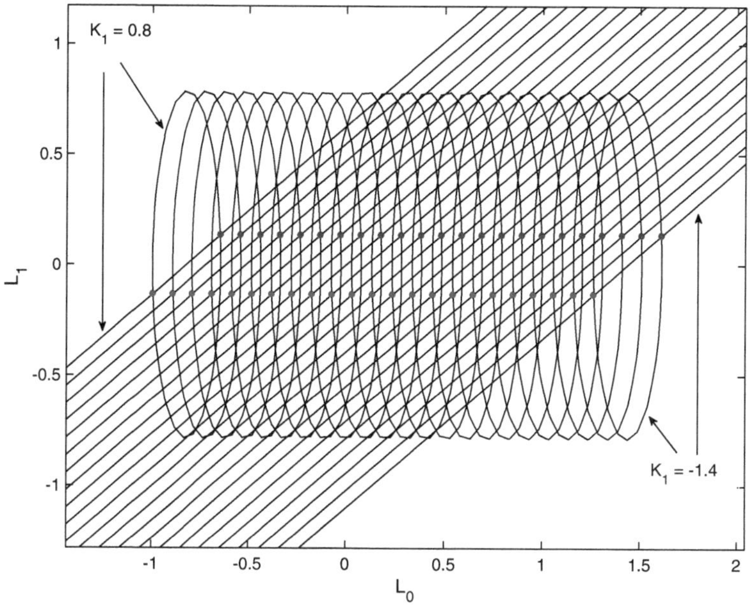

Fig. 7.9 Gain and phase loci for $K_1 \in [-1.4, 0.8]$ in Example 7.2. © 2015 IEEE. Reproduced from [3] with permission

$$\angle C(e^{j\theta_g}) = \tan^{-1}\left(\frac{\sin\theta_g\left(L_1(\cos\theta_g - 1) - (L_0\cos\theta_g + K_1)\right)}{(L_0\cos\theta_g + K_1)(\cos\theta_g - 1) + L_1\sin^2\theta_g}\right)$$
$$= \pi + \phi_g^* - \angle P(e^{j\theta_g}), \tag{7.50}$$

which implies that $L_1 = (0.7672)L_0 + 0.0787$.

We can consider a range of K_1 and desired phase margins so we have a set of ellipses and straight lines. For each ellipse there is a corresponding straight line with the intersection points. For $K_1 \in [-1.4, 0.8]$ we can see in Fig. 7.9 all the ellipses and straight lines for the different values. We choose the intersection point with $K_1 = 0.1$. Thus

$$K_0 = -0.0308, \quad K_1 = 0.1, \quad K_2 = 0.1041. \tag{7.51}$$

Then, our desired controller $C^*(z)$ to satisfy the specified phase margin is

$$C^*(z) = \frac{-0.0308 + 0.1z + 0.1041z^2}{z(z-1)}. \tag{7.52}$$

In Fig. 7.10, we can see the step response of the system with this controller. In classical control, it is known empirically that good phase margin leads to reduced overshoot;

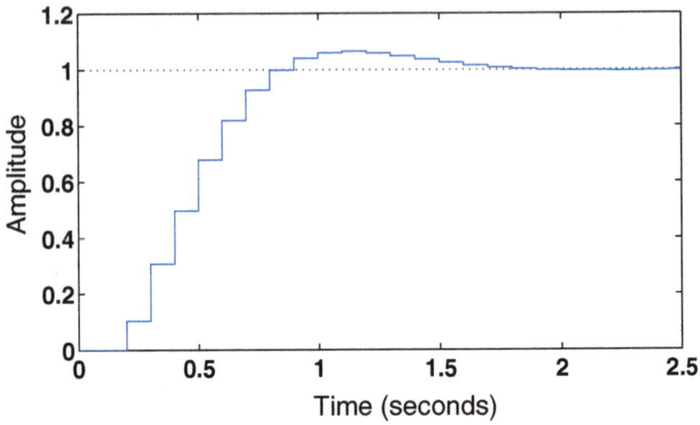

Fig. 7.10 Step response of the discrete-time system with $C^*(z)$ in Example 7.2. © 2015 IEEE. Reproduced from [3] with permission

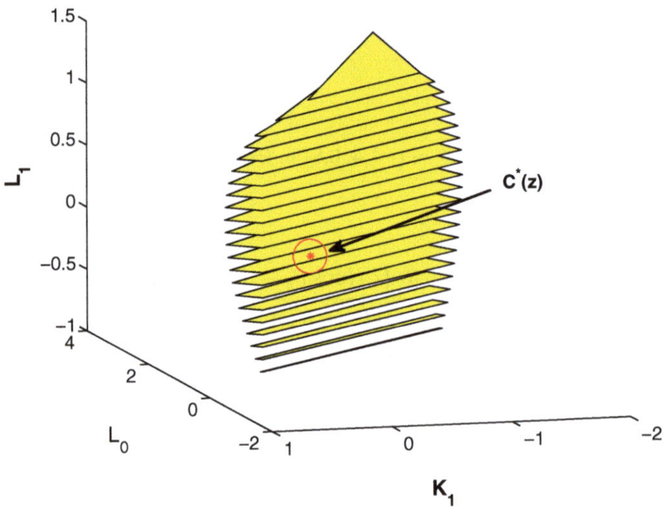

Fig. 7.11 $C^*(z)$ in red "*" contained in the stabilizing set in Example 7.2. © 2015 IEEE. Reproduced from [3] with permission

we observe that in this example. In Fig. 7.11, we can see that the controller is found to be contained in the stabilizing set shown previously.

7.4 Notes and References

The results given in this chapter were developed in the Ph.D. thesis [1] of Diaz-Rodriguez and are also presented in [2, 3].

References

1. Diaz Rodriguez, I.D.: Modern design of classical controllers. Ph.D. thesis. A&M University, Texas
2. Diaz-Rodriguez, I.D., Bhattacharyya, S.P.: Modern design of classical controllers: digital PID controllers. In: IEEE International Conference on Industrial Technology (ICIT), pp. 2112–2119 (2015)
3. Diaz-Rodriguez, I.D., Oliveira, V.A., Bhattacharyya, S.P.: Modern design of classical controllers: digital PID controllers. In: 2015 IEEE 24th International Symposium on Industrial Electronics (ISIE), pp. 1010–1015 (2015)

Chapter 8
PID Control of Multivariable Systems

Abstract In this chapter, we describe a novel approach to design multivariable controllers using single-input single-output theory. This method, based on the Smith-McMillan diagonal form of the plant, allows the scalar PID controller design methods, developed in earlier chapters, to be applied to multivariable systems, one loop at a time.

8.1 Introduction

The control of Multi-Input Multi-Output (MIMO) systems is more challenging compared to that of Single-Input Single-Output (SISO) systems because there is interaction between different inputs and outputs, making it impossible to design one loop independently without affecting the others. In particular, there are few results for the design of fixed and low order controllers for multivariable systems. This is a significant bottleneck in the field of control theory and control engineering.

In this chapter, we present an advanced tuning approach to design a controller for a MIMO plant, which takes advantage of the simplicity of SISO design methods. First, a transformation of the MIMO plant into multiple SISO systems is performed using the Smith-McMillan diagonalization procedure. Then, the SISO PI and PID controller design methods developed in earlier chapters are used to find the achievable performance regarding gain margin, phase margin, and gain crossover frequency for the independent SISO loops. After a diagonal controller is selected, based on achievable performance of each of the n SISO loops, a transformation to the $n \times n$ final controller is performed. The MIMO controller thus obtained is shown to satisfy gain and phase margins and time-delay tolerances that are the minimums achieved by the SISO loops.

Section 8.3 is reproduced from Iván D. Díaz-Rodríguez, Sangjin Han, Shankar P. Bhattacharyya, *Stability Margin Based Design of Multivariable Controllers. IEEE,* with permission © 2017 IEEE.

I. D. Díaz-Rodríguez et al., *Analytical Design of PID Controllers*,
https://doi.org/10.1007/978-3-030-18228-1_8

8.2 Design Methodology

We consider a MIMO continuous-time linear time-invariant system in Fig. 8.1.

$P(s)$ represents a given, strictly proper, n-input n-output plant transfer function matrix and $C(s)$ is an n-input n-output control transfer function matrix to be synthesized. The synthesis requirements are as follows:

(A) The closed loop should be stable.
(B) The controller should be proper and of low order.
(C) Asymptotic tracking of arbitrary step inputs must occur, that is, $y_i(t)$ must converge to the constant r_i, for $i = 1, \ldots, n =: \underline{n}$ as $t \to \infty$.
(D) The closed-loop system should be designed to have known gain margin, phase margin, and time-delay tolerances.

We describe, in this chapter, an approach to controller design that achieves objectives' (A) through (D) by exploiting the theory of SISO PID controller designs developed in earlier chapters.

A key element of the above approach is the Smith-McMillan diagonal form of the plant $P(s)$, denoted as $P_d(s)$, with diagonal elements $P_i(s)$.

8.2.1 *Transformation of the Multivariable Plant Transfer Function into its Smith-McMillan Form*

The $n \times n$ transfer function matrix $P(s)$ can be transformed into a unique diagonal transfer function matrix called the Smith-McMillan form. $P(s)$ can be written as

$$P(s) := \frac{1}{d(s)} N(s), \tag{8.1}$$

where $d(s)$ is the least common multiple of the denominators in $P(s)$ and $N(s)$ is a matrix of numerator polynomials. The Smith form of the polynomial matrix $N(s)$ is given by

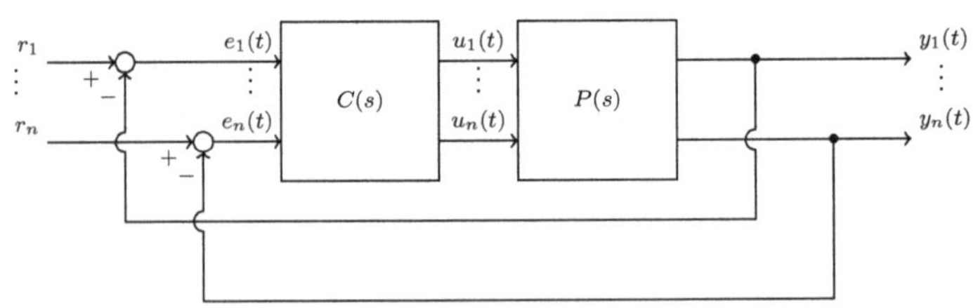

Fig. 8.1 Multivariable control system

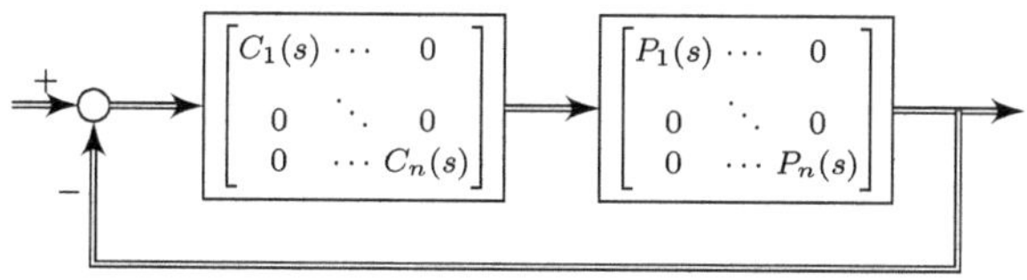

Fig. 8.2 Multiple SISO Unity Feedback Block Diagram. © 2017 IEEE. Reproduced from [1] with permission

$$S(s) = Y(s)N(s)U(s), \tag{8.2}$$

where $S(s)$ is a diagonal polynomial matrix, $Y(s)$ and $U(s)$ are unimodular polynomial matrices and

$$P_d(s) := \frac{S(s)}{d(s)} = \text{diag}\left[\frac{\epsilon_1(s)}{\psi_1(s)}, \ldots, \frac{\epsilon_n(s)}{\psi_n(s)}\right], \tag{8.3}$$

where $\epsilon_i(s)|\epsilon_{i+1}(s)$ and $\psi_{i+1}(s)|\psi_i(s)$, for $i = 1, \ldots, n-1$ and $\epsilon_i(s)$ and $\psi_i(s)$ are coprime $i \in \underline{n}$. $P_d(s)$ is the Smith-McMillan form of $P(s)$. The diagonal elements $P_i(s)$ are called the Smith-McMillan plants for $i \in \underline{n}$. Consider now a diagonal transfer function $C_d(s)$ with diagonal elements $C_i(s), i \in \underline{n}$. Then, the multivariable system in Fig. 8.1 can be considered as a "collection" of multiple SISO loops as shown in Fig. 8.2.

8.2.2 *Transformation of the Diagonal Controller Matrix Into the Corresponding MIMO Controller*

After designing the SISO controllers $C_i(s)$ for Smith-McMillan plants $P_i(s)$, we can transform the diagonal controller matrix $C_d(s)$ into the final MIMO controller matrix $C(s)$ via

$$C(s) = U(s)C_d(s)Y(s), \tag{8.4}$$

where $U(s)$ and $Y(s)$ are the unimodular matrices in (8.2). It is easy to see that

$$\begin{aligned} P(s)C(s) &= Y^{-1}(s)P_d(s)U^{-1}(s)U(s)C_d(s)Y(s) \\ &= Y^{-1}(s)P_d(s)C_d(s)Y(s). \end{aligned}$$

Let

$$e_d(s) := Y(s)e(s), \tag{8.5}$$
$$y_d(s) := Y(s)y(s), \tag{8.6}$$

where $e(s)$ and $y(s)$ are the Laplace transforms of the error $e(t)$ and output $y(t)$, respectively.

We show, in Lemma 8.1 below, that the multivariable control system in Fig. 8.1 is stable if and only if all n SISO unity feedback loops in Fig. 8.2 are stable. This satisfies the requirement (A). For (B), we show that $C(s)$ is guaranteed to be proper by requiring $C_i(s)$ for $i \in \underline{n}$ to have appropriate relative degrees. This is shown in Lemma 8.4. In the n SISO loops in Fig. 8.2, tracking requires an integrator in $C_i(s)$ for $i \in \underline{n}$. This guarantees that the outputs $y_d(t)$ track the references $r_d(t)$ provided that the n SISO loops are stable. We show that $y_d(t)$ tracks $r_d(t)$ if and only if $y(t)$ tracks $r(t)$ in Lemma 8.3. That is, the asymptotic tracking property of the multivariable control system in Fig. 8.1 is achieved by having integral action in $C_i(s)$ for $i \in \underline{n}$. For stability of the n SISO loops with integrators in $C_i(s)$, it is necessary that

$$s \nmid \epsilon_n(s), \tag{8.7}$$

that is, s is not a zero of $\epsilon_n(s)$. We assume that each $P_i(s) = \epsilon_i(s)/\psi_i(s)$ is stabilizable by some $C_i(s)$ containing a pole at $s = 0$. This satisfies (C). In particular, we show that gain margin, phase margin, and time-delay tolerance of the resulting multivariable system in Fig. 8.1 are the minimum gain and phase margins and time-delay tolerances of the n SISO loops.

We start with the statement of the stability of the multivariable system in terms of the stability of n SISO loops.

Lemma 8.1 *$C(s)$ stabilizes $P(s)$ in the mutivariable closed loop if and only if $C_i(s)$ stabilizes $P_i(s)$ in ith SISO loop for all $i \in \underline{n}$.*

Proof

$$\begin{aligned} det[I + P(s)C(s)] &= det[Y^{-1}(s)Y(s) + Y^{-1}(s)P_d(s)C_d(s)Y(s)] \\ &= det[Y^{-1}(s)(I + P_d(s)C_d(s))Y(s)] \\ &= det[Y^{-1}(s)]det[I + P_d(s)C_d(s)]det[Y(s)] \\ &= det[I + P_d(s)C_d(s)] \\ &= \prod_{i=1}^{n} (1 + P_i(s)C_i(s)) . \end{aligned}$$

Thus, the characteristic equation of the multivariable control system is Hurwitz if and only if the characteristic equations of the n SISO loops are Hurwitz. This completes the proof. □

We need a preliminary lemma to establish tracking property of the multivariable system. Write the unimodular matrix $Y(s)$ in (8.2)

$$Y(s) = Y_0 + Y_1 s + \cdots . \tag{8.8}$$

Lemma 8.2 *$Y_0 = Y(0)$ is a full rank matrix.*

Proof Since $Y^{-1}(s)$ is also a unimodular polynomial matrix and can be written as

$$Y^{-1}(s) = W_0 + W_1 s + \cdots , \tag{8.9}$$

we must have

$$\begin{aligned} I &= Y(s)Y^{-1}(s) \\ &= (Y_0 + Y_1 s + \cdots)(W_0 + W_1 s + \cdots) \\ &= Y_0 W_0 + (Y_0 W_1 + W_0 Y_1)s + \cdots . \end{aligned} \tag{8.10}$$

Therefore, (8.10) implies that

$$\begin{aligned} Y_0 W_0 &= I \\ Y_0 W_1 + W_0 Y_1 &= 0 \\ &\vdots \end{aligned}$$

which proves that $Y_0 = Y(0)$ is of full rank. □

Lemma 8.3 *$y(t)$ tracks $r(t)$ if and only if $y_d(t)$ tracks $r_d(t)$.*

Proof $y_d(t)$ tracks $r_d(t)$ if and only if

$$\begin{aligned} &\lim_{t\to\infty} e_d(t) = 0 \\ \Leftrightarrow &\lim_{s\to 0} s e_d(s) = 0 \text{ by the Final Value Theorem} \\ \Leftrightarrow &\lim_{s\to 0} sY(s)e(s) = 0 \text{ by (8.5).} \end{aligned}$$

Since Y_0 is full rank by Lemma 8.2, the above is equivalent to

$$\begin{aligned} &\lim_{s\to 0} se(s) = 0 \\ \Leftrightarrow &\lim_{t\to\infty} e(t) = 0 \text{ by the Final Value Theorem.} \end{aligned}$$

Thus, $e(t)$ goes to 0 as $t \to \infty$ and this completes the proof. □

It is necessary that $C(s)$ be proper and of low order. The following lemma asserts that $C_i(s)$ for $i \in \underline{n}$ has to have certain relative degrees in order for $C(s)$ to be proper.

Lemma 8.4 *Let r_k be the relative degree of the controller $C_k(s)$ and let d_{ik}^U and d_{kj}^Y be the degrees of the $(i, k)^{th}$ and $(k, j)^{th}$ polynomial elements of the unimodular polynomial matrices $U(s)$ and $Y(s)$, respectively. If r_k, $k \in \underline{n}$ satisfies*

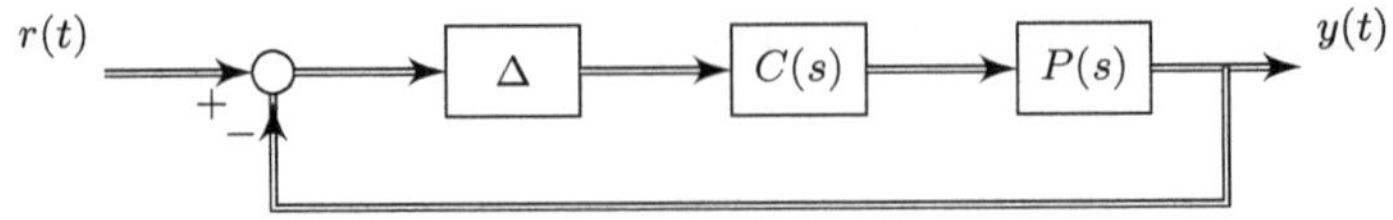

Fig. 8.3 The multivariable control system with perturbation Δ.

$$\min_{k=1,2,\ldots,n} \left\{ r_k - d_{ik}^U - d_{kj}^Y \right\} \geq 0, \quad \forall i, j = 1, 2, \ldots, n \tag{8.11}$$

then the multivariable controller $C(s)$ is proper.

Proof The $(i, j)^{th}$ element of $C(s)$ can be written as

$$C_{ij}(s) = \sum_{k=1}^{n} u_{ik}(s) C_k^d(s) y_{kj}(s), \tag{8.12}$$

which has the relative degree r_{ij}^c.

$$r_{ij}^c = \min_{k=1,2,\ldots,n} \left\{ r_k^d - d_{ik}^u - d_{kj}^y \right\}. \tag{8.13}$$

If for a specific $k \in \underline{n}$, $u_{ik}(s) C_k^d(s) y_{kj}(s) = 0$, then the corresponding $r_k^d - d_{ik}^u - d_{kj}^y$ term in (8.13) needs to be neglected. If $r_{ij}^c \geq 0, \quad \forall i, j = 1, 2, \ldots, n$, $C(s)$ is proper. □

Let us define the perturbation matrix Δ as

$$\Delta := \begin{bmatrix} \delta & & 0 \\ & \ddots & \\ 0 & & \delta \end{bmatrix}. \tag{8.14}$$

If we insert Δ at the usual loop breaking point as shown in Fig. 8.3, we can define the multivariable stability margins, namely, gain and phase margins and time-delay tolerances for the multivariable control system. For gain margin, we replace δ by k in (8.14) and find the smallest k called k^*, such that the loop in Fig. 8.3 becomes just unstable. A similar definition applies to the phase margin and time-delay tolerances by replacing δ with $e^{-j\theta}$ and e^{-sT}, respectively.

Theorem 8.1 *Suppose $C(s)$ in* (8.4) *is a proper controller such that $C_i(s)$ stabilizes the Smith-McMillan plant $P_i(s)$ with gain margin g_i, phase margin ϕ_i, and time-delay tolerance τ_i for the ith SISO loop for each $i \in \underline{n}$. Then, $C(s)$ stabilizes $P(s)$ with a gain marginG, phase marginΦ, and time-delay tolerance T, where*

$$G = \min_{i=1,2,\ldots,n} \{g_i\}, \tag{8.15}$$

$$\Phi = \min_{i=1,2,\ldots,n} \{\phi_i\}, \tag{8.16}$$

$$T = \min_{i=1,2,\ldots,n} \{\tau_i\}, . \tag{8.17}$$

Proof The proof of this theorem follows from the following observations. Substituting $\delta = k$ in (8.14),

$$\Delta := \begin{bmatrix} k & & 0 \\ & \ddots & \\ 0 & & k \end{bmatrix}. \tag{8.18}$$

By Lemma 8.1, we obtain

$$det[I + \Delta P(s)C(s)] = \prod_{i=1}^{n} (1 + P_i(s)C_i(s)). \tag{8.19}$$

Substituting $\delta = e^{j\theta}$ in (8.14), we have

$$\Delta := \begin{bmatrix} e^{j\theta} & & 0 \\ & \ddots & \\ 0 & & e^{j\theta} \end{bmatrix}. \tag{8.20}$$

We obtain, in the same way,

$$det[I + \Delta P(s)C(s)] = \prod_{i=1}^{n} det[1 + e^{j\theta} P_i(s)C_i(s)]. \tag{8.21}$$

Substituting $\delta = e^{-sT}$ in (8.14), we have

$$\Delta := \begin{bmatrix} e^{-sT} & & 0 \\ & \ddots & \\ 0 & & e^{-sT} \end{bmatrix}. \tag{8.22}$$

We obtain likewise

$$det[I + \Delta P(s)C(s)] = \prod_{i=1}^{n} det[1 + e^{-sT} P_i(s)C_i(s)]. \tag{8.23}$$

Evaluating (8.19), (8.21), and (8.23) at $s = j\omega$, we conclude (8.15), (8.16), and (8.17), respectively. □

This theorem shows that the gain margin, phase margin, and time-delay tolerance for the multivariable system are the minimum of the gain margins, the minimum of the phase margins, and the minimum of the time-delay tolerances, respectively, of the n SISO loops in Fig. 8.2.

8.3 Example: Multivariable PI Controller Design

For (D) in Sect. 8.2, the following steps are needed:

Step 1. Compute a stabilizing set for each SISO loop.
Step 2. Parametrize constant gain and constant phase loci for PI or PID controllers.
Step 3. Construct the achievable Gain–Phase margin design curves.
Step 4. Select achievable gain margin, phase margin, and gain crossover frequency and retrieve the controller $C_i(s)$.

The methods for the steps above are developed in Chap. 6 with examples. The example below illustrates our approach to design $C(s)$ with predesigned gain and phase margins and time-delay tolerance specifications for a given multivariable plant $P(s)$.

Example 8.1 Let us consider a two-input two-output system $P(s)$ in Fig. 8.1 as given below:

$$P(s) = \begin{bmatrix} \frac{4}{(s+1)(s+2)} & \frac{-1}{s+1} \\ \frac{2}{s+1} & -\frac{6s+7}{2(s^2+3s+2)} \end{bmatrix}. \tag{8.24}$$

The objective is to find the controller $C(s)$ such that it satisfies the predesigned gain margin, phase margin, and gain crossover frequency.

A. The Smith-McMillan form of $P(s)$ and structure of $C_d(s)$

The least common multiple of the denominators in $P(s)$ is

$$d(s) = (s+1)(s+2). \tag{8.25}$$

$P(s)$ can be written as

$$P(s) = \frac{1}{(s+1)(s+2)} \underbrace{\begin{bmatrix} 4 & -1(s+2) \\ 2(s+2) & -(3s+3.5) \end{bmatrix}}_{=:N(s)}. \tag{8.26}$$

The Smith form of $N(s)$ is expressed as

$$S(s) = \underbrace{\begin{bmatrix} \frac{1}{4} & 0 \\ -(s+2) & 2 \end{bmatrix}}_{=:Y(s)} \underbrace{\begin{bmatrix} 4 & -1(s+2) \\ 2(s+2) & -(3s+3.5) \end{bmatrix}}_{N(s)} \underbrace{\begin{bmatrix} 1 & \frac{1}{4}(s+2) \\ 0 & 1 \end{bmatrix}}_{=:U(s)}$$
$$= \begin{bmatrix} 1 & 0 \\ 0 & s^2 - 2s - 3 \end{bmatrix}. \tag{8.27}$$

Dividing $S(s)$ by $d(s)$, we get the Smith-McMillan form

$$P_d(s) = \begin{bmatrix} \frac{1}{(s+1)(s+2)} & 0 \\ 0 & -\frac{s-3}{s+2} \end{bmatrix}. \tag{8.28}$$

Let us consider $C_d(s)$ as

$$C_d(s) = \begin{bmatrix} \frac{k_{p_1}s+k_{i_1}}{s} & 0 \\ 0 & \frac{k_{p_2}s+k_{i_2}}{s(s+2)^2} \end{bmatrix}. \tag{8.29}$$

There are two additional poles included in $C_2(s)$. The relative degree must be $r_2 = 2$ for the controller $C(s)$ to be proper. The location of poles can be considered as additional design variable since it will affect the achievable performance of the system.

B. Computation of the stabilizing sets for Smith-McMillan plants

We can find the stabilizing set and the achievable performance in terms of gain margin, phase margin, and ω_g for each SISO loop. For the first SISO loop, the admissible range of k_{p_1} for stability was determined to be $k_{p_1} \in (-2, \infty)$. For the second SISO loop, the admissible range of k_{p_2} for stability was determined to be $k_{p_2} \in (-9.2702, 2.6667)$. By sweeping k_{p_1} and k_{p_2} over these intervals, we can generate the stabilizing sets in (k_{p_1}, k_{i_1}) and (k_{p_2}, k_{i_2}) spaces, respectively. The stabilizing sets are shown in Figs. 8.4 and 8.5.

C. Gain–Phase margin design curves

The achievable gain–phase margin curves for $P_1(s)$ are generated as shown in Figs. 8.6, 8.7. Similarly, the achievable gain–phase margin curves for $P_2(s)$ are generated as shown in Figs. 8.8, 8.9.

D. Retrieval of the PI controller gains

Figure 8.6 shows that the gain margin increases without bound as the phase margin increases. For $P_1(s)$, it can be observed from Fig. 8.6 that the achievable phase margin is 83° at $\omega_g = 0.1$ rad/s.

For $P_2(s)$, we see that the achievable gain margin is bounded by 17.69 with the phase margin of 83° at $\omega_g = 0.1$ rad/s as shown in Fig. 8.8. The achievable gain margins for different ω_g values are also shown in Fig. 8.8. The designer has the liberty to choose different values of gain margin, phase margin, and ω_g that best suits the design needs from the generated curves. In this example, we notice, by

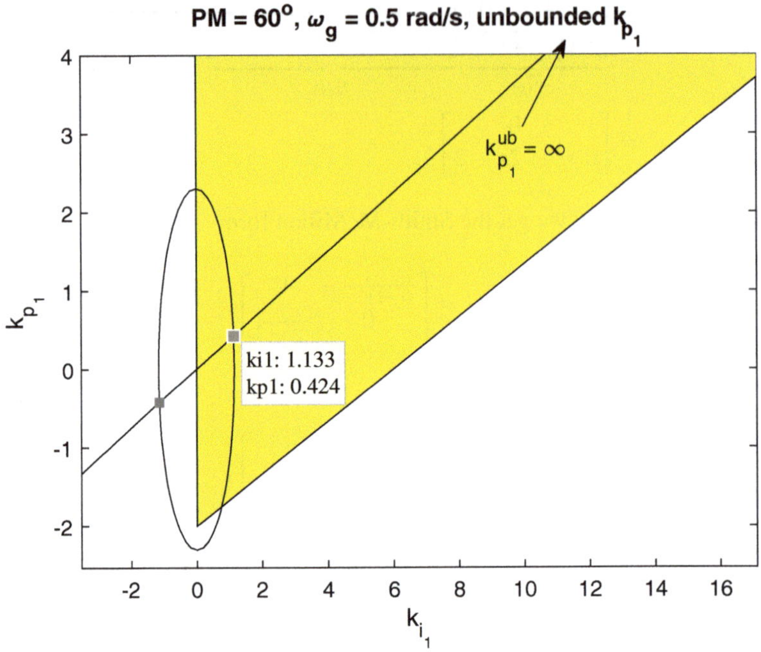

Fig. 8.4 Stabilizing set for $P_1(s)$, intersection of an ellipse and a straight line, and the PI controller $C_1(s)$ (inside the stabilizing set) achieving 60° degree of phase margin at $\omega_g = 0.5$ rad/s. $k_{p1}^{ub} = \infty$ indicates that the upper gain margin is infinity. © 2017 IEEE. Reproduced from [1] with permission

Theorem 8.1, that the achievable gain margin for $P(s)$ is equal to the minimum of the margins. For illustration purposes, we select $GM = \infty$, $PM = 60°$, and $\omega_g = 0.5$ rad/s for $P_1(s)$. For $P_2(s)$, we select $GM = 3.518$, $PM = 60°$, and $\omega_g = 0.5$ rad/s. The time-delay tolerances are $\tau_1 = 2.094$ and $\tau_2 = 2.094$, respectively. After the selection of the margins from the achievable gain–phase margin curves, the designer can retrieve the controller gains corresponding to these points. The controller gains are found to be $k_{p_1}^* = 0.424$ and $k_{i_1}^* = 1.133$ for $P_1(s)$ and $k_{p_2}^* = -1.059$ and $k_{i_2}^* = -1.34$ for $P_2(s)$.

E. Synthesis of $C(s)$ from $C_d(s)$ and design verification

The final step is to obtain $C(s)$ from $C_d(s)$ using (8.4) and (8.27):

$$C(s) = \begin{bmatrix} \frac{0.371s+0.618}{s} & \frac{-0.53s-0.67}{s(s+2)} \\ \frac{1.061s+1.34}{s(s+2)} & \frac{-2.12s-2.68}{s(s+2)^2} \end{bmatrix}. \tag{8.30}$$

We can verify the results by computing the gain margin and phase margin of the multivariable system. The characteristic polynomial of the multivariable system is given by

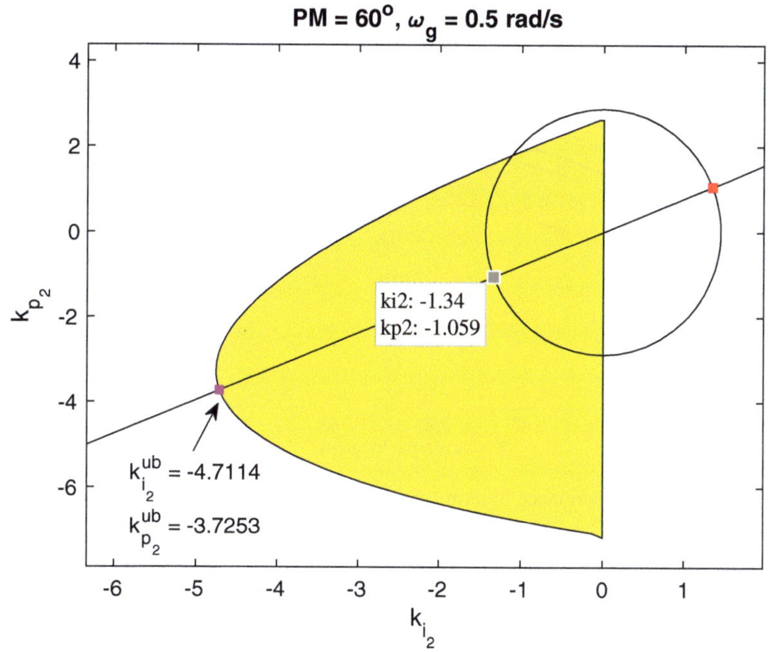

Fig. 8.5 Stabilizing set for $P_2(s)$, intersection of an ellipse and a straight line, and the PI controller $C_2(s)$ (inside the stabilizing set) achieving 60° degree of phase margin at $\omega_g = 0.5$ rad/s. ($k_{p2}^{ub} = -3.7253$, $k_{i_2}^{ub} = -4.7114$) is a boundary point in the stabilizing set for the computation of the upper gain margin. © 2017 IEEE. Reproduced from [1] with permission

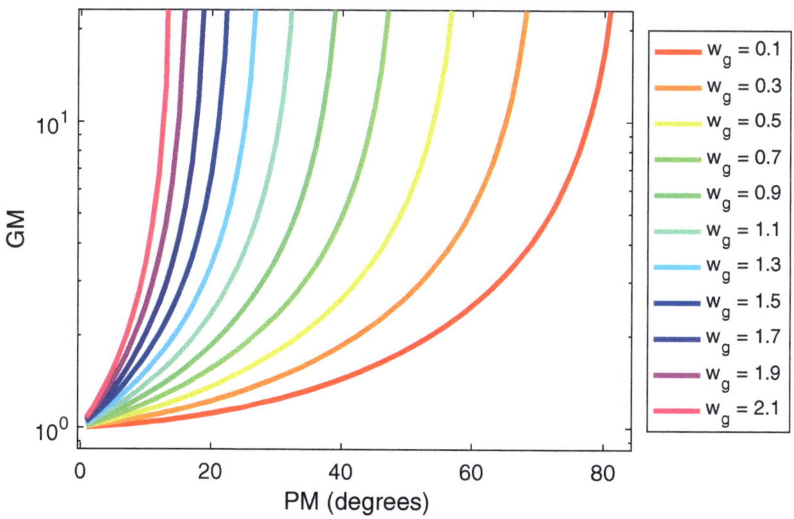

Fig. 8.6 Achievable performance in terms of gain margin(GM), phase margin(PM), and gain crossover frequency ω_g for $P_1(s)$. © 2017 IEEE. Reproduced from [1] with permission

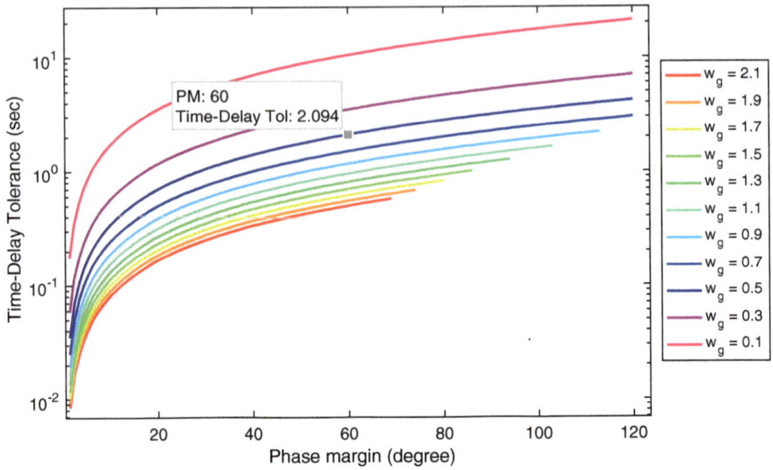

Fig. 8.7 Achievable performance in terms of time-delay tolerance τ_{max}, phase margin(PM), and gain crossover frequency ω_g for $P_1(s)$. © 2017 IEEE. Reproduced from [1] with permission

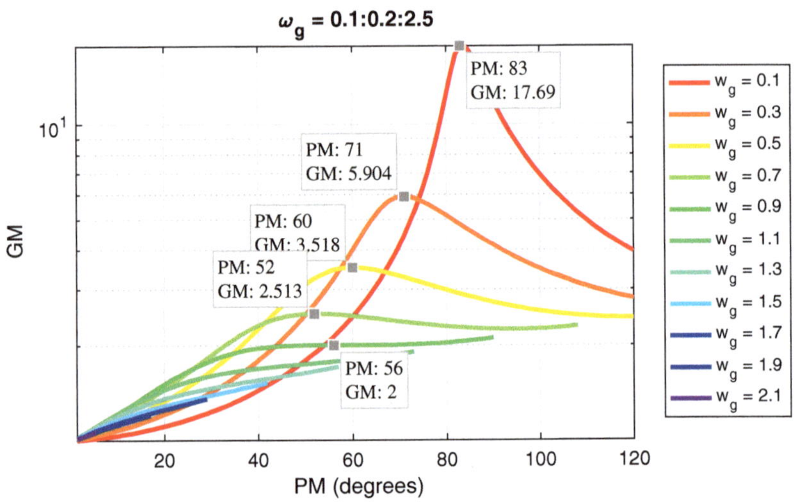

Fig. 8.8 Achievable performance in terms of gain margin(GM), phase margin(PM), and gain crossover frequency ω_g for $P_2(s)$. © 2017 IEEE. Reproduced from [1] with permission

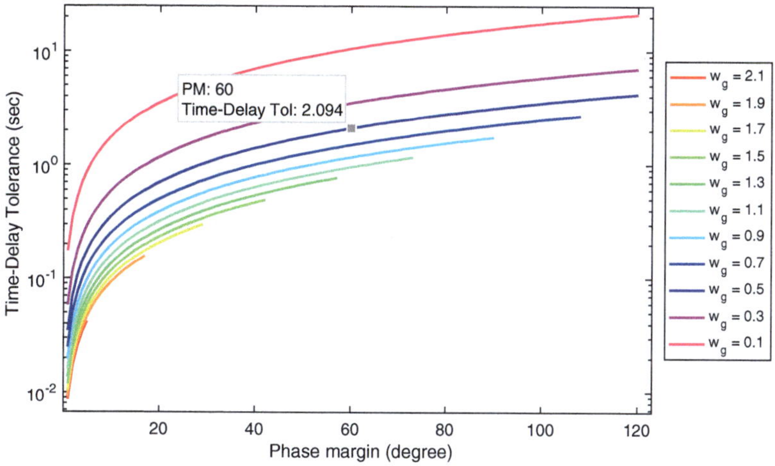

Fig. 8.9 Achievable performance in terms of time-delay tolerance τ_{max}, phase margin(PM), and gain crossover frequency ω_g for $P_2(s)$. © 2017 IEEE. Reproduced from [1] with permission

$$det[I + \Delta P(s)C(s)], \tag{8.31}$$

where Δ is defined as in (8.14). For the gain margin, substitute (8.18) into (8.31):

$$\begin{aligned} det[I + \Delta P(s)C(s)] = & s^7 + 9s^6 + (32 - 0.637k)s^5 + (56 + 2.33k)s^4 \\ & + (48 - 0.4484k^2 + 19.288k)s^3 \\ & + (16 - 0.4216k^2 + 32.708k)s^2 \\ & + (3.7833k^2 + 17.096k)s + 4.5506k^2. \end{aligned} \tag{8.32}$$

The range of k for the closed-loop stability is determined to be

$$0 < k < 3.518, \tag{8.33}$$

and the roots of (8.32) at $k = 3.518$ are

$$\begin{cases} 0.00030784192 & -j1.5532363 \\ 0.00030784192 & +j1.5532363 \\ -0.38174169 & +j1.2786136 \\ -0.38174169 & -j1.2786136 \\ -1.2283752 & \\ -2.2365012 & \\ -4.7722559. & \end{cases} \tag{8.34}$$

The real part of two of the roots just crossed the imaginary axis. Thus, the gain margin is $k^* = 3.518$ for the MIMO control system and is the minimum value of the gain margins that we selected in the design steps.

For the phase margin, substitute (8.20) into (8.31):

$$\begin{aligned} det[I + \Delta P(s)C(s)] = {} & s^7 + 9s^6 + (32 - 0.637e^{-j\theta})s^5 + (2.33e^{-j\theta} + 56.0)s^4 \\ & + (19.29e^{-j\theta} - 0.4484e^{-j2\theta} + 48.0)s^3 \\ & + (32.7e^{-j\theta} - 0.4218e^{-j2\theta} + 16.0)s^2 \\ & + (e^{-j\theta} + 3.784e^{-j2\theta} + 17.1)s + 4.551e^{-j2\theta}. \end{aligned} \tag{8.35}$$

The range of θ for the closed-loop stability is determined to be

$$0 < \theta < 60°, \tag{8.36}$$

and the roots of (8.35) for $\theta = 60°$ are

$$\begin{cases} -3.5159631 & -j0.67153706 \\ -2.0790212 & -j0.081079624 \\ -1.2506782 & -j0.1211673 \\ -1.2333837 & +j1.2928996 \\ -0.92099468 & +j0.58077106 \\ 0.000015906289 & -j0.49969144 \\ 0.000025099732 & -j0.50019525. \end{cases} \tag{8.37}$$

The real part of two of the roots just crossed the imaginary axis. Therefore, the phase margin is $\theta^* = 60°$ for the MIMO control system and is the minimum value of the phase margins that we selected in the design steps.

For the time-delay tolerance, substitute (8.22) into (8.31):

$$\begin{aligned} det[I + \Delta P(s)C(s)] = {} & s^7 + 9s^6 + s^5(0.423e^{-Ts} + 32.0) \\ & + s^4(4.45e^{-Ts} + 56.0) + s^3(18.23e^{-Ts} + 48.0) \\ & + s^2(30.59e^{-Ts} + 0.3299e^{-2Ts} + 16.0) \\ & + s(17.1e^{-Ts} + 2.583e^{-2Ts}) + 4.551e^{-2Ts}. \end{aligned}$$

The range of T for the closed-loop stability is determined to be

$$0 < T < 2.094 \text{ s}. \tag{8.38}$$

The time-delay tolerance is $T = 2.094$ for the MIMO control system and is the minimum value of the time-delay tolerances that we have selected in the design steps.

8.4 Notes and References

In [5], the Smith-McMillan form is considered as a design method of multivariable controllers. The main results in this chapter were first developed in [1]. In [7–9], an equivalent representation is presented to design multi-loop PI or PID controllers mainly using Internal Model Control approach. In [2, 4, 6], a design of MIMO systems is presented as individual channel designs where SISO design methods are exploited for MIMO systems. The transformation of a multivariable system into a scalar equivalent system is presented in [3].

References

1. Diaz-Rodriguez, I., Han, S., Bhattacharyya, S.P.: Stability margin based design of multivariable controllers. In: IEEE Conference on Control Technology and Applications (CCTA), IEEE, pp. 1661–1666 (2017)
2. Kallakuri, P., Keel, L.H., Bhattacharyya, S.P.: Multivariable controller design with integrity. In: 2013 American Control Conference, IEEE, pp. 5159–5164 (2013)
3. Keel, L.H., Bhattacharyya, S.P.: On the stability of multivariable feedback systems. In: 2015 IEEE 54th Annual Conference on Decision and Control (CDC), IEEE, pp. 4627–4631 (2015)
4. Keel, L.H., Bhattacharyya, S.P. Exact multivariable control design using siso methods: recent results. In: 2016 IEEE Conference on Control Applications (CCA), IEEE, pp. 215–223 (2016)
5. Mohsenizadeh, D.N., Keel, L.H., Bhattacharyya, S.P.: Multivariable controller synthesis using siso design methods. In: 2015 54th IEEE Conference on Decision and Control (CDC), IEEE, pp. 2680–2685 (2015)
6. O'Reilly, J., Leithead, W.E.: Multivariable control by individual channel design. Int. J. Control **54**(1), 1–46 (1991)
7. Rajapandiyan, C., Chidambaram, M.: Controller design for MIMO processes based on simple decoupled equivalent transfer functions and simplified decoupler. Ind. Eng. Chem. Res. **51**(38), 12398–12410 (2012)
8. Vu, T.N.L., Lee, M.: Independent design of multi-loop PI/PID controllers for interacting multivariable processes. J. Process Control **20**(8), 922–933 (2010)
9. Zanwar, S.R., Sankeshwari, S.S., Scholar, P.G.: Design of multi loop PI/PID controller for interacting multivariable process with effective open loop transfer function. Int. J. Eng. Sci. **7603** (2016)

Part III
H_∞ Optimal PID Control

Chapter 9
H_∞ Optimal Synthesis for Continuous-Time Systems

Abstract The H_∞ norm has proved itself to be a useful criterion for control system design. This chapter presents a constructive determination of a set of stabilizing PI and PID controllers, for a given plant, achieving an H_∞ norm bound of γ on the error transfer function. The results in this chapter utilize the computation of the complete stabilizing set $\mathcal{S}$. We point out connections between the H_∞ design and gain and phase margin designs. We show that the design criterion is expressed as the intersection of the stabilizing set and the outside of a family of ellipses in control parameter space.

9.1 Introduction

The Nyquist stability criterion entails the frequency response of the open-loop transfer function to stay away from the critical point $-1 + j0$ in the complex plane. The stability margins such as gain and phase margins are a measure of "robustness" of a given system as they represent how far the frequency response is away from the critical point at the crossover frequencies. An H_∞ norm specification on the error transfer function measures the closest distance to the critical point from all frequencies. In this chapter, we consider the H_∞ norm on the error transfer function as a design criterion. The complete stabilizing set $\mathcal{S}$ of PI and PID controllers is computed in Chap. 2. Having the set $\mathcal{S}$ in hand, we find the subset $\mathcal{S}_\gamma$ of PI and PID controllers satisfying the H_∞ norm less than γ.

In the next section, we develop a useful relationship between H_∞ norm specification on the error transfer function and guaranteed gain and phase margins. Following this, we present our constructive calculation of $\mathcal{S}_\gamma$ for PI or PID controller sets satisfying the given H_∞ norm specification.

I. D. Díaz-Rodríguez et al., *Analytical Design of PID Controllers*,
https://doi.org/10.1007/978-3-030-18228-1_9

9.2 H_∞ Optimal Control and Stability Margins

Consider the unity feedback system in Fig. 9.1 with the error transfer function

$$\frac{e(s)}{r(s)} = \frac{1}{1+G(s)}. \tag{9.1}$$

Suppose that $G(s)$ includes a controller designed to make the H_∞ norm of (9.1) less than γ, a prescribed real positive number. Then

$$\frac{1}{|1+G(j\omega)|} < \gamma, \qquad \text{for all } \omega \geq 0 \tag{9.2}$$

and (9.2) is equivalent to

$$|1+G(j\omega)| > \frac{1}{\gamma}, \qquad \text{for all } \omega \in [0, \infty). \tag{9.3}$$

We will now establish that (9.3) implies guaranteed gain and phase margins at the loop breaking point "m" in Fig. 9.1.

Remark 9.1 Let γ^* denote the infimum value of γ satisfying (9.3). When $G(s)$ is strictly proper, $\gamma^* \geq 1$. When $G(s)$ is proper, $\gamma^* > 1/\,|1+G(j\infty)|$.

Case 1: $\gamma > 1$
The condition (9.3) implies that the Nyquist plot $G(j\omega)$ stays out of the circle CEDB centered at $-1+j0$ and of radius $1/\gamma$. In Fig. 9.2, we have the limiting case in which $G(j\omega)$ passes through B, the phase margin is ϕ and

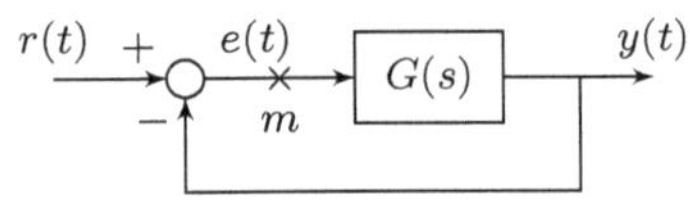

Fig. 9.1 Unity feedback control loop

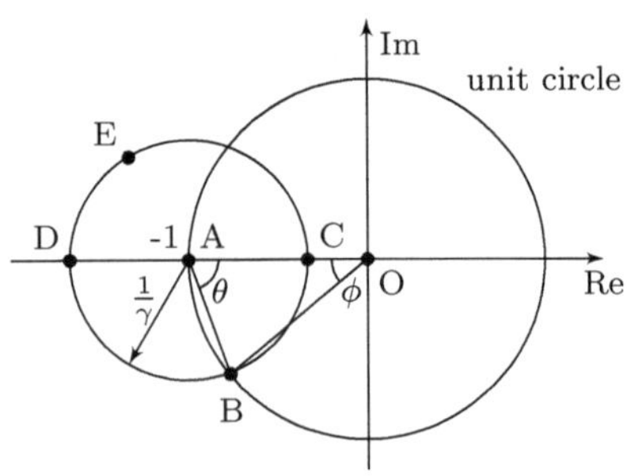

Fig. 9.2 $\gamma > 1$. © 2018 IFAC. Reproduced from the original publication "PID Controller Design with an H_∞ criterion", IFAC-PapersOnline, Volume 51, Issue 4, June 2018, Pages 400–405 with permission

$$G(j\omega) = \overrightarrow{\text{OB}}, \tag{9.4}$$

$$-1 + j0 = \overrightarrow{\text{OA}}, \tag{9.5}$$

$$1 + G(j\omega) = \overrightarrow{\text{AB}}. \tag{9.6}$$

Since $\overrightarrow{\text{OA}} + \overrightarrow{\text{AB}} = \overrightarrow{\text{OB}}$, we have

$$-1 + j0 + \frac{1}{\gamma}e^{-j\theta} = -1e^{j\phi}. \tag{9.7}$$

Also

$$2\theta + \phi = \pi \tag{9.8}$$

from the triangle $\overrightarrow{\text{OAB}}$.

From (9.7) and (9.8),

$$-1 + \frac{1}{\gamma}\sin\left(\frac{\phi}{2}\right) = -\cos\phi \tag{9.9}$$

$$\sin\phi = \frac{1}{\gamma}\cos\left(\frac{\phi}{2}\right). \tag{9.10}$$

From (9.10),

$$\phi = 2\sin^{-1}\left(\frac{1}{2\gamma}\right), \tag{9.11}$$

which is the guaranteed minimum phase margin for the H_∞ controller with norm less than γ.

The guaranteed gain margin is the interval:

$$\left[\frac{1}{\text{OD}}, \frac{1}{\text{OC}}\right] = \left[\frac{\gamma}{\gamma+1}, \frac{\gamma}{\gamma-1}\right]. \tag{9.12}$$

Case 2: $\gamma = 1$
In this case, Fig. 9.2 is replaced by Fig. 9.3. It is easy to see that the guaranteed phase margin is $\phi = \pi/3$ and the guaranteed gain margin is $\left[\frac{1}{2}, \infty\right]$. These also follow from formulas (9.11) and (9.12) evaluated at $\gamma = 1$.
Case 3: $\gamma < 1$
The geometry corresponding to this case is shown in Fig. 9.4. In this case, it also follows that the guaranteed phase margin is

$$\phi = 2\sin^{-1}\left(\frac{1}{2\gamma}\right) \tag{9.13}$$

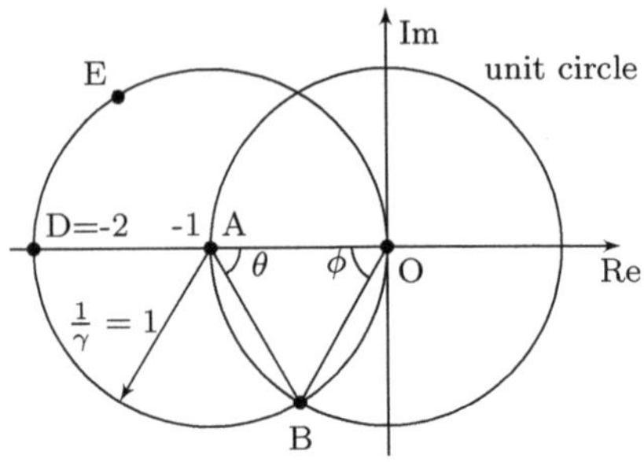

Fig. 9.3 $\gamma = 1$. © 2018 IFAC. Reproduced from the original publication "PID Controller Design with an H_∞ criterion", IFAC-PapersOnline, Volume 51, Issue 4, June 2018, Pages 400–405 with permission

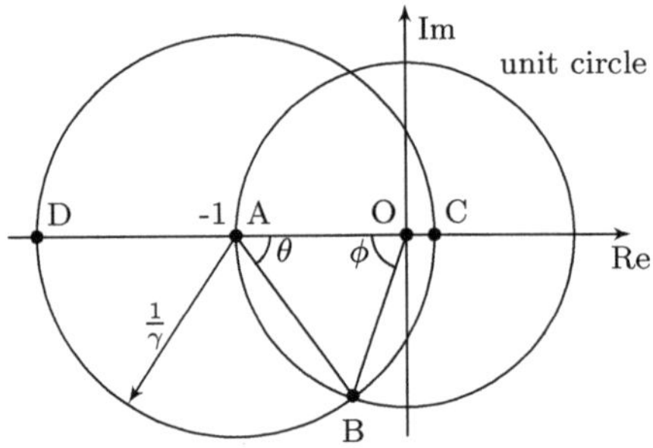

Fig. 9.4 $\gamma < 1$. © 2018 IFAC. Reproduced from the original publication "PID Controller Design with an H_∞ criterion", IFAC-PapersOnline, Volume 51, Issue 4, June 2018, Pages 400–405 with permission

and the guaranteed gain margin is

$$\left[\frac{1}{\text{OD}}, \infty\right] = \left[\frac{\gamma}{1+\gamma}, \infty\right]. \tag{9.14}$$

Combining the above cases, we have the following result.

Theorem 9.1 *Consider the unity feedback system in Fig. 9.1. If the H_∞ norm of the error transfer function is less than γ:*

$$\left\| \frac{1}{1+G(s)} \right\|_\infty < \gamma, \tag{9.15}$$

then the guaranteed phase margin at the loop breaking point "m" is

$$\phi = 2\sin^{-1}\left(\frac{1}{2\gamma}\right). \tag{9.16}$$

The guaranteed gain margin is

$$g_m = \begin{cases} \left[\frac{\gamma}{\gamma+1}, \frac{\gamma}{\gamma-1}\right], & \textit{for } \gamma > 0 \\ \left[\frac{\gamma}{\gamma+1}, \infty\right], & \textit{for } \gamma \le 0. \end{cases} \tag{9.17}$$

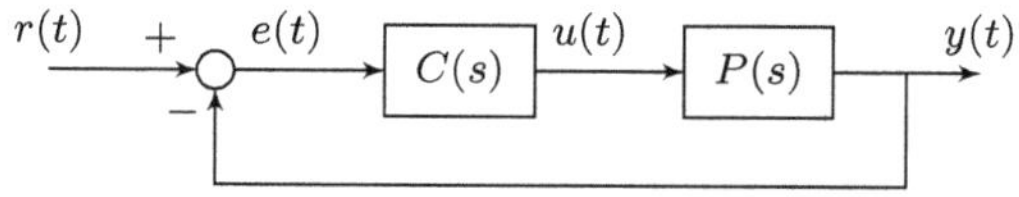

Fig. 9.5 Unity feedback control loop

Now, consider the control system in Fig. 9.5 where $r(t)$ is the reference signal, $e(t)$ the error signal, $u(t)$ the input signal (to the plant), $y(t)$ the output signal, $P(s)$ is the plant transfer function, and $C(s)$ is the controller transfer function which we will consider to be either PI or PID.

The problem to be solved in this chapter is: Find the set $\mathcal{S}_\gamma$ of all stabilizing PI or PID controllers satisfying

$$\left\| \frac{1}{1+P(s)C(s)} \right\|_\infty < \gamma. \tag{9.18}$$

In the following two sections, we develop the computation of $\mathcal{S}_\gamma$ for PI and PID controllers. Note that (9.18) is equivalent to

$$|1+P(j\omega)C(j\omega)| > \frac{1}{\gamma}, \quad \forall\, \omega \in [0,\infty). \tag{9.19}$$

9.3 Computation of $\mathcal{S}_\gamma$ for PI Controllers

PI controllers have the following form:

$$C(s) = k_p + \frac{k_i}{s}. \tag{9.20}$$

Write

$$P(j\omega) = P_r(\omega) + j\omega P_i(\omega), \tag{9.21}$$

$$C(j\omega) = k_p - j\frac{k_i}{\omega}. \tag{9.22}$$

Substituting (9.21) and (9.22) in (9.19), we get

$$|\underbrace{1+k_pP_r(\omega)+k_iP_i(\omega)}_{L_0(\omega)}+j\underbrace{(\omega k_pP_i(\omega)-\frac{k_i}{\omega}P_r(\omega))}_{L_1(\omega)}| > \frac{1}{\gamma}, \tag{9.23}$$

which can be rewritten as

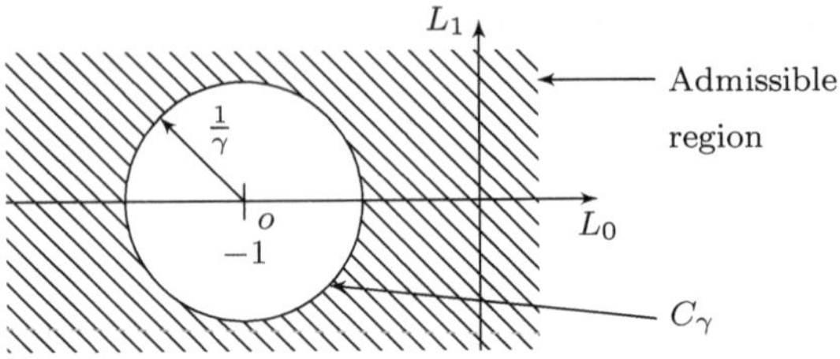

Fig. 9.6 The C_γ circle. © 2018 IFAC. Reproduced from the original publication "PID Controller Design with an H_∞ criterion", IFAC-PapersOnline, Volume 51, Issue 4, June 2018, Pages 400–405 with permission

$$(1 + L_0(\omega))^2 + L_1^2(\omega) > \frac{1}{\gamma^2} \tag{9.24}$$

$$\begin{bmatrix} P_r(\omega) & P_i(\omega) \\ \omega P_i(\omega) & -\frac{P_r(\omega)}{\omega} \end{bmatrix} \begin{bmatrix} k_p \\ k_i \end{bmatrix} = \begin{bmatrix} L_0(\omega) \\ L_1(\omega) \end{bmatrix}. \tag{9.25}$$

Equation (9.25) has a unique solution if

$$|P(j\omega)| \neq 0, \tag{9.26}$$

that is, the plant has no $j\omega$ axis zeros.

Assuming (9.26), (9.25) can be solved:

$$\begin{bmatrix} k_p \\ k_i \end{bmatrix} = \underbrace{\frac{1}{|P(j\omega)|^2} \begin{bmatrix} P_r(\omega) & \omega P_i(\omega) \\ -\omega^2 P_i(\omega) & -\omega P_r(\omega) \end{bmatrix}}_{T(\omega)} \begin{bmatrix} L_0(\omega) \\ L_1(\omega) \end{bmatrix}. \tag{9.27}$$

Equation (9.24) represents the outside of a circle C_γ of radius $\frac{1}{\gamma}$ in the (L_0, L_1) plane centered at $(-1, 0)$ (Figs. 9.6 and 9.7):

Lemma 9.1 *Condition* (9.19) *at a fixed* ω *is equivalent to* k_p, k_i *lying in the exterior of the axis-parallel ellipse* $E_\gamma(\omega)$ *with center* o' *at* $(\frac{-\omega^2 P_i(\omega)}{|P(j\omega)|^2}, \frac{-P_r(\omega)}{|P(j\omega)|^2})$, *and major and minor axes of lengths* $\frac{2}{\gamma|P(j\omega)|}$, $\frac{2\omega}{\gamma|P(j\omega)|}$.

Proof For each $\omega \geq 0$, (9.23) is

$$\begin{aligned} &\left|1 + (P_r(j\omega) + j\omega P_i(j\omega))(k_p - j\frac{k_i}{\omega})\right| > \frac{1}{\gamma} \\ \Leftrightarrow\ &\left(1 + P_r(j\omega)k_p + P_i(j\omega)k_i\right)^2 + \left(\omega P_i(j\omega)k_p - P_r(j\omega)\frac{k_i}{\omega}\right)^2 > \frac{1}{\gamma^2} \\ \Leftrightarrow\ &\frac{(k_i - c_1)^2}{a^2} + \frac{\left(k_p - c_2\right)^2}{b^2} > 1, \end{aligned} \tag{9.28}$$

where

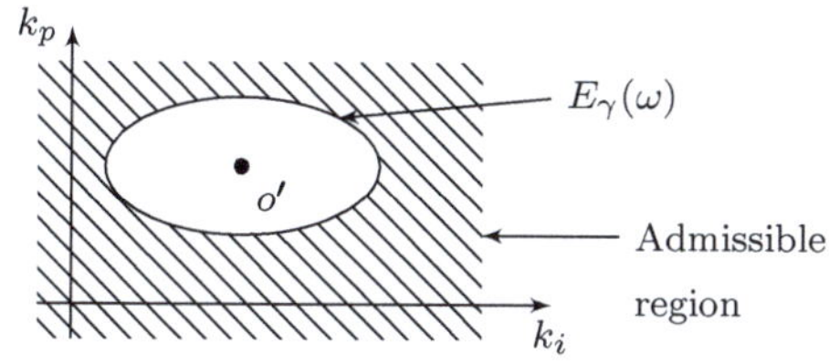

Fig. 9.7 The $E_\gamma(\omega)$ ellipse. © 2018 IFAC. Reproduced from the original publication "PID Controller Design with an H_∞ criterion", IFAC-PapersOnline, Volume 51, Issue 4, June 2018, Pages 400–405 with permission

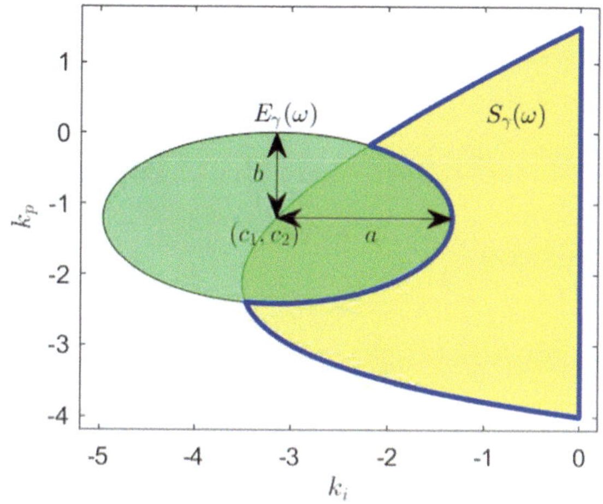

Fig. 9.8 $E_\gamma(\omega)$ and $S_\gamma(\omega)$. © 2018 IFAC. Reproduced from the original publication "PID Controller Design with an H_∞ criterion", IFAC-PapersOnline, Volume 51, Issue 4, June 2018, Pages 400–405 with permission

$$c_1 = \frac{-\omega^2 P_i(\omega)}{|P(j\omega)|^2}, \; c_2 = \frac{-P_r(\omega)}{|P(j\omega)|^2}, \; a = \frac{\omega/\gamma}{|P(j\omega)|}, \; b = \frac{1/\gamma}{|P(j\omega)|}. \tag{9.29}$$

□

For a fixed ω, let $\mathcal{S}_\gamma(\omega)$ denote the intersection of the stabilizing set S with the exterior of the ellipse $E_\gamma(\omega)$ as shown in Fig. 9.8. In other words,

$$\mathcal{S}_\gamma(\omega) = \mathcal{S} \setminus E_\gamma(\omega) \;\; \forall\, \omega \in [0, \infty). \tag{9.30}$$

Since (9.19) must hold for all ω,

$$\mathcal{S}_\gamma = \bigcap_{\omega=0}^{\infty} \mathcal{S}_\gamma(\omega) \tag{9.31}$$

as shown in Fig. 9.9.

We state this result in the following theorem.

Theorem 9.2 *In the unity feedback control loop, suppose that the plant $P(s)$ has no $j\omega$ axis zeros. All stabilizing PI controllers $C(s)$ satisfying the H_∞ norm bound of γ on the error transfer function is the set $\mathcal{S}_\gamma$:*

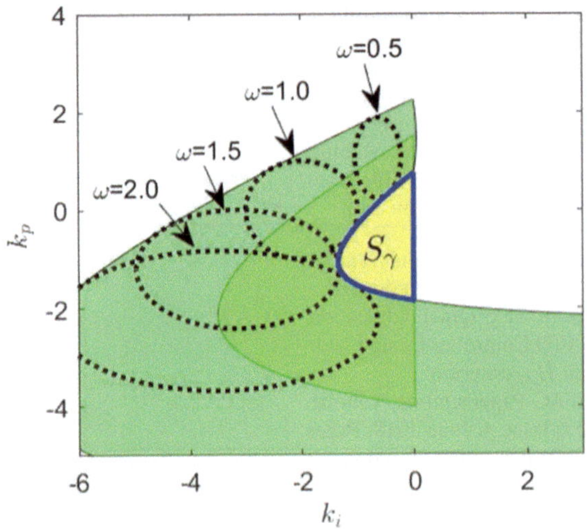

Fig. 9.9 $\mathcal{S}_\gamma$. © 2018 IFAC. Reproduced from the original publication "PID Controller Design with an H_∞ criterion", IFAC-PapersOnline, Volume 51, Issue 4, June 2018, Pages 400–405 with permission

$$\mathcal{S}_\gamma = \bigcap_{\omega=0}^{\infty} \mathcal{S}_\gamma(\omega). \tag{9.32}$$

Proof $\mathcal{S}_\gamma(\omega)$ is the admissible set for each ω and the controller must satisfy the H_∞ norm for all frequencies. Hence, we have the set $\mathcal{S}_\gamma$ by intersecting the admissible sets $\mathcal{S}_\gamma(\omega)$ for all ω. □

Note that $\mathcal{S}$ can be determined using the concept of signature developed in Chap. 2. If $E_\gamma(\omega)$ is outside of $\mathcal{S}$, then $\mathcal{S}_\gamma(\omega) = \mathcal{S}$. If $\mathcal{S} \subset E_\gamma(\omega)$, then $\mathcal{S}_\gamma$ is empty.

Remark 9.2 We can determine the minimum achievable γ for a given plant under PI or PID control. The minimum γ denoted γ^* is the value for which the union of family of ellipses eclipses the stabilizing set $\mathcal{S}$.

Remark 9.3 The computation of $\mathcal{S}_\gamma$ would not be possible without knowing the stabilizing set $\mathcal{S}$.

Example 9.1 Consider the second-order plant and the PI controller:

$$P(s) = \frac{s-2}{s^2+4s+3}, \quad C(s) = k_p + \frac{k_i}{s}. \tag{9.33}$$

The stabilizing set was first computed for the plant and the PI controller given in (9.33). The family of ellipses $E_\gamma(\omega)$ were drawn by sweeping over ω and $\mathcal{S}_\gamma$ was found accordingly for $\gamma = 1.6$, 2.0, 4.0, and 8.0. In Fig. 9.10, we observed that $\mathcal{S}_\gamma$ were contained in the stabilizing set $\mathcal{S}$ and $\mathcal{S}_{\gamma_1} \subset \mathcal{S}_{\gamma_2}$ if $\gamma_1 < \gamma_2$. So, $\mathcal{S}_\gamma$ for $\gamma \in [1, \infty)$ is the telescoping series of sets shown. If k_p, k_i were chosen from sets $\mathcal{S}_\gamma$, the Nyquist plot must stay outside of a circle centered at the critical point $-1 + j0$ with

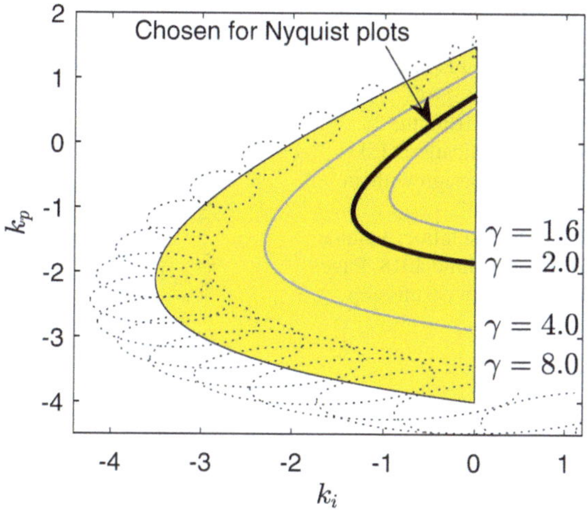

Fig. 9.10 $\mathcal{S}_\gamma$ for $\gamma = 1.6$, 2.0, 4.0, 8.0 with the stabilizing set. © 2018 IFAC. Reproduced from the original publication "PID Controller Design with an H_∞ criterion", IFAC-PapersOnline, Volume 51, Issue 4, June 2018, Pages 400–405 with permission

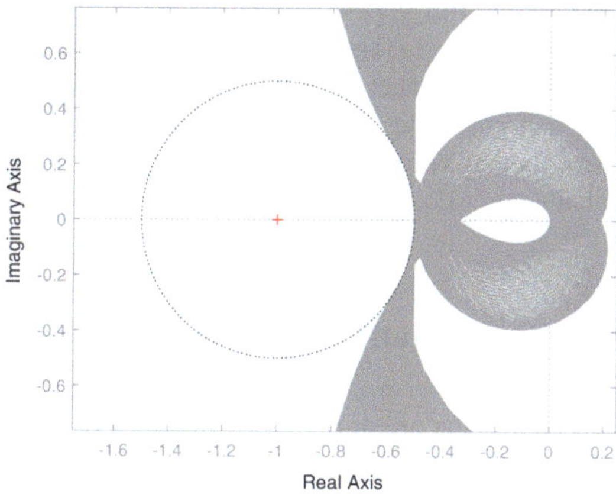

Fig. 9.11 Nyquist plots with k_p, k_i along the curve of $\gamma = 2$. © 2018 IFAC. Reproduced from the original publication "PID Controller Design with an H_∞ criterion", IFAC-PapersOnline, Volume 51, Issue 4, June 2018, Pages 400–405 with permission

radius of $1/\gamma$. We chose some boundary points in $\mathcal{S}_\gamma$ that were inside the stabilizing set $\mathcal{S}$ where $\gamma = 2$ and drew the Nyquist plots in Fig. 9.11. Each Nyquist plot was at least 0.5 away from the critical point.

Following Theorem 9.1, the guaranteed gain margin was

$$\left[\frac{\gamma}{\gamma+1},\ \frac{\gamma}{\gamma-1}\right] = \left[\frac{2}{3},\ 2\right], \tag{9.34}$$

and the guaranteed phase margin ϕ was

$$\phi = 2\sin^{-1}\left(\frac{1}{2\gamma}\right) = 28.955^o, \tag{9.35}$$

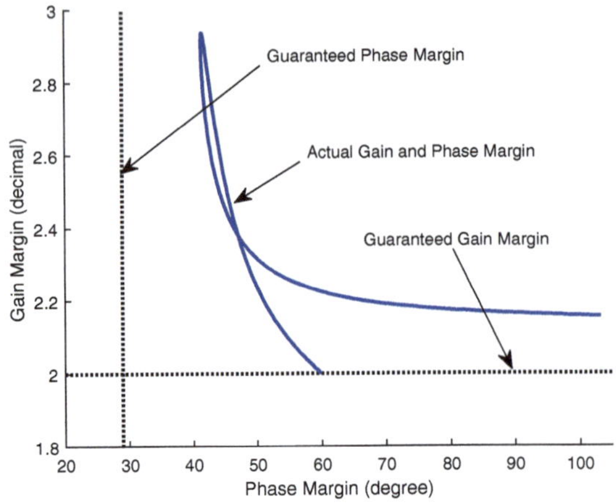

Fig. 9.12 Guaranteed gain and phase margin of the boundary points of $\mathcal{S}_\gamma$ for $\gamma = 2$. © 2018 IFAC. Reproduced from the original publication "PID Controller Design with an H_∞ criterion", IFAC-PapersOnline, Volume 51, Issue 4, June 2018, Pages 400–405 with permission

for $\gamma = 2$. Figure 9.12 shows the guaranteed gain and phase margins when we choose k_p and k_i from S_γ for $\gamma = 2$. For all controllers achieving the same H_∞ norm at the boundary of S_γ, there is a trade-off between gain and phase margins. When higher gain margin is desired, one should sacrifice some phase margin and vice versa. Nevertheless with the H_∞ norm, we get the guaranteed gain and phase margins calculated in Eqs. (9.34) and (9.35).

9.4 Computation of $\mathcal{S}_\gamma$ for PID Controllers

PID controllers are of the following form (Fig. 9.13):

$$C(s) = k_p + \frac{k_i}{s} + k_d s. \tag{9.36}$$

Substituting $s = j\omega$, we have

$$C(j\omega) = k_p - j\frac{1}{\omega}\left(k_i - \omega^2 k_d\right). \tag{9.37}$$

Notice that (9.37) is equal to (9.22) if we replace k_i in (9.22) with $k_i' = k_i - \omega^2 k_d$. By analysis similar to the PI case, it is easy to show that (9.19) implies that the controller parameters k_p, k_i, k_d must lie in the exterior of $E_\gamma(\omega)$ described by

$$\frac{\left(k_i - \omega^2 k_d - c_1\right)^2}{a^2} + \frac{\left(k_p - c_2\right)^2}{b^2} > 1, \tag{9.38}$$

where $E_\gamma(\omega)$ is an elliptic cylinder with the center lying on the line

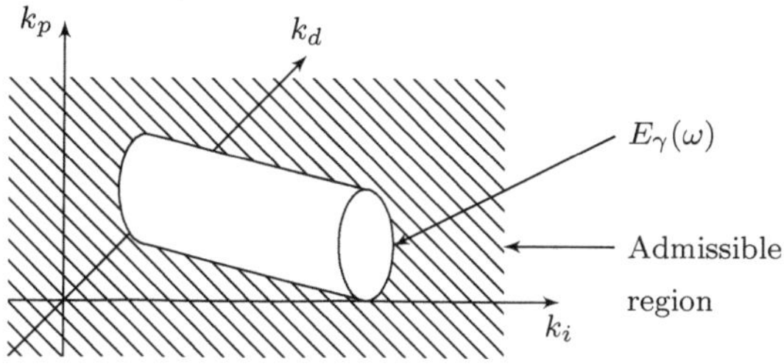

Fig. 9.13 The $E_\gamma(\omega)$ elliptic cylinder. © 2018 IFAC. Reproduced from the original publication "PID Controller Design with an H_∞ criterion", IFAC-PapersOnline, Volume 51, Issue 4, June 2018, Pages 400–405 with permission

$$\begin{cases} k_i - \omega^2 k_d = \frac{-\omega^2 P_i(\omega)}{|P(j\omega)|^2}, \\ k_p = \frac{-P_\gamma(\omega)}{|P(j\omega)|^2}, \end{cases} \tag{9.39}$$

and major and minor axes $\frac{2}{\gamma|P(j\omega)|}$ and $\frac{2\omega}{\gamma\sqrt{\omega^4+1}|P(j\omega)|}$.

As before,

$$\mathcal{S}_\gamma(\omega) = \mathcal{S} \setminus E_\gamma(\omega) \ \forall\, \omega \in [0, \infty) \tag{9.40}$$

and

$$\mathcal{S}_\gamma = \bigcap_{\omega=0}^{\infty} \mathcal{S}_\gamma(\omega). \tag{9.41}$$

Remark 9.4 We can also consider the H_∞ norm with a weighting function $W(s)$ multiplied by the error transfer function in (9.18). In this case, we may replace γ by γ', where $\gamma' = \frac{\gamma}{|W(j\omega)|}$. Then, the major and minor axes of the axis-parallel ellipse $E_\gamma(\omega)$ are subject to change with ω in accordance to the frequency response of the weighting function. However, the rest of the derivation of the equations in this section remains the same.

Remark 9.5 If $C(s)$ is replaced by

$$C_\tau(s) = \frac{k_p s + k_i + k_d s^2}{s(\tau s + 1)}, \tag{9.42}$$

then

$$C_\tau(s)P(s) = C(s)\frac{1}{\tau s + 1}P(s). \tag{9.43}$$

Since τ can be fixed a priori, replace $P_r(j\omega)$ and $P_i(j\omega)$ by

$$P_r'(j\omega) = \frac{P_r(j\omega) + \tau\omega^2 P_i(j\omega)}{1 + \tau^2\omega^2}$$
$$P_i'(j\omega) = \frac{P_i(j\omega) - \tau P_r(j\omega)}{1 + \tau^2\omega^2}.$$

Then, the controller design can be carried out as before.

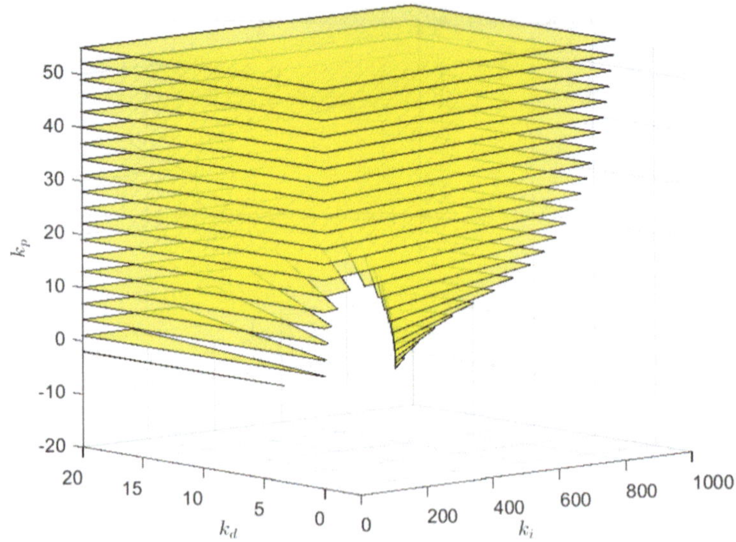

Fig. 9.14 The stabilizing set in k_p, k_i, k_d space using the signature method. © 2018 IFAC. Reproduced from the original publication "PID Controller Design with an H_∞ criterion", IFAC-PapersOnline, Volume 51, Issue 4, June 2018, Pages 400–405 with permission

Example 9.2 Consider a rational plant transfer function and the PID controller:

$$P(s) = \frac{10s^3 + 9s^2 + 362.4s + 36.16}{2s^5 + 2.7255s^4 + 138.4292s^3 + 156.471s^2 + 637.6472s + 360.1779}, \tag{9.44}$$

$$C(s) = k_p + \frac{k_i}{s} + k_d s. \tag{9.45}$$

The stabilizing set was computed using the signature method and is shown in Fig. 9.14. We chose $k_d = 9$ and computed $\mathcal{S}_\gamma$ for $\gamma = 1$ in the k_p, k_i plane. Figure 9.15 shows S_γ and the family of ellipses, $E_\gamma(\omega)$.

We observed that the stabilizing set with $k_d = 9$ was unbounded in the k_p, k_i plane. However, $\mathcal{S}_\gamma$ for $\gamma = 1$ in the same plane was bounded. For high values of ω, the major and minor axes of the ellipses grow as the centers c_1 and c_2 in (9.29) go away from the origin. So, we suggest that the family of ellipses be computed for high enough values of ω to get the exact set $\mathcal{S}_\gamma$.

Clearly, in this case, $\mathcal{S}_\gamma$ is not empty and the H_∞ norm condition less than $\gamma = 1$ provides very good robustness, namely, $[0.5, \infty]$ gain margin and 60° phase margin. All of the points in $\mathcal{S}_\gamma$ guarantee such good robustness. In fact, since the open-loop transfer function $P(s)C(s)$ is strictly proper, the Nyquist plot of $P(j\omega)C(j\omega)$ goes to 0 as $\omega \to \infty$ and so every point in $\mathcal{S}_\gamma$ achieves the same H_∞ norm.

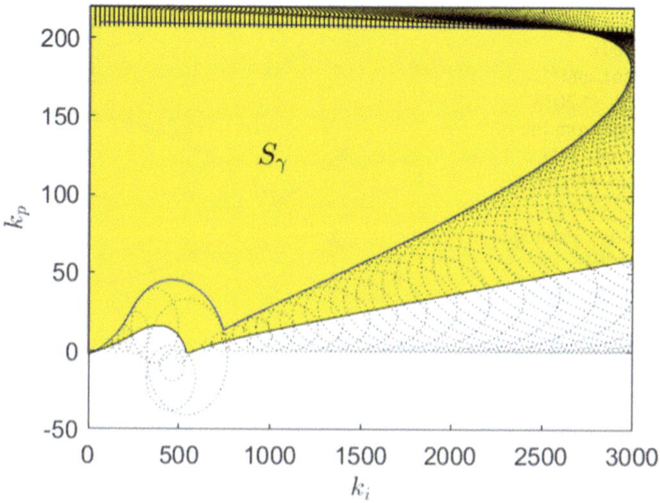

Fig. 9.15 $\mathcal{S}_\gamma$ and family of ellipses for $\gamma = 1$ in k_p, k_i plane with $k_d = 9$. © 2018 IFAC. Reproduced from the original publication "PID Controller Design with an H_∞ criterion", IFAC-PapersOnline, Volume 51, Issue 4, June 2018, Pages 400–405 with permission

Time Response Considerations

So far we have discussed stability and robustness. However, the design of a controller should pay attention to the time response. In order to demonstrate this, we chose the following three design points:

$$\begin{cases} C_1(s) = 185 + \dfrac{2986}{s} + 9s, \\ C_2(s) = 20 + \dfrac{800}{s} + 9s, \\ C_3(s) = 19 + \dfrac{200}{s} + 9s. \end{cases} \tag{9.46}$$

The first point has the maximum k_i value in $\mathcal{S}_\gamma$, the second and the third are arbitrary points from the boundary of $\mathcal{S}_\gamma$.

The Nyquist plots in Fig. 9.16 confirm that all three design points satisfy the robustness condition. The step responses in Fig. 9.17 show that the three controller designs result in different time responses in terms of overshoot and settling time. While $C_1(s)$ and $C_2(s)$ have highest and intermediate integral gains, $C_3(s)$ provides much shorter settling time and lower overshoot than the other two controllers do.

The integrator in the controller provided zero steady-state error and we found all stabilizing controllers achieving prescribed H_∞ norm of the error transfer function. While the robustness and zero steady-state error could be achieved by the proposed method, one should also consider the quality of the transient response when tuning

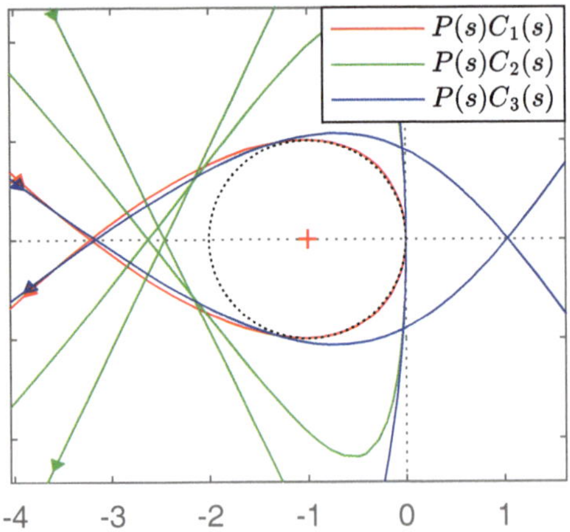

Fig. 9.16 Nyquist diagram for $P(s)C_1(s)$ (red), $P(s)C_2(s)$ (green), and $P(s)C_3(s)$ (blue). © 2018 IFAC. Reproduced from the original publication "PID Controller Design with an H_∞ criterion", IFAC-PapersOnline, Volume 51, Issue 4, June 2018, Pages 400–405 with permission

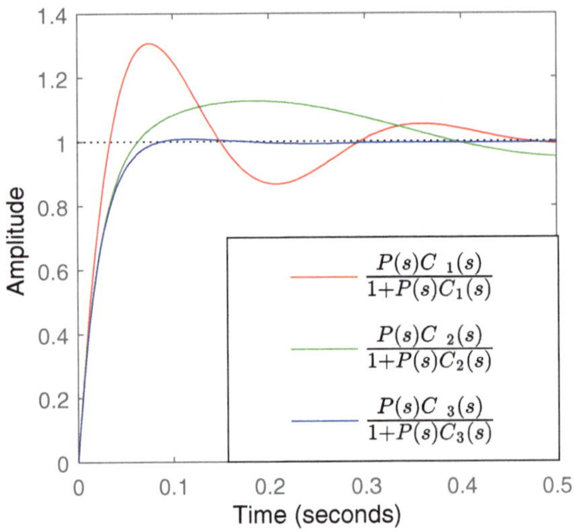

Fig. 9.17 Step responses for the closed-loop systems of $P(s)C_1(s)$ (red), $P(s)C_2(s)$ (green), and $P(s)C_3(s)$ (blue). © 2018 IFAC. Reproduced from the original publication "PID Controller Design with an H_∞ criterion", IFAC-PapersOnline, Volume 51, Issue 4, June 2018, Pages 400–405 with permission

the PID parameters within the set $\mathcal{S}_\gamma$. Thus, PID controller design for better transient response within the same degree of robustness is an important area of research.

9.5 Notes and References

The main results of this chapter are taken from Han, Keel, and Bhattacharyya [3]. In [2], the 2D regions of stabilizing PID controllers achieving the H_∞ norm bound of γ on the sensitivity and complementary sensitivity functions with weightings were found by using Neimark's D-decomposition. The difference is that our approach explicitly uses the *stabilizing set*. A similar approach was adopted in [5] for first-order controllers and in this case, the stability region was computed a priori. In [4], it was shown that at a fixed frequency (and for a fixed k_d, the derivative gain) the L_2 norm of the error transfer function being equal to γ was represented by an ellipse in (k_p, k_i) space. An H_∞ optimal PID design using a frequency loop-shaping approach was reported in [1, 6].

References

1. Ashfaque, B.S., Tsakalis, K.: Discrete-time PID controller tuning using frequency loop-shaping. IFAC Proc. Vol. **45**(3), 613–618 (2012)
2. Emami, T., Watkins, J.M.: Robust performance characterization of PID controllers in the frequency domain. WSEAS Trans. Syst. Control **4**(5), 232–242 (2009)
3. Han, S., Keel, L.H., Bhattacharyya, S.P.: PID controller design with an H_∞ criterion. IFAC-PapersOnLine 51(4), 400–405 (2018). 3rd IFAC Conference on Advances in Proportional-Integral-Derivative Control PID 2018
4. Krajewski, W., Viaro, U.: On robust PID control for time-delay plants. In: 2012 17th International Conference on Methods and Models in Automation and Robotics (MMAR), pp. 540–545. IEEE (2012)
5. Tantaris, R.N., Keel, L.H., Bhattacharyya, S.P.: H_∞ design with first-order controllers. IEEE Trans. Autom. Control **51**(8), 1343–1347 (2006)
6. Tsakalis, K.S., Dash, S.: Approximate H_∞ loop shaping in PID parameter adaptation. Int. J. Adapt. Control Signal Process. **27**(1–2), 136–152 (2013)

Chapter 10
H_∞ Optimal Synthesis for Discrete-Time Systems

Abstract In this chapter, we consider the H_∞ optimal synthesis of digital PI and PID controllers. We present the computation of $\mathcal{S}_\gamma$, the set of stabilizing digital PI or PID controllers for a given plant, satisfying an H_∞ norm bound of γ on the error transfer function.

10.1 Introduction

In digital control, dynamic systems are often represented in terms of z-transforms of discrete-time signals and systems. In this chapter, we extend the H_∞ norm approach for the design of continuous-time systems developed in Chap. 9 to discrete-time systems. The H_∞ norm criterion on the error transfer function is written in terms of the z-transforms of discrete-time signals and systems. We compute the stabilizing set $\mathcal{S}$ of digital PI or PID controllers based on the results in Chap. 4. Having computed the stabilizing set $\mathcal{S}$, we constructively determine the subset $\mathcal{S}_\gamma$ of digital PI or PID controllers for which the error transfer function of the closed-loop system has an H_∞ norm less than a prescribed real value $\gamma > 0$.

10.2 Computation of $\mathcal{S}_\gamma$ for Digital PI Controllers

Consider the unity feedback control loop with a discrete-time controller and plant in Fig. 10.1, where

$$P(z) = \frac{N(z)}{D(z)}, \tag{10.1}$$

$$C(z) = \frac{K_1 z + K_0}{z - 1}, \tag{10.2}$$

I. D. Díaz-Rodríguez et al., *Analytical Design of PID Controllers*,
https://doi.org/10.1007/978-3-030-18228-1_10

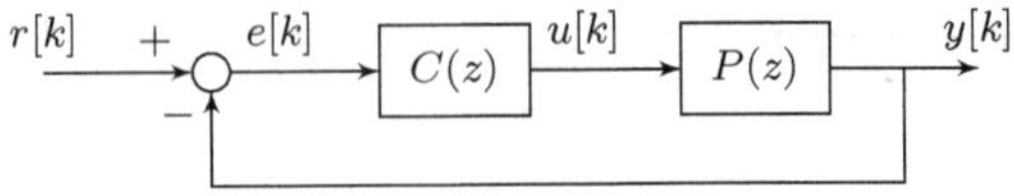

Fig. 10.1 Unity feedback control loop in the discrete-time domain

and $D(z)$, $N(z)$ are polynomials in z with real coefficients and $C(z)$ is the transfer function of a PI controller. We assume that $P(1) \neq 0$ since the closed loop cannot be stabilized otherwise. The error transfer function is

$$\frac{E(z)}{R(z)} = \frac{1}{1 + P(z)C(z)}, \tag{10.3}$$

where $E(z)$ and $R(z)$ are the z-transforms of $e[k]$ and $r[k]$, respectively.

The H_∞ norm criterion on the error transfer function is

$$\left\| \frac{E(z)}{R(z)} \right\|_\infty < \gamma \tag{10.4}$$

for a prescribed $\gamma > 0$. This is equivalent to

$$\sup_\theta \left| \frac{1}{1 + P(e^{j\theta})C(e^{j\theta})} \right| < \gamma, \quad \forall \theta \in [0, 2\pi)\,. \tag{10.5}$$

Since $N(z)$ and $D(z)$ have real coefficients, it is sufficient to consider $\theta \in [0, \pi)$ as in Chap. 4. Substituting $z = e^{j\theta}$, we have

$$N(z)|_{z=e^{j\theta}} = N(e^{j\theta}), \tag{10.6}$$

$$D(z)|_{z=e^{j\theta}} = D(e^{j\theta}). \tag{10.7}$$

Writing $u := -\cos\theta$, we decompose (10.6) and (10.7) into real and imaginary parts:

$$N(e^{j\theta})\big|_{u=-\cos\theta} =: R_N(u) + j\sqrt{1-u^2}T_N(u), \tag{10.8}$$

$$D(e^{j\theta})\big|_{u=-\cos\theta} =: R_D(u) + j\sqrt{1-u^2}T_D(u). \tag{10.9}$$

Denoting $v := \sqrt{1-u^2}$,

$$P(e^{j\theta})\big|_{u=-\cos\theta} := \frac{R_N(u) + jvT_N(u)}{R_D(u) + jvT_D(u)}. \tag{10.10}$$

Define $R_P(u)$ and $T_P(u)$ for $P(e^{j\theta})$ similar to those in (10.8) and (10.9). Then,

$$P(e^{j\theta})\,|_{u=-\cos\theta} = R_P(u) + jvT_P(u) =: P(u), \tag{10.11}$$

where

$$R_P(u) = \frac{R_N(u)R_D(u) + v^2 T_N(u)T_D(u)}{R_D(u)R_D(u) + v^2 T_D(u)T_D(u)}, \tag{10.12}$$

$$T_P(u) = \frac{T_N(u)R_D(u) - R_N(u)T_D(u)}{R_D(u)R_D(u) + v^2 T_D(u)T_D(u)}. \tag{10.13}$$

Now, for $C(z)$, we have

$$C(z)|_{z=e^{j\theta}} = C(e^{j\theta}) = \frac{K_1 e^{j\theta} + K_0}{e^{j\theta} - 1}. \tag{10.14}$$

By substituting $u = -\cos\theta$ and denoting $v = \sqrt{1-u^2}$, we have

$$\begin{aligned}
C(e^{j\theta}) =& \frac{K_1 e^{j\theta} + K_0}{e^{j\theta} - 1}, \\
=& \frac{K_1(-u + jv) + K_0}{(-u + jv) - 1}, \\
=& \frac{K_0 - K_1 u + jvK_1}{-(u+1) + jv}, \\
=& \frac{(K_0 - K_1 u + jvK_1)(u + 1 + jv)}{-\{(u+1)^2 + (1-u^2)\}}, \\
=& \frac{(u+1)(K_0 - K_1 u) - v^2 K_1 + jv(K_1(u+1) + K_0 - K_1 u)}{-\{u^2 + 2u + 1 + 1 - u^2\}}, \\
=& \frac{-1}{2(u+1)} \left[(u+1)\{(K_0 - K_1 u) - (1-u)K_1\} + jv(K_0 + K_1)\right], \\
=& \frac{-1}{2}(K_0 - K_1) - j\frac{v}{2(1+u)}(K_0 + K_1), \\
=& \underbrace{\frac{1}{2}(K_1 - K_0)}_{=:L_0} - j\frac{v}{u+1}\underbrace{\frac{1}{2}(K_0 + K_1)}_{=:L_1},
\end{aligned}$$

where

$$\begin{aligned}
L_0 &= -\frac{1}{2}K_0 + \frac{1}{2}K_1, \\
L_1 &= \frac{1}{2}K_0 + \frac{1}{2}K_1.
\end{aligned} \tag{10.15}$$

or

$$\begin{bmatrix} L_0 \\ L_1 \end{bmatrix} = \underbrace{\begin{bmatrix} -\frac{1}{2} & \frac{1}{2} \\ \frac{1}{2} & \frac{1}{2} \end{bmatrix}}_{=:W} \begin{bmatrix} K_0 \\ K_1 \end{bmatrix}. \tag{10.16}$$

Notice that the mapping $W : (K_0, K_1) \rightarrow (L_0, L_1)$ is invertible and

$$W^{-1} = \begin{bmatrix} -1 & 1 \\ 1 & 1 \end{bmatrix}. \tag{10.17}$$

Thus, we can represent $C(z)|_{z=e^{j\theta}}$ as a function of u, L_0, and L_1:

$$C(u, L_0, L_1) = L_0 - j\frac{v}{u+1}L_1, \tag{10.18}$$

and we simply denote

$$\begin{aligned} C(u) &:= C(u, L_0, L_1), \\ P(u) &:= R_P(u) + j\,v\,T_P(u). \end{aligned}$$

Condition (10.5) is then equivalent to

$$\max_{-1\le u<1} \left| \frac{1}{1+P(u)C(u)} \right| < \gamma. \tag{10.19}$$

For a fixed $u \in [-1, 1)$,

$$\begin{aligned} &\left| \frac{1}{1+P(u)C(u)} \right| < \gamma, \\ \Leftrightarrow &\left| 1 + (R_P(u) + jvT_P(u))(L_0 - j\frac{v}{u+1}L_1) \right| > \frac{1}{\gamma}, \\ \Leftrightarrow &\left(1 + R_P(u)L_0 + \frac{v^2}{u+1}T_P(u)L_1 \right)^2 \\ &+ v^2 \left(T_P(u)L_0 - \frac{1}{u+1}R_P(u)L_1 \right)^2 > \frac{1}{\gamma^2}. \end{aligned} \tag{10.20}$$

Since $|P(u)|^2 = R_P^2(u) + v^2T_P^2(u)$, it is easy to show that (10.20) is equivalent to

$$\frac{\left(L_0 + \frac{R_P(u)}{|P(u)|^2} \right)^2}{\frac{1}{\gamma^2|P(u)|^2}} + \frac{\left(L_1 + \frac{(u+1)T_P(u)}{|P(u)|^2} \right)^2}{\frac{(u+1)^2}{\gamma^2 v^2 |P(u)|^2}} > 1. \tag{10.21}$$

Thus, condition (10.21) represents the outside of an axis-parallel ellipse in (L_0, L_1) space. We use W^{-1} in (10.17) to map the ellipse back to (K_0, K_1) space.

Example 10.1 Consider the second-order plant and the PI controller:

$$P(z) = \frac{0.5}{z^2 - z + 0.5}, \quad C(z) = \frac{K_1 z + K_0}{z - 1}. \tag{10.22}$$

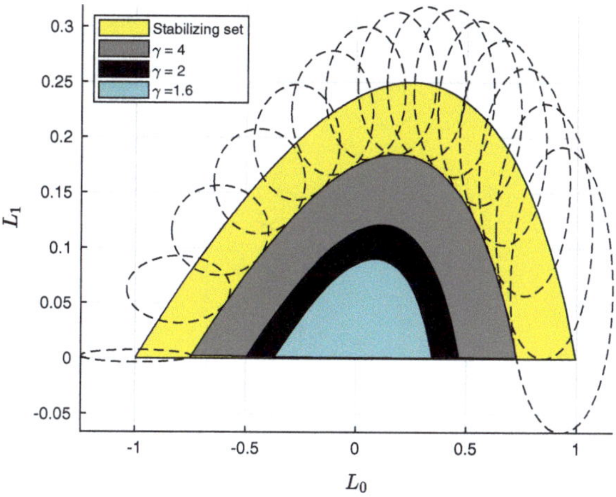

Fig. 10.2 $\mathcal{S}_\gamma$ for $\gamma = 1.6, 2.0$ and 4.0 with the stabilizing set in (L_0, L_1) space

The stabilizing set is first computed for the plant and the PI controller given in (10.22). Figure 10.3 shows the stabilizing set (yellow) in (K_0, K_1) space. The stabilizing set is mapped to (L_0, L_1) space in order to compute $\mathcal{S}_\gamma$ sets for $\gamma = 1.6$, 2.0, and 4.0 in Fig. 10.2. Then, the $\mathcal{S}_\gamma$ sets in (L_0, L_1) space given by (10.21) are mapped back to (K_0, K_1) space to obtain the PI controllers achieving the H_∞ norm less than γ specification as shown in Fig. 10.3.

Fixing $\gamma = 2$, some of the boundary points of the $\mathcal{S}_\gamma$ set are chosen for displaying the Nyquist plots. It is important to verify that the Nyquist plots stay outside of a circle centered at the critical point $-1 + j0$ with radius of $1/\gamma$. In Fig. 10.4, the Nyquist plots are at least 0.5 away from the critical point.

10.3 Computation of $\mathcal{S}_\gamma$ for Digital PID Controllers

Consider a unity feedback control loop with a plant and a controller in Fig. 10.1, where

$$P(z) = \frac{N(z)}{D(z)}, \tag{10.23}$$

$$C(z) = \frac{K_2 z^2 + K_1 z + K_0}{z(z-1)}, \tag{10.24}$$

$D(z)$, $N(z)$ are polynomials in z with real coefficients and $C(z)$ is the same transfer function representation for a PID controller. Again, it is assumed that $P(1) \neq 0$, since stabilization is impossible otherwise.

Now, for $C(z)$, we have

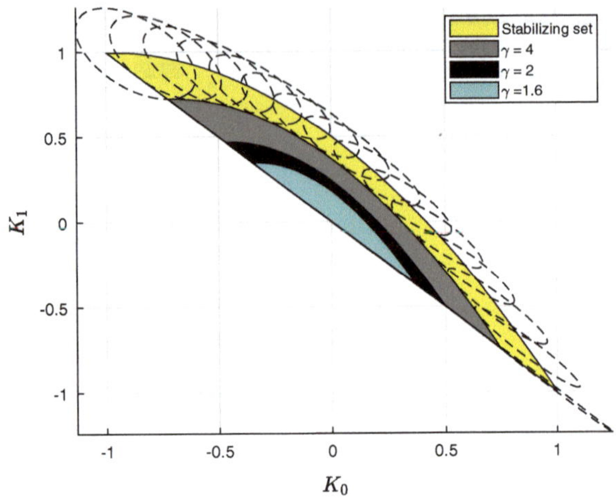

Fig. 10.3 $\mathcal{S}_\gamma$ for $\gamma = 1.6, 2.0$ and 4.0 with the stabilizing set in (K_0, K_1) space

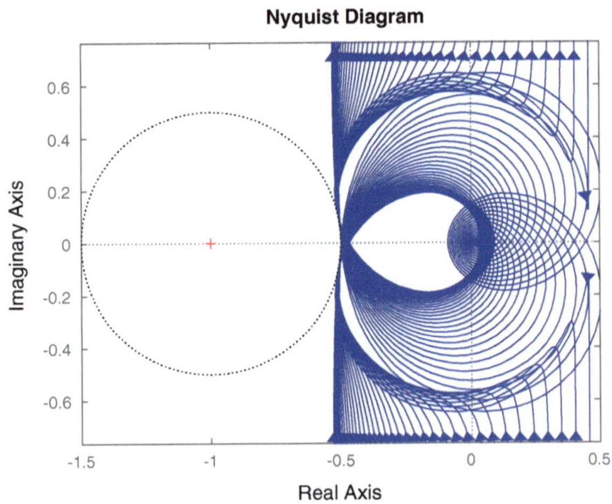

Fig. 10.4 Nyquist plots with (K_0, K_1) in the stabilizing set along the curve of $\gamma = 2$

$$C(z)|_{z=e^{j\theta}} = C(e^{j\theta}) = \frac{K_2 e^{2j\theta} + K_1 e^{j\theta} + K_0}{e^{j\theta}(e^{j\theta} - 1)} = \frac{K_2 e^{j\theta} + K_1 + K_0 e^{-j\theta}}{e^{j\theta} - 1}. \quad (10.25)$$

By substituting $u = -\cos\theta$ and $v = \sqrt{1 - u^2}$ as before,

$$\begin{aligned} &\frac{K_2 e^{j\theta} + K_1 + K_0 e^{-j\theta}}{e^{j\theta} - 1} \\ &= \frac{K_2(-u + jv) + K_1 + K_0(-u - jv)}{(-u + jv) - 1}, \\ &= \frac{-u(K_2 + K_0) + K_1 + jv(K_2 - K_0)}{-(u + 1) + jv}, \end{aligned}$$

$$
\begin{aligned}
&= \frac{\{-u(K_2+K_0)+K_1+jv(K_2-K_0)\}\,(u+1+jv)}{\{-(u+1)+jv\}\,(u+1+jv)},\\
&= \frac{-(u+1)\,\{K_2-K_1-(1-2u)K_0\}+jv\,\{K_2+K_1-(1+2u)K_0\}}{-(u+1)^2-(1-u^2)},\\
&= \frac{1}{2}\,\{K_2-K_1-(1-2u)K_0\}-j\frac{v}{1+u}\frac{1}{2}\,\{K_2+K_1-(1+2u)K_0\}\\
&=: C(u). \qquad (10.26)
\end{aligned}
$$

Let

$$
\begin{aligned}
W_0 &:= \frac{1}{2}K_2-\frac{1}{2}K_1\\
W_1 &:= \frac{1}{2}K_2+\frac{1}{2}K_1.
\end{aligned} \qquad (10.27)
$$

By substituting (10.27) into (10.26), we obtain

$$
C(u)=\left(W_0-\frac{(1-2u)}{2}K_0\right)-j\frac{v}{(1+u)}\left(W_1-\frac{(1+2u)}{2}K_0\right). \qquad (10.28)
$$

Let

$$
\tilde{W}_0 := W_0-\frac{(1-2u)}{2}K_0, \qquad (10.29)
$$

$$
\tilde{W}_1 := W_1-\frac{(1+2u)}{2}K_0, \qquad (10.30)
$$

so that (10.19) can be rewritten as follows:

$$
|1+P(u)C(u)|^2>\frac{1}{\gamma^2} \qquad (10.31)
$$

$$
\Leftrightarrow |1+(R+jvT)(\tilde{W}_0-j\frac{v}{(1+u)}\tilde{W}_1)|^2>\frac{1}{\gamma^2} \qquad (10.32)
$$

$$
\Leftrightarrow \left(1+R\tilde{W}_0+\frac{v^2}{(1+u)}T\tilde{W}_1\right)^2+v^2\left(T\tilde{W}_0-\frac{1}{(1+u)}R\tilde{W}_1\right)^2>\frac{1}{\gamma^2} \qquad (10.33)
$$

$$
\begin{aligned}
\Leftrightarrow &\left(R^2\tilde{W}_0^2+\frac{v^4}{(1+u)^2}T^2\tilde{W}_1^2+2R\tilde{W}_0\right.\\
&\left.+\frac{2v^2}{(1+u)}T\tilde{W}_1+\frac{2v^2}{(1+u)}R\tilde{W}_0T\tilde{W}_1+1\right)\\
&+v^2\left(T^2\tilde{W}_0^2+\frac{1}{(1+u)^2}R^2\tilde{W}_1^2-\frac{2}{(1+u)}T\tilde{W}_0R\tilde{W}_1\right)>\frac{1}{\gamma^2} \qquad (10.34)
\end{aligned}
$$

$$
\Leftrightarrow \left(R^2\tilde{W}_0^2+v^2T^2\tilde{W}_0^2+2R\tilde{W}_0+1\right)
$$

$$+ v^2\left(\frac{1}{(1+u)^2}R^2\tilde{W}_1^2 + \frac{v^2}{(1+u)^2}T^2\tilde{W}_1^2 + \frac{2}{(1+u)}T\tilde{W}_1\right) > \frac{1}{\gamma^2} \tag{10.35}$$

$$\Leftrightarrow \left(|P|^2\tilde{W}_0^2 + 2R\tilde{W}_0 + 1\right)$$

$$+ v^2\left(\frac{1}{(1+u)^2}|P|^2\tilde{W}_1^2 + \frac{2}{(1+u)}T\tilde{W}_1\right) > \frac{1}{\gamma^2} \tag{10.36}$$

$$\Leftrightarrow |P|^2\left(\tilde{W}_0 + \frac{R}{|P|^2}\right)^2 - \frac{R^2}{|P|^2} + 1$$

$$+ v^2|P|^2\left(\frac{1}{(1+u)}\tilde{W}_1 + \frac{T}{|P|^2}\right)^2 - \frac{v^2T}{|P|^2} > \frac{1}{\gamma^2} \tag{10.37}$$

$$\Leftrightarrow |P|^2\left(\tilde{W}_0 + \frac{R}{|P|^2}\right)^2 + \frac{v^2|P|^2}{(1+u)^2}\left(\tilde{W}_1 + \frac{(1+u)T}{|P|^2}\right)^2 > \frac{1}{\gamma^2} \tag{10.38}$$

$$\Leftrightarrow \frac{\left(\tilde{W}_0 + \frac{R}{|P|^2}\right)^2}{1/|P|^2} + \frac{\left(\tilde{W}_1 + \frac{(1+u)T}{|P|^2}\right)^2}{(1+u)^2/v^2|P|^2} > \frac{1}{\gamma^2} \tag{10.39}$$

$$\Leftrightarrow \frac{\left(\tilde{W}_0 + \frac{R}{|P|^2}\right)^2}{\frac{1}{\gamma^2|P|^2}} + \frac{\left(\tilde{W}_1 + \frac{(1+u)T}{|P|^2}\right)^2}{\frac{(1+u)^2}{\gamma^2v^2|P|^2}} > 1 \tag{10.40}$$

$$\Leftrightarrow \frac{\left(W_0 - \frac{(1-2u)}{2}K_0 + \frac{R}{|P|^2}\right)^2}{\frac{1}{\gamma^2|P|^2}} + \frac{\left(W_1 - \frac{(1+2u)}{2}K_0 + \frac{(1+u)T}{|P|^2}\right)^2}{\frac{(1+u)^2}{\gamma^2v^2|P|^2}} > 1 \tag{10.41}$$

$$\Leftrightarrow \frac{\left(W_0 - \frac{(1-2u)}{2}K_0 + \frac{R_P(u)}{|P|^2}\right)^2}{\frac{1}{\gamma^2|P|^2}} + \frac{\left(W_1 - \frac{(1+2u)}{2}K_0 + \frac{(1+u)T_P(u)}{|P|^2}\right)^2}{\frac{(1+u)^2}{\gamma^2v^2|P|^2}} > 1. \tag{10.42}$$

Equation (10.42) represents the exterior of axis-parallel ellipse in (W_0, W_1) space for each $u \in (-1, 1)$, when we fix $K_0 := K_0^*$.

Algorithm:

1. Compute $\mathcal{S}(K_0, K_1, K_2)$ using the signature method for a given plant.
2. Choose $\gamma \geq 1$.
3. Fix $K_0 = K_0^*$ and find the subset $\mathcal{S}(K_0^*)$.
4. For each $u \in (-1, 1)$ find the ellipse corresponding to (10.42) in (W_0, W_1) space.
5. Map the ellipses from (W_0, W_1) space to (K_0^*, K_1, K_2) space. Since it is a linear map, the resulting boundary for the inequality is also an ellipse, rotated 45° (or −45°).
6. Intersect the outside of the ellipse with $\mathcal{S}(K_0^*)$ to find the subset $\mathcal{S}_\gamma(K_0^*, u)$.
7. Sweep over $u \in (-1, 1)$ to get $\mathcal{S}_\gamma(K_0^*) = \bigcap_{u\in(-1,1)} \mathcal{S}_\gamma(K_0^*, u)$.
8. Sweep over K_0 to get $\mathcal{S}_\gamma = \bigcup_{K_0^*} \mathcal{S}_\gamma(K_0^*)$.

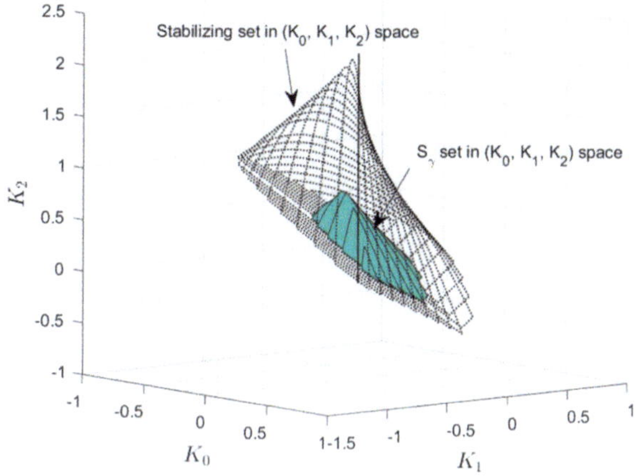

Fig. 10.5 The stabilizing set in (K_0, K_1, K_2) space

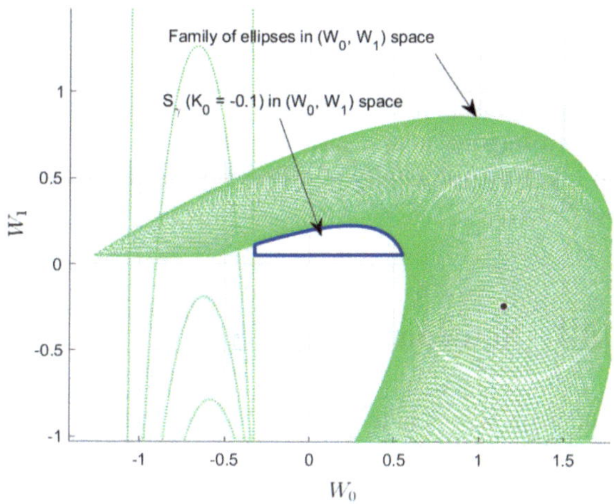

Fig. 10.6 The family of axis-parallel ellipses in (W_0, W_1) space as u runs from -1 to 1

Example 10.2 Consider the second-order plant and the PID controller:

$$P(z) = \frac{1}{z^2 - 0.25}, \quad C(z) = \frac{K_2 z^2 + K_1 z + K_0}{z(z-1)}. \tag{10.43}$$

The stabilizing set $\mathcal{S}$ is computed for the plant $P(z)$ and the PID controller $C(z)$ in (10.43). Figure 10.5 shows $\mathcal{S}$ in (K_0, K_1, K_2) space. We fix $\gamma = 2$ and $\mathcal{S}_\gamma$ is overlapped with $\mathcal{S}$.

Fix $K_0 = -0.1$. In (W_0, W_1) space, the H_∞ criterion in (10.19) is represented by the outside of an axis-parallel ellipse for a fixed u. Since the criterion is the worst-case condition—the maximum magnitude is less than γ, the family of ellipses is drawn and the intersection of all outsides of ellipses is the admissible region. The family of ellipses are drawn in Fig. 10.6. $\mathcal{S}_\gamma$ in (W_0, W_1) space is thus the intersection of $\mathcal{S}(K_0 = -0.1)$ and the outside of ellipses.

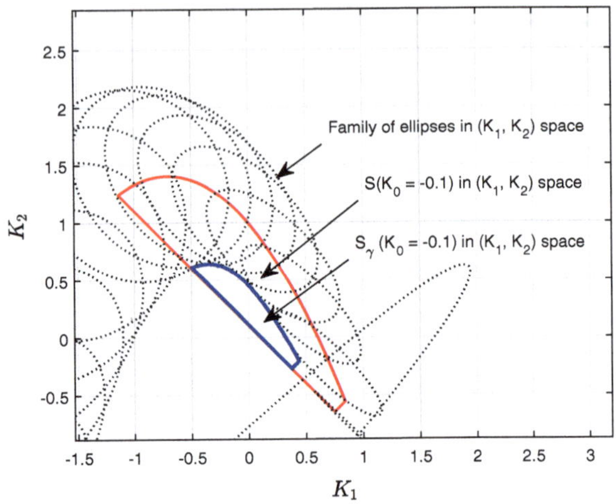

Fig. 10.7 $\mathcal{S}_\gamma(K_0)$ and the family of ellipses in (K_1, K_2) space with $K_0 = -0.1$

Now, we map the set $\mathcal{S}_\gamma$ in (W_0, W_1) space to (K_1, K_2) space for $K_0 = -0.1$. Sweeping over K_0 gives the set $\mathcal{S}_\gamma$ as seen in Fig. 10.5. Since the mapping in (10.27) is not a diagonal matrix, each mapped ellipse is not necessarily an axis-parallel ellipse. In Fig. 10.7, the family of ellipses is displayed and overlapped with the stabilizing set $\mathcal{S}_\gamma(K_0)$.

10.4 Notes and References

The main results in this chapter were first obtained in [1].

Reference

1. Han, S.: Robust and optimal PID controller synthesis for linear time invariant systems. Ph.D. Thesis, Texas A & M University (2019)

Appendix A
Application Examples

In this appendix, we present several examples to which various results developed in earlier chapters apply. The examples are largely from practical applications where PI and PID controllers are used.

A.1 Introduction

In this appendix, we present some examples that are related to practical applications. Section A.2 deals with AC drives which utilizes the σ-Hurwitz design discussed in Chap. 2. Section A.3 deals with a Half-Bridge Voltage Source Inverter which utilizes gain and phase margin design for continuous-time systems discussed in Chap. 6. Section A.4 deals with harmonic distortion which utilizes the H_∞ design for continuous-time systems discussed in Chap. 9. Section A.5 deals with position control of a metal plate which utilizes the H_∞ design for discrete-time systems discussed in Chap. 10.

A.2 AC Drives

Three-phase systems can be modeled as complex transfer functions. The complex transfer function representation assumes that the input and output signals are also of a complex form. For current control, in particular, the Surface-Mounted Permanent-Magnet (SMPM) synchronous machine is modeled by a first-order complex transfer function $P(s)$, and the tracking PI controller $C(s)$ is tuned to cancel the plant pole which is stable

I. D. Díaz-Rodríguez et al., *Analytical Design of PID Controllers*,
https://doi.org/10.1007/978-3-030-18228-1

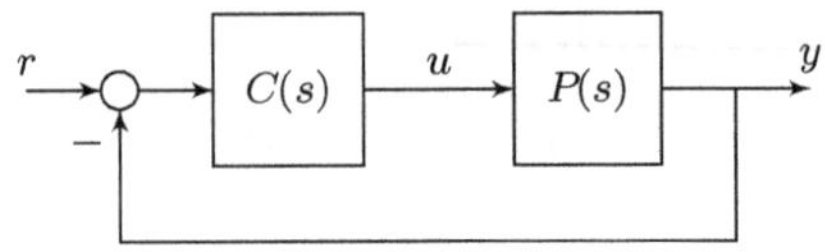

Fig. A.1 Unity feedback control loop

$$P(s) := \frac{1}{Ls + j\omega_e L + R} e^{-sT_d} \tag{A.1}$$

$$C(s) := \frac{k(Ls + j\omega_e L + R)}{s}. \tag{A.2}$$

In Fig. A.1, r is the reference current input, u is the voltage input to the plant $P(s)$, and y is the measured output current. All complex signals and plant and the controller transfer functions are in the rotating dq frame. ω_e is the synchronous frequency, L and R are stator inductance and resistance, T_d is the computation and modulation time delay, and k is a design parameter of the controller.

We set $\omega_e = 50\,\text{Hz}$, $L = 17.6\,\text{mH}$, and $R = 2.8\ \Omega$. We replaced the delay term e^{-sT_d} with the second-order Padé approximation. Then, the resulting closed-loop system is the same as in Fig. 2.1, where

$$C(s) = \frac{k}{s}, \quad P(s) = \frac{1 - \frac{1}{2}T_d s + \frac{1}{12}T_d^2 s^2}{1 + \frac{1}{2}T_d s + \frac{1}{12}T_d^2 s^2}. \tag{A.3}$$

Setting $k_p = 0$, $k_i = k$, and $k_d = 0$ in (2.165), we have

$$\begin{aligned}\delta'(s') &= (s' - \sigma)\left(\frac{1}{12}T_d^2(s' - \sigma)^2 + \frac{1}{2}T_d(s' - \sigma) + 1\right)\\ &\quad + k\left(\frac{1}{12}T_d^2(s' - \sigma)^2 - \frac{1}{2}T_d(s' - \sigma) + 1\right)\end{aligned}$$

$$N'(-s') = \frac{1}{12}T_d^2(s' + \sigma)^2 + \frac{1}{2}T_d(s' + \sigma) + 1$$

and

$$\begin{aligned}\nu'(s') &= \delta'(s')N'(-s')\\ \nu'(j\omega) &= p_1(\omega, \sigma, T_d) + kp_2(\omega, \sigma, T_d) + jq(\omega, \sigma, T_d),\end{aligned}$$

where

$$\begin{aligned}p_1(\omega, \sigma, T_d) &= -\frac{1}{12}\left(\frac{1}{12}\sigma T_d^4 - T_d^3\right)\omega^4\\ &\quad - \left(\frac{1}{72}T_d^4\sigma^3 - \frac{1}{12}T_d^3\sigma^2 - \frac{5}{12}T_d^2\sigma + T_d\right)\omega^2\end{aligned}$$

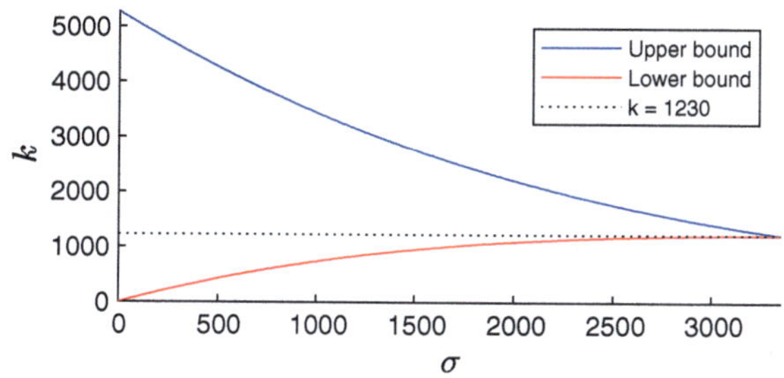

Fig. A.2 Stabilizing k bounds versus σ. © 2018 IEEE. Reproduced from [6] with permission

$$- \frac{1}{144} T_d^4 \sigma^5 + \frac{1}{12} T_d^2 \sigma^3 - \sigma,$$

$$p_2(\omega, \sigma, T_d) = \frac{1}{144} T_d^4 \omega^4 + \frac{1}{12} \left(\frac{1}{6} T_d^4 \sigma^2 + T_d^3 \sigma + T_d^2 \right) \omega^2 + \frac{1}{144} T_d^4 \sigma^4 + \frac{1}{12} T_d^3 \sigma^3 + \frac{5}{12} T_d^2 \sigma^2 + T_d \sigma + 1,$$

$$q(\omega, \sigma, T_d) = \frac{1}{144} T_d^4 \omega^5 + \frac{1}{12} \left(\frac{1}{6} T_d^4 \sigma^2 + T_d^3 \sigma - 5 T_d^2 \right) \omega^3 + \frac{1}{12} \left(\frac{1}{12} T_d^4 \sigma^4 + T_d^3 \sigma^3 - T_d^2 \sigma^2 \right) \omega + \left(-T_d \sigma + 1 \right) \omega.$$

The sets $S(\sigma)$ are intervals on the k axis and the upper and lower bounds for each interval with a prescribed σ are plotted in Fig. A.2. $S(\sigma)$ just becomes empty at around $\sigma = 3356$ and the corresponding k is around 1230.

A.3 Half-Bridge Voltage Source Inverter

In this example, we apply the special case of the first-order controller $C(s) = (x_1 s + x_2)/(s + x_3)$ when $x_3 = 0$ (PI controller) design approach to a single-phase voltage source inverter application. Figure A.3 shows the Half-Bridge Voltage Source Inverter. In this application, an ideal voltage source V_{DC} is considered. Also, the power switches plus diode couple are assumed to behave like an ideal switch, i.e., one whose voltage is zero in the "on" state and whose current is zero in the "off" state. Moreover, it is assumed that the change from the "on" state to the "off" state and vice versa takes place in zero time. The load will be described as the series connection of a resistor R_S, an inductor L_S, and a voltage source E_S, which can be either DC or AC.

The control problem considered in this application is the linear regulation of the output current I_O of the voltage source inverter. In Fig. A.4, we have a block

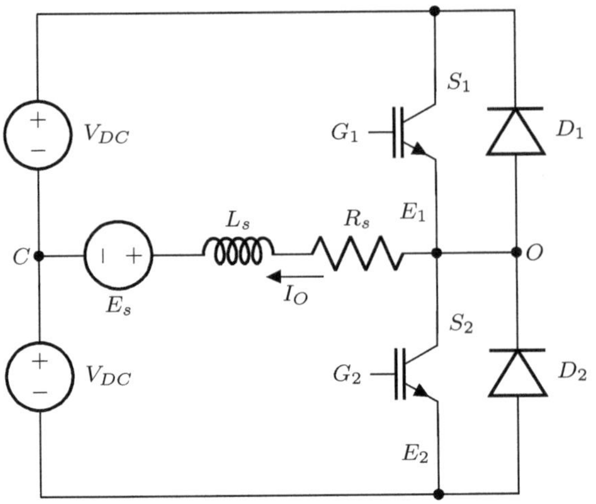

Fig. A.3 Half-bridge source inverter. © 2006 Morgan and Claypool Publishers. Reproduced from [3] with permission

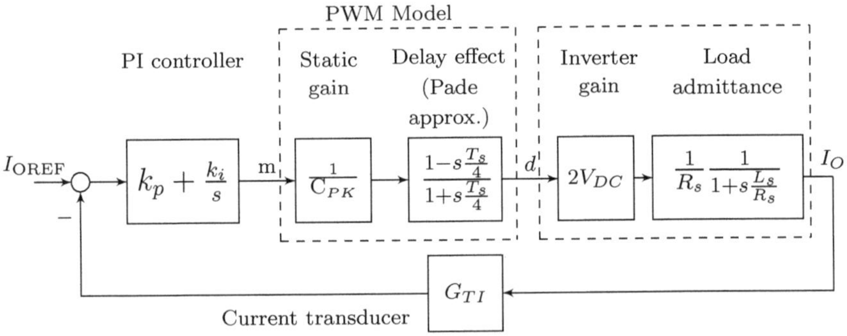

Fig. A.4 Control block diagram. © 2006 Morgan and Claypool Publishers. Reproduced from [3] with permission

diagram of the system to be considered. In this example, the purpose of the voltage source inverter is to deliver a given amount of output power P_O to the load of the inverter inductor. This can be difficult in typical ac motor drive applications, where a sinusoidal current of suitable amplitude and given frequency, f_O, must be generated on each motor phase. Consequently, it is customary to use a current transducer, whose gain, G_{TI}, is given. In the block diagram, the controller to be considered is a Proportional–Integral controller.

The output of the regulator represents the modulating signal that drives the Pulse Width Modulator. In this PWM block, a time delay has been considered; this has been replaced with a Pade approximation. Then, we have the inverter and load models and a typical implementation of a transducer gain.

A. Computation of the Stabilizing Set

First, the open-loop transfer function is as follows:

$$G_{OL}(s) = C(s)P(s)$$
$$G_{OL}(s) = \left(k_p + \frac{k_i}{s}\right)\frac{2V_{DC}}{C_{PK}}\frac{1 - s\frac{T_S}{4}}{1 + s\frac{T_S}{4}}\frac{G_{TI}}{R_S}\frac{1}{1 + s\frac{L_S}{R_S}}, \tag{A.4}$$

where $V_{DC} = 250\,(\text{V})$, $C_{PK} = 4\ (V)$, $Ts = 0.00002\,(\text{s})$, $G_{TI} = 0.1\ (V/A)$, $R_S = 1\ (\Omega)$, $L_S = 1.5\,(\text{mH})$. We will consider the notation of the PI controller in (A.4) to be $x_1 := k_p$, $x_2 := k_i$, and $x_3 = 0$ for our special case of first-order controller.

We have

$$G_{OL}(s) = \left(\frac{x_1 s + x_2}{s}\right)\frac{-6.25 \times 10^{-5}s + 12.5}{7.5 \times 10^{-9}s^2 + 0.0015\,s + 1}. \tag{A.5}$$

The closed-loop characteristic polynomial is

$$\begin{aligned}\delta(s, x_1, x_2) =& 7.5 \times 10^{-9}s^3 + (0.0015 - 6.25 \times 10^{-5}x_1)s^2\\ &+ (12.5x_1 - 6.25 \times 10^{-5}x_2 + 1)s + 12.5x_2.\end{aligned} \tag{A.6}$$

Here, $n = 2$, $m = 1$, and $N(-s) = 12.5 + 6.25 \times 10^{-5}s$. Therefore, we obtain

$$\begin{aligned}\nu(s) &= \delta(s, x_1, x_2)N(-s)\\ &= 4.69 \times 10^{-13}s^4 + (1.87 \times 10^{-7} - 3.91 \times 10^{-9}x_1)s^3\\ &+ (0.0188 - 3.91 \times 10^{-9}x_2)s^2 + (156x_1 + 12.5)s + 156x_2\end{aligned} \tag{A.7}$$

so that

$$\begin{aligned}\nu(j\omega, x_1, x_2) &= 4.69 \times 10^{-13}\omega^4 + (3.91 \times 10^{-9}x_2 - 0.0188)\omega^2 + 156x_2\\ &+ j[(3.91 \times 10^{-9}x_1 - 1.87 \times 10^{-7})\omega^3 + (156x_1 + 12.5)\omega]\\ &= p(\omega) + jq(\omega).\end{aligned} \tag{A.8}$$

We find that $z^+ = 1$ so that the signature requirement on $\nu(s)$ for stability is

$$n - m + 1 + 2z^+ = 4. \tag{A.9}$$

Since the degree of $\nu(s)$ is even, we see from the signature formulas that $q(\omega)$ must have at least one positive real root of odd multiplicity. The range of x_1 such that $q(\omega, x_1)$ has at least one real, positive, distinct, finite zero with odd multiplicities was determined to be $x_1 \in (-0.08, \infty)$. However, the range for which we can get a nonempty region is given by $x_1 \in (-0.08, 24)$ the allowable range for x_1. By sweeping over different x_1 values within the interval $(-0.08, 24)$, we can generate the set of stabilizing (x_1, x_2) values. This set is shown in Fig. A.5.

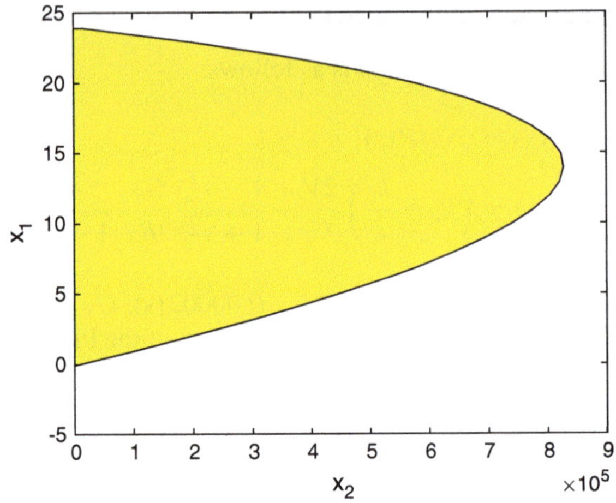

Fig. A.5 Stability region for $-0.08 \leq x_1 \leq 24$

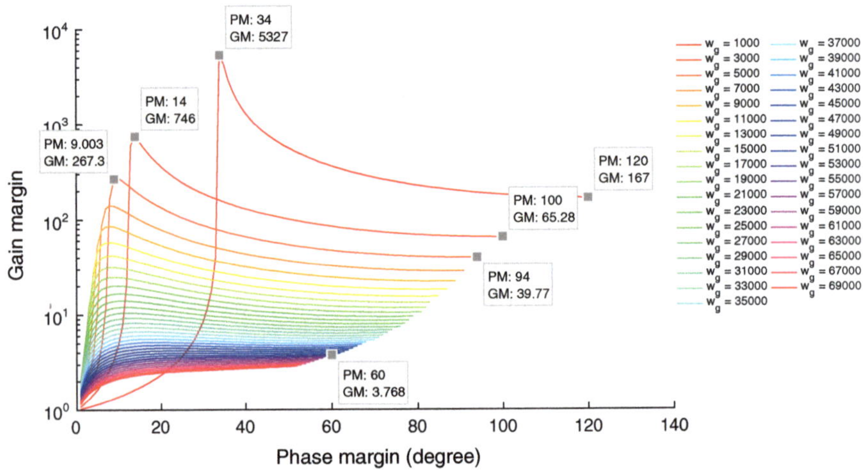

Fig. A.6 Achievable performance in terms of GM, PM, and ω_g for PI controller design

B. Construction of Achievable Gain–Phase Margin Design Curves

For the construction of the achievable gain–phase margin set for the PI controller design case, the evaluated range of ω_g is [1000, 69000] and the range for PM is from 1° to 120°. For the PI case, using the constant gain and constant phase loci equations, (6.13) and (6.14), we now get an ellipse and a straight line in the (x_1, x_2) space, respectively. The intersection (x_1, x_2) superimposed on the stabilizing set represents the PI controller gains that satisfy the PM and ω_g. Evaluating the range of PM and ω_g, we can construct the achievable gain–phase margin curves represented in Fig. A.6.

C. Simultaneous Performance Specifications and Retrieval of Controller Gains

In Fig. A.6, we display the achievable gain–phase margin set of curves indexed by a fixed ω_g in different colors. Notice that we can get more GM and PM for lower values of ω_g. For example, for $\omega_g = 1000$ rad/s, the maximum GM that we can get is 5327 with a PM of 34°. For $\omega_g = 3000$ rad/s, the maximum GM is 746 with a $PM = 14°$. Using Fig. A.6, the designer has the liberty to choose values for GM, PM, and ω_g that best suits his design needs.

After the selection of simultaneous GM, PM, and ω_g from the achievable gain–phase margin set, the designer can retrieve the controller gains corresponding to this point. For illustration purposes, let us say that the desired performance values chosen for this example are a PM of 60°, GM of 3.768, and a ω_g of 53000 rad/s (see Fig. A.6). Then, taking these values and the constant gain and constant phase loci for PI controllers presented in the methodology, we can find the intersection of the ellipse and the straight line in the (x_1, x_2) space shown in Fig. A.7. The controller gains are

$$x_1^* = 6.34 \quad \text{(A.10)}$$
$$x_2^* = 5812. \quad \text{(A.11)}$$

In Fig. A.8, we can see the Nyquist plot for the controller gains selected. Here, we can see that those controller gains satisfy the desired performance specifications, $PM = 60°$, $GM = 3.7681$ (11.5 dB).

We can also compute the time-delay tolerance design curves. Following equation (6.43) and taking the values from Fig. A.6, we get Fig. A.9.

In Fig. A.9, we can see the achievable time-delay tolerance for the system using the proposed controller. We can select any point from these curves and retrieve the controller gains following the same procedure as taking a point from the gain–phase margin design curves. In this case, we selected the same $PM = 60°$ and $\omega_g = 53000$ rad/s. The time-delay tolerance is

$$\tau = 1.976 \text{ x } 10^{-5} \text{ s.} \quad \text{(A.12)}$$

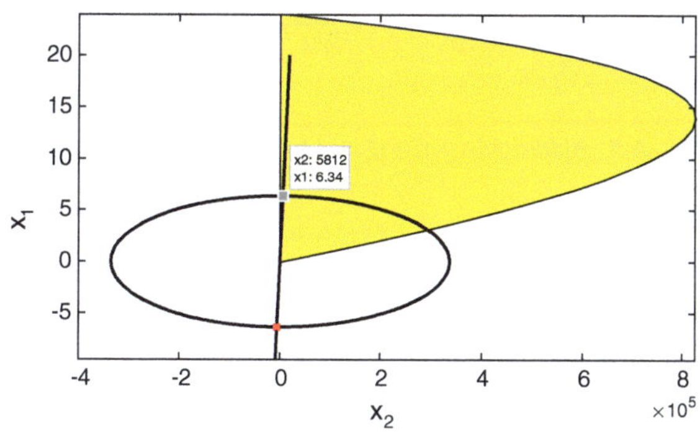

Fig. A.7 Ellipse and straight line superimposed in the PI controller stabilizing set

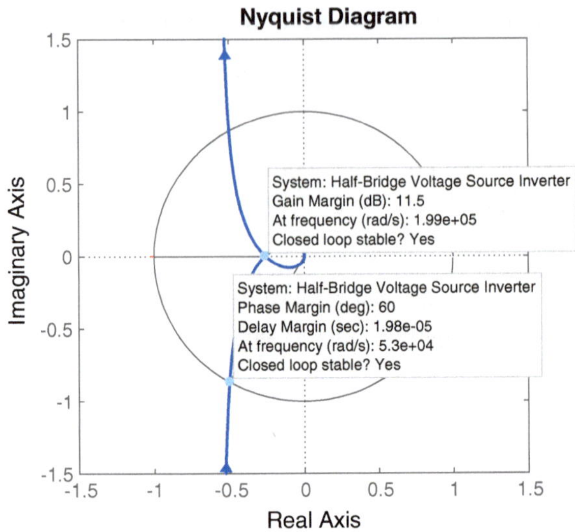

Fig. A.8 System Nyquist plot for $x_1^* = 6.34$, $x_2^* = 5812$ in the PI controller design

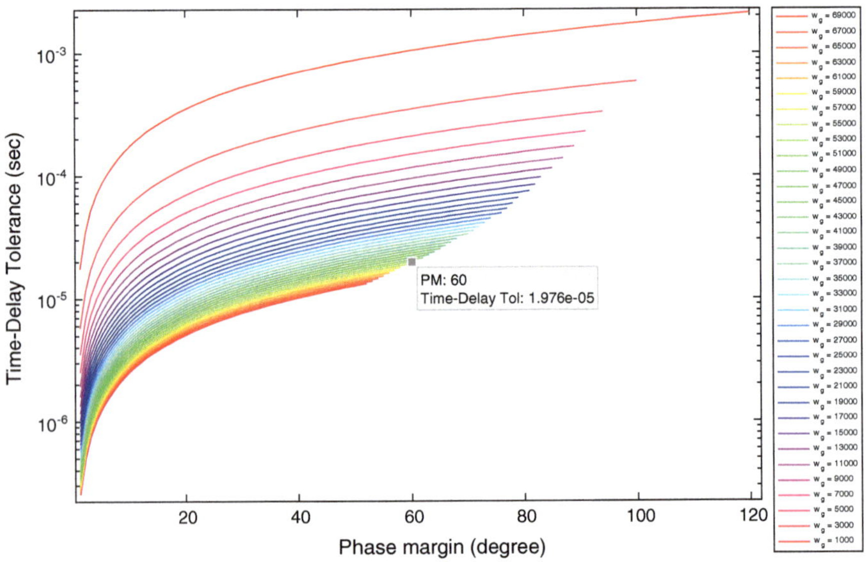

Fig. A.9 Achievable performance in terms of time-delay tolerance, PM, and ω_g for PI controller design

A.4 Selective Mitigation of Harmonic Distortion

Consider a unity feedback control loop with a disturbance input to the plant in Fig. A.10 where $r(t)$ is the reference input, $u(t)$ is the input to the plant, $\xi(t)$ is the disturbance, $y(t)$ is the output of the plant, and $e(t)$ is the error between the reference and the output. The control objective is to make the output $y(t)$ track the reference $r(t)$. However, the system is subject to the disturbance $\xi(t)$ and the output deviates from the reference depending on the disturbance signals. Let the Laplace transforms of the reference and disturbance signals be

$$R(s) := \frac{n_r(s)}{d_r(s)}, \quad \Xi(s) := \frac{n_\xi(s)}{d_\xi(s)}, \tag{A.13}$$

where $n_r(s), d_r(s), n_\xi(s), d_\xi(s)$ are polynomials in s with real coefficients.

We assume that (see Fig. A.11)

$$Y(s) = \underbrace{\begin{bmatrix} P_u(s) & P_\xi(s) \end{bmatrix}}_{=:P(s)} \begin{bmatrix} U(s) \\ \Xi(s), \end{bmatrix} \tag{A.14}$$

where

$$P_u(s) = \frac{n_{P_u}(s)}{d_{P_u}(s)}, \quad P_\xi(s) = \frac{n_{P_\xi}(s)}{d_{P_\xi}(s)}. \tag{A.15}$$

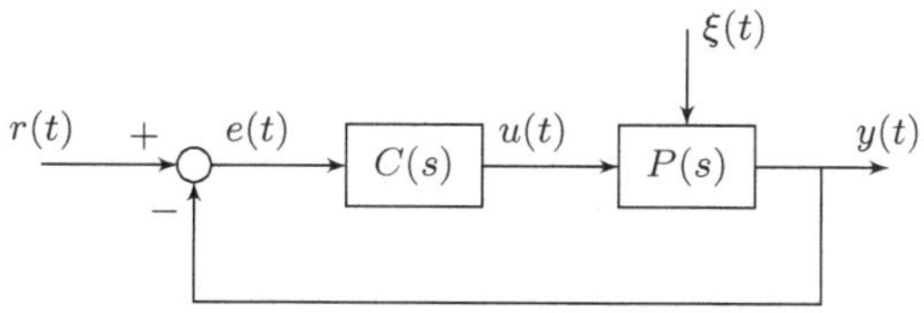

Fig. A.10 Unity feedback under disturbance

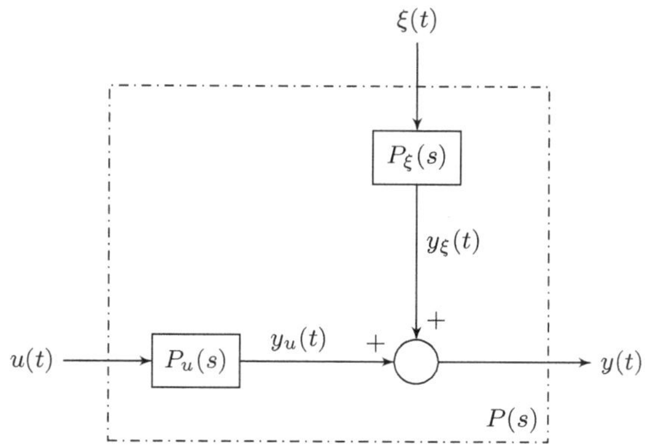

Fig. A.11 A detailed look of $P(s)$

Assumption A.1 $d_{P_u}(s) = d_{P_\xi}(s)$.

Since

$$E(s) = R(s) - Y(s), \tag{A.16}$$

$$U(s) = C(s)E(s), \tag{A.17}$$

where

$$C(s) = \frac{n_C(s)}{d_C(s)} \tag{A.18}$$

we have by Assumption A.1 that

$$E(s) = \frac{d_{P_u}(s)d_C(s)}{d_{cl}(s)}R(s) - \frac{d_{P_u}(s)d_C(s)n_{P_\xi}(s)}{d_{cl}(s)}\Xi(s), \tag{A.19}$$

where the closed-loop characteristic polynomial is

$$d_{cl}(s) = d_C(s)d_P(s) + n_C(s)n_P(s). \tag{A.20}$$

Assumption A.2 $C(s)$ is of a PID type.

By Assumption A.2, $y_u(t)$ tracks the arbitrary constant reference input $r(t)$ in the steady state provided that the closed loop is stable. Since $e(t) = r(t) - y(t)$, $e(t) \to y_\xi(t)$ as $t \to \infty$. Thus, the error signal only has the closed-loop response to the disturbance $\xi(t)$ in the steady state. Let $Y_\xi(s)$ be the Laplace transform of $y_\xi(t)$ and the disturbance $\xi(t)$ be a linear combination of sinusoidal signals of fixed frequencies, say $\omega_1, \ldots, \omega_q$

$$\xi(t) = \sum_{i=1}^{q} \xi_i(t) = \sum_{i=1}^{q} A_i sin(\omega_i t), \tag{A.21}$$

where A_i is the amplitude of ith sinusoidal. If the frequencies of signals are multiples of a fundamental frequency, say ω_1, we call each signal $\xi_i(t)$ a *harmonic* of $\xi_1(t)$.

$$\left|Y_\xi(j\omega_i)\right| = \left|\frac{P_\xi(j\omega_i)}{1 + C(j\omega_i)P_u(j\omega_i)}\right| A_i \tag{A.22}$$

Let $\tilde{\gamma}_i$ be a prescribed upper bound for the amplitude of $y_\xi(t)$ at frequency ω_i, that is,

$$\left|Y_\xi(j\omega_i)\right| < \tilde{\gamma}_i. \tag{A.23}$$

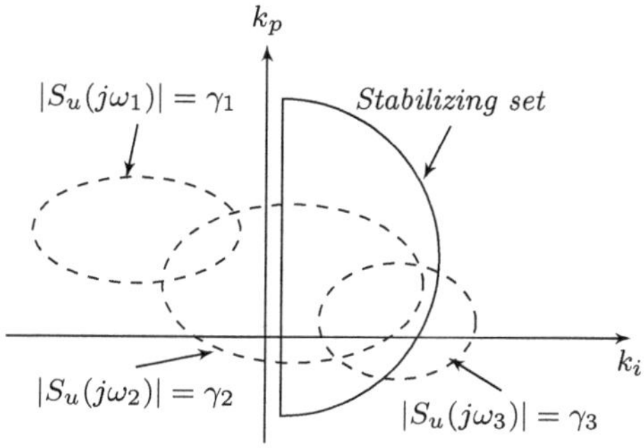

Fig. A.12 The stabilizing controllers satisfying the harmonic mitigation are the intersection of the outside of the ellipses and inside the stabilizing set

Then,

$$\begin{aligned} \left|Y_\xi(j\omega_i)\right| < \tilde{\gamma}_i \Leftrightarrow \underbrace{\left|\frac{1}{1+C(j\omega_i)P_u(j\omega_i)}\right|}_{=:S_u(j\omega_i)} &< \underbrace{\frac{\tilde{\gamma}_i}{\left|P_\xi(j\omega_i)\right| A_i}}_{=:\gamma_i} \\ \Leftrightarrow |1+C(j\omega_i)P_u(j\omega_i)| &> \frac{1}{\gamma_i}. \end{aligned} \tag{A.24}$$

Thus, it is possible to design a stabilizing controller that compensates each harmonic present in the disturbance.

In the sequel, we graphically design a PI controller that renders the closed-loop system to have the frequency response of the error signal less than a prescribed value γ_i at each fixed frequency ω_i. This is illustrated in Fig. A.12. The stabilizing controllers are simultaneously outside of the ellipses (dashed lines) and inside of the stabilizing set (solid line).

Let us consider the following continuous-time LTI plant:

$$P_u(s) = \frac{s-5}{10\,s^2+16\,s+2}, \tag{A.25}$$

and the disturbance transfer function:

$$P_\xi(s) = \frac{20\,s^2+8\,s+1}{10\,s^2+16\,s+2}. \tag{A.26}$$

Suppose that the controller to be designed is a PI controller given by

$$C(s) = k_p + \frac{k_i}{s}, \tag{A.27}$$

the reference input for this system is a unit step and it is subject to a disturbance signal $\xi(t) = \xi_1(t) + \xi_2(t)$ with

$$\xi_1(t) = 0.6\,sin(2t) \tag{A.28}$$
$$\xi_2(t) = 0.4\,sin(3t). \tag{A.29}$$

Suppose also that the specifications to be attained are

- Amplitude of the harmonic at 2 rad/s is less than 2.
- Amplitude of the harmonic at 3 rad/s is less than 2.

The stabilizing set of PI controllers for the given plant is computed first. From the disturbance signal characterization, we identify that $A_1 = 0.6$, $A_2 = 0.4$, $\omega_1 = 2$ rad/s, and $\omega_2 = 3$ rad/s. From the specifications, we have $\tilde{\gamma}_1 = 2$ and $\tilde{\gamma}_2 = 2$. Substituting (9.28) and (9.29) into (A.24), we obtain that $\gamma_1 \approx 2.0544$, $\gamma_2 \approx 2.7752$, and the set of all possible controllers in the stabilizing set attaining the specifications. This is shown in Fig. A.13. We select four sample points, shown in Table A.1, and examine whether or not the chosen controllers achieve the specifications.

The controller ■ lies outside of the ellipse $|S_u(j\omega_1)| = \gamma_1$ but inside of the ellipse $|S_u(j\omega_2)| = \gamma_2$. The controller guarantees $y_\xi(t)$ to have the harmonic amplitude less than 2 at ω_1 but not for ω_2 as shown in Fig. A.14. The controller ◆ lies inside of the ellipse $|S_u(j\omega_1) = \gamma_1|$ and the ellipse $|S_u(j\omega_2)| = \gamma_2$. This means that the controller does not guarantee any of the amplitude specifications. This is shown in Fig. A.15. The controller ▲ lies inside of the ellipse $|S_u(j\omega_1)| = \gamma_1$ and outside of the ellipse $|S_u(j\omega_2)| = \gamma_2$. This controller achieves the amplitude specification for the harmonic at ω_2 but not for ω_1 as shown in Fig. A.16. The controller • lies outside of both

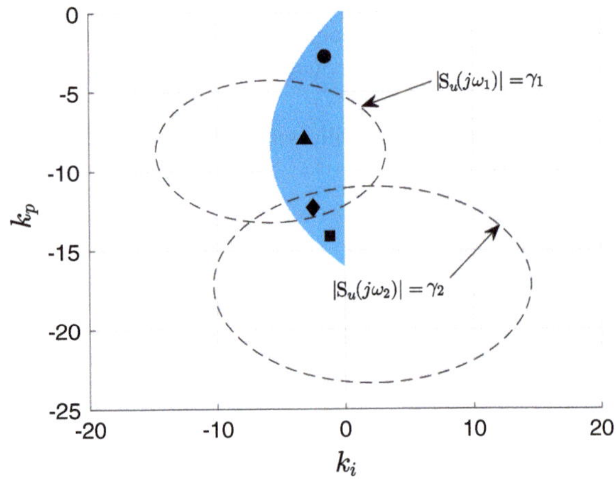

Fig. A.13 Stabilizing set with harmonic mitigation specifications

Table A.1 Selected controllers

	■	◆	▲	•
k_p	−14.0861	−12.2734	−7.9683	−2.7568
k_i	−1.1521	−2.4424	−3.0876	−1.5207

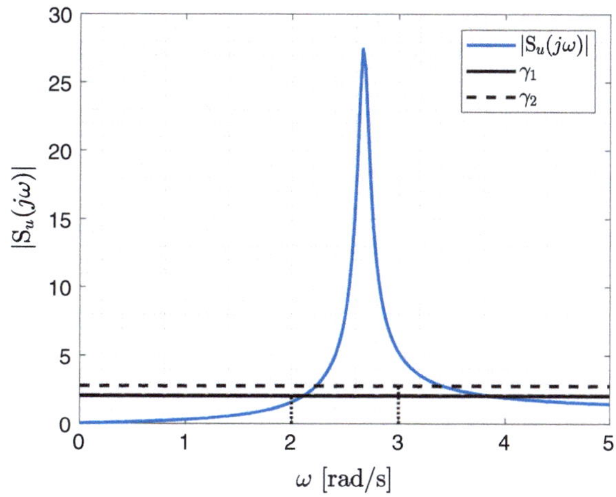

Fig. A.14 Magnitude of the frequency response of $S_u(j\omega)$ for controller ■

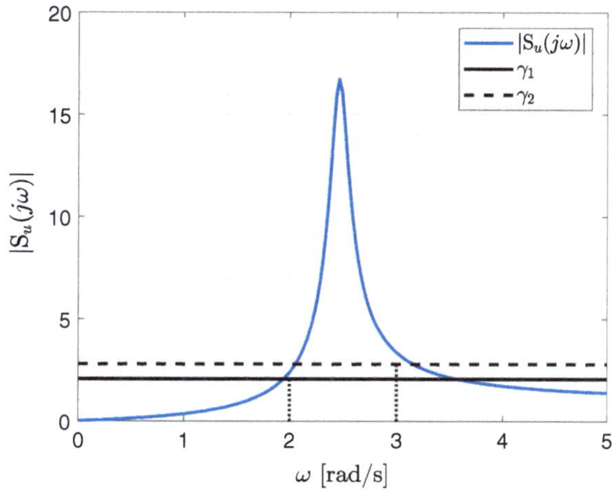

Fig. A.15 Magnitude of the frequency response of $S_u(j\omega)$ for controller ♦

the ellipses. Thus, the controller achieves the amplitude specifications as shown in Fig. A.17. The output time response for the point • is shown in Fig. A.18 and its Fourier transform is shown in Fig. A.19. It verifies that the controller • achieves the amplitude specifications γ_1 and γ_2 for both the harmonics at ω_1 and ω_2, respectively.

A.5 Position Control of a Metal Plate

In this section, we design digital PID controllers for a metal plate position system. Dincel and Söylemez implemented a laboratory experiment system where a motor rotated a fan to blow a metal plate such that the angle of the metal was maintained at a specific angle. The system is described in Fig. A.20.

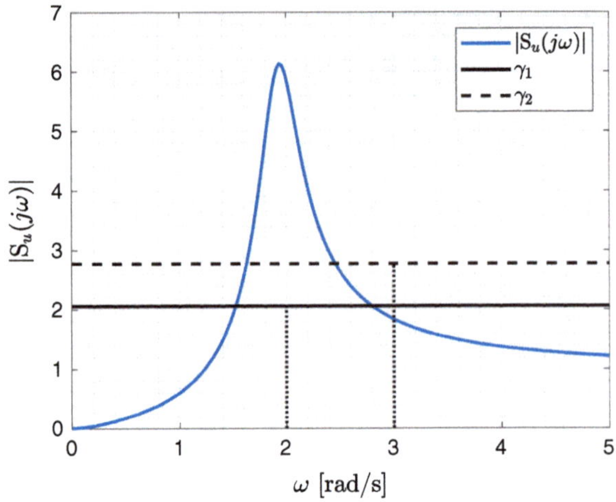

Fig. A.16 Magnitude of the frequency response of $S_u(j\omega)$ for controller ▲

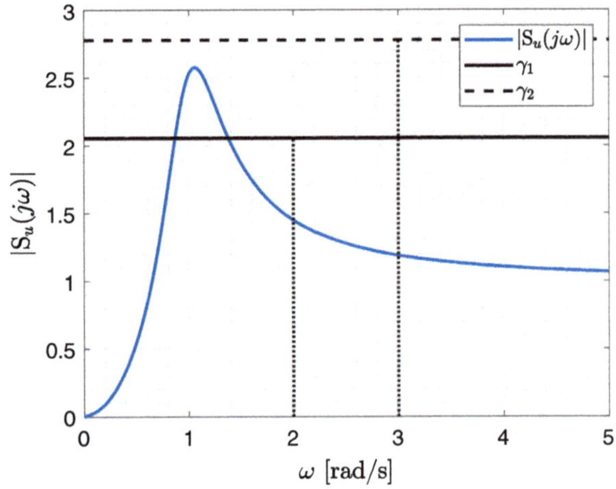

Fig. A.17 Magnitude of the frequency response of $S_u(j\omega)$ for controller •

The open-loop linear model of the system was obtained by system identification and was given by

$$P(z) = \frac{0.016546 + 0.017457z}{0.85206 - 1.7348z + z^2} \tag{A.30}$$

with a sampling time $t_s = 0.05$ s. By using the so-called "dominant pole placement", a PI-PD controller was designed so that it achieved a settling time of 2.3 s with an overshoot of 5%. The PI-PD controller is

$$\begin{aligned} C_{PI}(z) &= 0.05 + 0.1665\frac{z}{z-1}, \\ C_{PD}(z) &= -0.87 - 0.5016\frac{z-1}{z} \end{aligned} \tag{A.31}$$

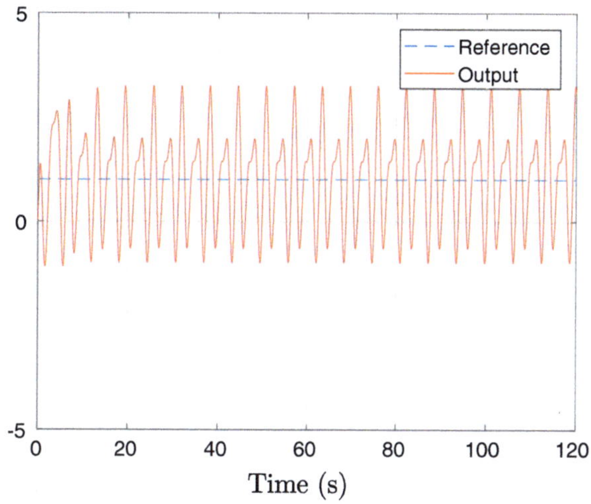

Fig. A.18 Output time response $y(t)$ for the designed controller •

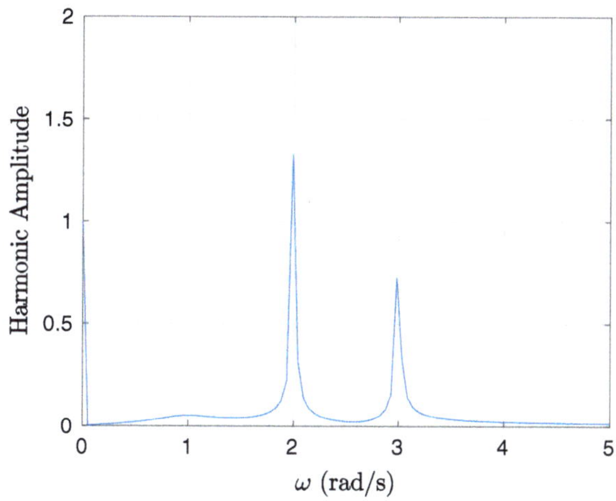

Fig. A.19 FFT of the output response for the designed controller •

and achieved the H_∞ norm of γ less than 1.6553 on the error transfer function.

In Fig. A.21, two structures are compared. We consider the same system and design digital PID controllers with some prescribed γ values.

We consider the PID controller system where $P(z)$ is as in (A.30) and

$$C(z) = \frac{K_0 + K_1 z + K_2 z^2}{z(z-1)}. \tag{A.32}$$

We follow the algorithm and notations described in Sect. 10.3, Chap. 10.

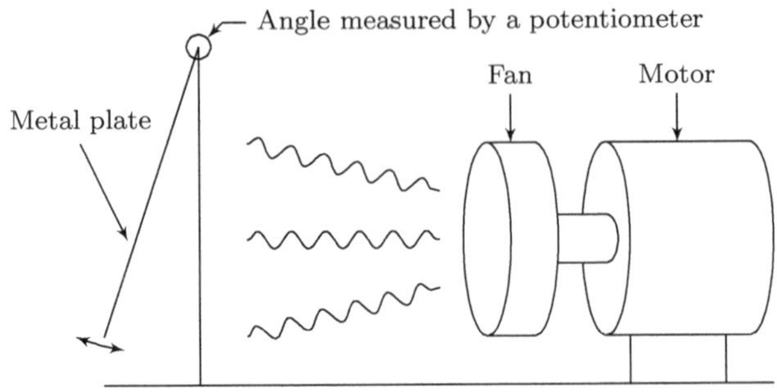

Fig. A.20 A metal plate system

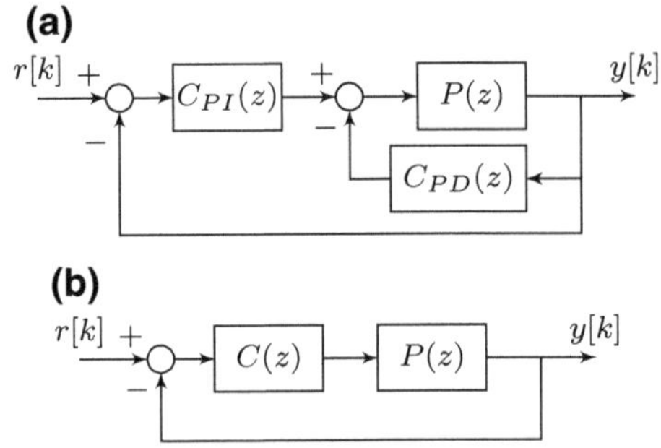

Fig. A.21 **a** Digital PI-PD controller structure, **b** Digital PID controller structure

A. Computation of the Stabilizing Set $\mathcal{S}$

The stabilizing set $\mathcal{S}$ for PID controllers in (K_1, K_2, K_3) space is described in Chap. 4. Since $K_3 := K_2 - K_0$, we can convert the stabilizing set $\mathcal{S}(K_1, K_2, K_3)$ into $\mathcal{S}(K_0, K_1, K_2)$ using the following relationship

$$\begin{bmatrix} K_0 \\ K_1 \\ K_2 \end{bmatrix} = \begin{bmatrix} 0 & 1 & -1 \\ 1 & 0 & 0 \\ 0 & 1 & 0 \end{bmatrix} \begin{bmatrix} K_1 \\ K_2 \\ K_3 \end{bmatrix}. \tag{A.33}$$

Two stabilizing sets are shown in Figs. A.22 and A.23.

B. Computation of Family of Ellipses in (W_0, W_1) Space

Let us first choose $\gamma = 1.6553$ and $K_0 = -0.5016$ since the PI-PD controller in (A.31) renders the H_∞ norm less than 1.6533 and the constant term in the numerator of the PD controller is equal to -0.5016. Since the ellipse in (10.43) is an axis-parallel ellipse, the family of ellipses must be a collection of axis-parallel ellipses in (W_0, W_1) space as shown in Fig. A.24. The stabilizing set $\mathcal{S}$ at $K_0 = -0.5016$ and the subset $\mathcal{S}_\gamma$ with $\gamma = 1.6553$ space are to be determined in the next step.

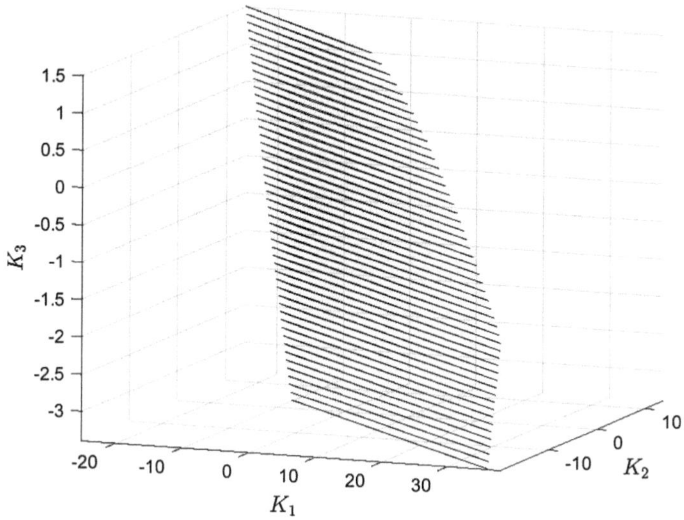

Fig. A.22 $\mathcal{S}(K_1, K_2, K_3)$

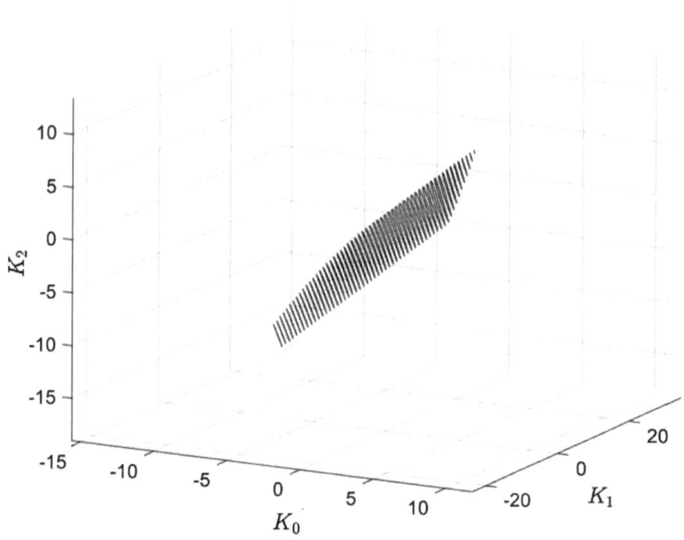

Fig. A.23 $\mathcal{S}(K_0, K_1, K_2)$

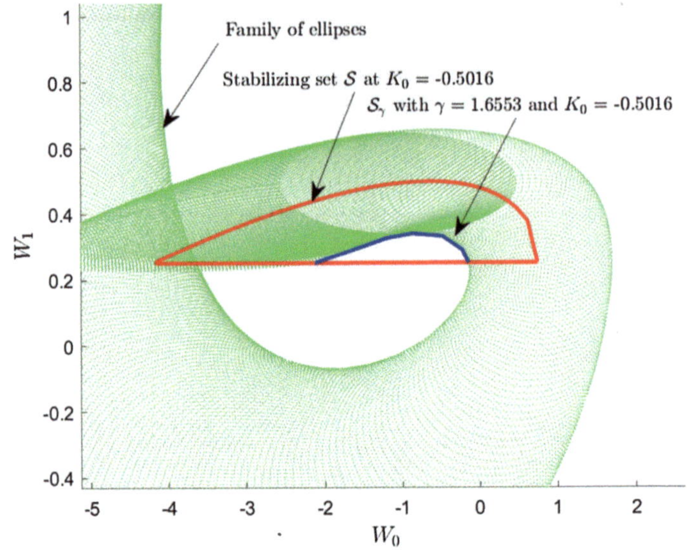

Fig. A.24 Family of ellipses in (W_0, W_1) space. Stabilizing set $\mathcal{S}$ at $K_0 = -0.5016$ and the subset $\mathcal{S}_\gamma$ with $\gamma = 1.6553$ in (W_0, W_1) space are overlaid

C. Mapping Ellipses to (K_0, K_1, K_2) Space and Computation of $\mathcal{S}_\gamma$ for a Fixed K_0

Due to the mapping in (10.27), the mapped ellipses are not axis-parallel. As u varies from -1 to 1, the mapped ellipses encompass the stabilizing set $\mathcal{S}(K_0 = -0.5016)$ in (K_1, K_2) space and carve the stabilizing set $\mathcal{S}(K_0 = -0.5016)$. The subset $\mathcal{S}_\gamma$ with $\gamma = 1.6553$ is then the leftover set as shown in Fig. A.25.

D. Sweeping Over K_0 and Computation of $\mathcal{S}_\gamma$

By sweeping K_0 over a range of values at each of which the stabilizing set $\mathcal{S}(K_0)$ is not empty, we can compute the set $\mathcal{S}_\gamma$ as shown in Fig. A.26.

E. Controller Selection with Performance Specifications

Using the procedure described above, we collect all controllers achieving overshoot of less than 5% and settling time of less than 2.3 s for $\gamma = 1.5, 1.6, 1.7$, and 1.8. In particular, for each $\mathcal{S}_\gamma$ set, we pick such a controller that provides the least settling time. The $\mathcal{S}_\gamma$ sets are shown in Fig. A.27. Table A.2 shows controller gains which give the "best" performance corresponding to γ values. The step responses and Nyquist plots of the "best" controllers for each γ value are shown in Figs. A.28 and A.29, respectively.

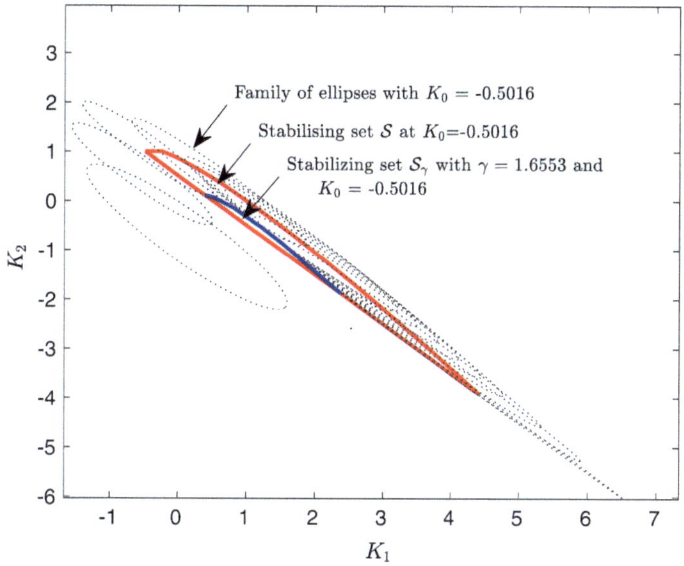

Fig. A.25 Family of ellipses in (K_1, K_2) space. Stabilizing set $\mathcal{S}$ at $K_0 = -0.5016$ and the subset $\mathcal{S}_\gamma$ with $\gamma = 1.6553$ are overlaid

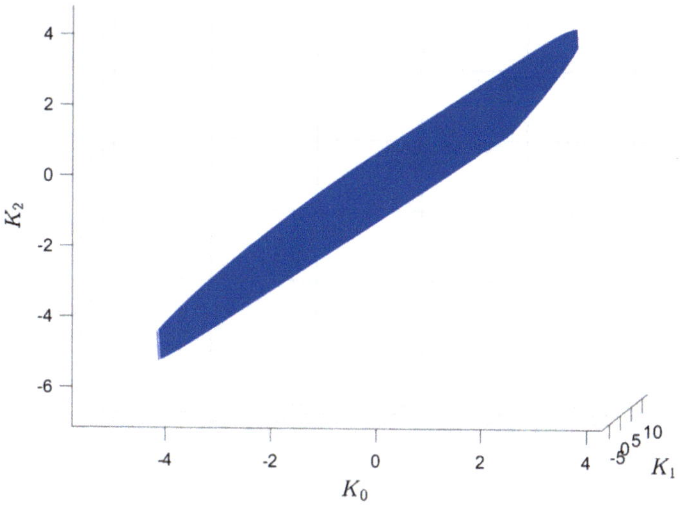

Fig. A.26 $\mathcal{S}_\gamma$ set with $\gamma = 1.6553$

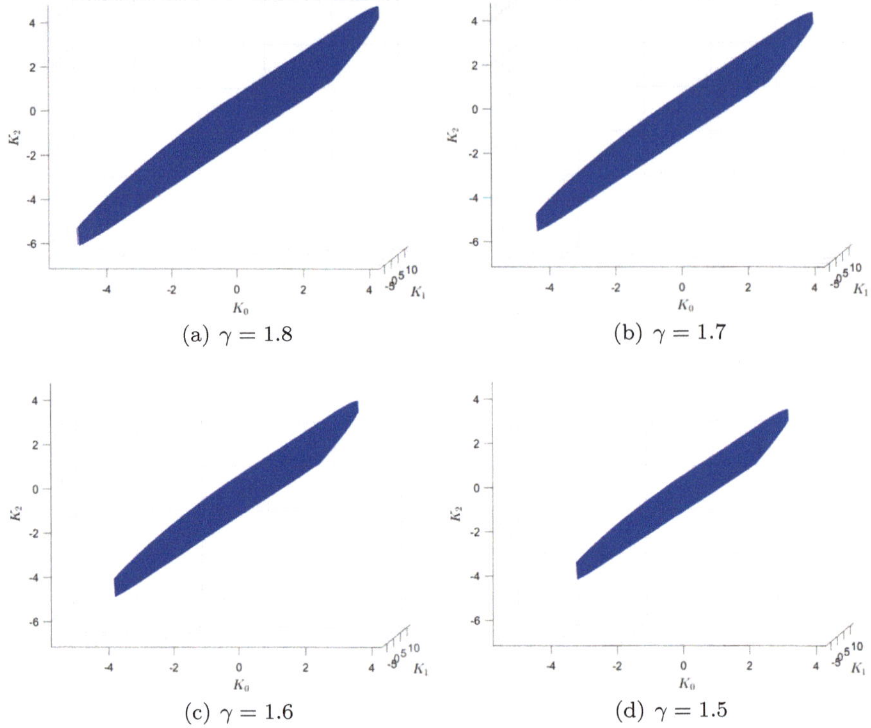

Fig. A.27 $\mathcal{S}_\gamma$ sets for different values of γ

Table A.2 Best controller gains for various γ values

	K_2	K_1	K_0	Overshoot (%)	Settling time
$\gamma = 1.8$	−0.1504	0.4767	−0.11	1.6	1.14
$\gamma = 1.7$	2.3725	−3.8903	1.82	1.99	0.883
$\gamma = 1.6$	2.1939	−3.7020	1.79	1.54	1.04
$\gamma = 1.5$	2.2254	−3.9765	2.00	1.77	1.25

A.6 Notes and References

The complex transfer function for three-phase AC drives is established in [2, 7, 11]. The pole-zero cancelation for the surface-mounted permanent-magnet synchronous machine using PI controllers is introduced in [2, Fig. 9] and in [11, Fig. 2b]. The k value in Fig. A.2 matches the result in [11] at the sampling frequency f_s of 5 kHz.

For more information concerning the controller design case in Sect. A.3, the reader can refer to Sect. 2 of the book by Buso and Mattavelli [3].

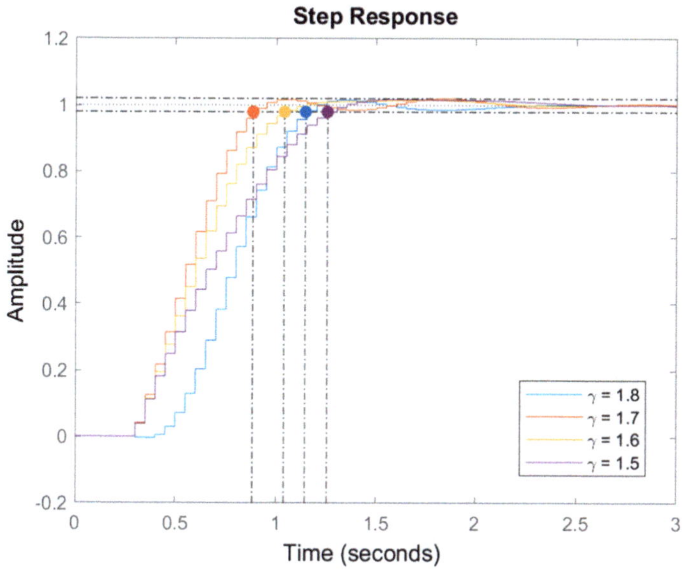

Fig. A.28 Step responses of the selected controllers in Table A.2

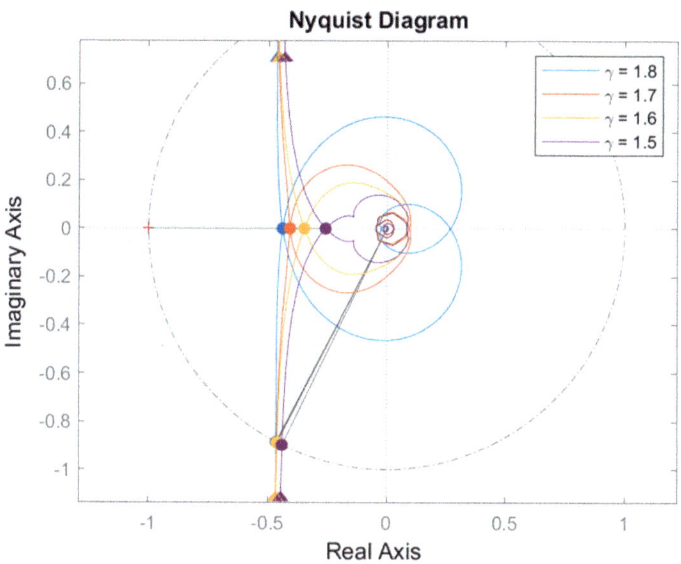

Fig. A.29 Nyquist plots of the selected controllers in Table A.2

The PI-PD controller example was implemented by Dincel and Söylemez [4]. The main results in Sect. A.5 were taken from [5]. The results on selective harmonic mitigation were developed by Magossi, Han, Oliveira, and Bhattacharyya [8]. In Power Electronics, it is often desirable to attenuate the amplitude of selected harmonics in a system [1, 9, 10].

References

1. Aguiar, C.R., Bastos, R.F., Gonalves, A.F.Q., Neves, R.V.A., Reis, G.B., Machado, R.Q.: Frequency fuzzy anti-islanding for grid-connected and islanding operation in distributed generation systems. IET Power Electron. **8**(7), 1255–1262 (2015)
2. Briz, F., Degner, M.W., Lorenz, R.D.: Analysis and design of current regulators using complex vectors. IEEE Trans. Ind. Appl. **36**(3), 817–825 (2000)
3. Buso, S., Mattavelli, P.: Digital Control in Power Electronics. Morgan and Claypool Publishers, San Rafael (2006)
4. Dincel, E., Söylemez, M.T.: Digital PI-PD controller design for arbitrary order systems: dominant pole placement approach. ISA Trans. (2018).
5. Han, S.: Robust and optimal PID controller synthesis for linear time invariant control systems. Ph.D. thesis, Texas A&M University (2019)
6. Han, S., Bhattacharyya, S.: PID controller synthesis using a σ-Hurwitz stability criterion. IEEE Control Syst. Lett. **2**(3), 525–530 (2018)
7. Harnefors, L.: Modeling of three-phase dynamic systems using complex transfer functions and transfer matrices. IEEE Trans. Ind. Electron. **54**(4), 2239–2248 (2007)
8. Magossi, R.F.Q., Han, S., Oliveira, V.A., Bhattacharyya, S.P.: Proportional-Integral controller design for selective harmonic mitigation. In: Congreso Latinoamericano de Control Automàtico, CLCA 2018, Quito, Ecuador, Octubre 24–26 (2018)
9. Teodorescu, R., Blaabjerg, F., Borup, U., Liserre, M.: A new control structure for grid-connected LCL PV inverters with zero steady-state error and selective harmonic compensation. In: Applied Power Electronics Conference and Exposition, vol. 1, pp. 580–586 (2004)
10. Teodorescu, R., Blaabjerg, F., Liserre, M., Loh, P.C.: Proportional-resonant controllers and filters for grid-connected voltage-source converters. IEE Proc. Electr. Power Appl. **153**(5), 750–762 (2006)
11. Yepes, A.G., Vidal, A., Malvar, J., López, O., Doval-Gandoy, J.: Tuning method aimed at optimized settling time and overshoot for synchronous proportional-integral current control in electric machines. IEEE Trans. Power Electron. **29**(6), 3041–3054 (2014)

Appendix B
Sample MATLAB Codes

We present sample MATLAB codes and the associated figures. The purpose is to help the readers to understand the results developed and used in this book. Moreover, they can readily apply the codes to their research problems as a starting point. We recommend readers to use the latest version of the MATLAB software.

B.1 Sample MATLAB Codes for Continuous-Time Systems

We present how to graphically visualize the stabilizing set of PI and PID controllers for a given continuous-time system. The mathematical account of the stabilizing set is discussed in Part I. The computation of the stabilizing set is a foundation of all of the design methods we developed in the rest of this book. Following the stabilizing set, we provide how to get the design curves for gain and phase margin design in Part II. Next, we show the MATLAB codes for all stabilizing PI and PID controllers satisfying an H_∞ criterion in Part III.

B.1.1 PI Controller Stabilizing Set

The following MATLAB sample code is used to compute the PI stabilizing set of the following plant:

$$P(s) = \frac{s-5}{s^2+1.6s+0.2}. \tag{B.1}$$

I. D. Díaz-Rodríguez et al., *Analytical Design of PID Controllers*,
https://doi.org/10.1007/978-3-030-18228-1

```
% PI stabilizing set
clear all;
syms s;
kp = sym('kp','real');
ki = sym('ki','real');
w = sym('w','real');
N = [1 -5];                        % Numerator
D = [1 1.6 0.2];                   % Denominator
P = tf(N,D);                       % Plant
tam_N = size(N);
tam_D = size(D);
n = tam_D(2)-1;
m = tam_N(2)-1;
ze = roots(N);                     % Zeros of N(s)
l = 1;
nz = 0;
for k = 1:m
    if real(ze(k)) > 0
        nz = 1;                    % RHP Zeros of N(s)
        l = l + 1;
    end
end
signature = n - m + 1 + 2*nz;             % Signature number
D_s = poly2sym(D,s);                      % D(s)
N_s = poly2sym(N,s);                      % N(s)
N_ms = subs(N_s, -s);                     % N(-s)
Delta = s*D_s + (ki)*N_s + kp*s*N_s;      % Characteristic ...
    equation
Delta = collect(Delta,s);                 % simplify ...
    expression in terms of s
V_s = collect(Delta*N_ms,s);              % V = Delta*N(-s)
V = subs(V_s, 1i*w);                      % V(jw)
Vr = real(V);                             % Real part of V
Vi = imag(V);                             % Imaginary part ...
    of V
z = 1;
e = 1;
Range = -1.6:0.01:0.04;
% This is the range considered for fixed Kp values. This ...
    needs to be changed depending the Plant
for kp_f = Range
    f_kp = subs(Vi, kp, kp_f);      % substitute value of Kp
    f_Vi = sym2poly(f_kp);          % convert to polynomial
    r = roots(f_Vi);                % find the roots
    tam3 = size(r);
    l = 1;
    r2 = 0;
    for k = 1:tam3
        if imag(r(k)) == 0 && real(r(k)) > 0  % select ...
            real, positive roots
            r2(l) = real(r(k));
            l = l + 1;
        end
    end
    wt = [0 r2];
    tam_w = size(wt);
    R(1) = subs(Vr, w, wt(1));
    C1(1,:) = [coeffs(R(1)),0];
    Te(1,:) = [ki,0];
    R(tam_w(2)) = subs(Vr, w, wt(tam_w(2)));
    for k=2:tam_w(2)
        R(k) = subs(Vr, w, wt(k));
        C1(k,:) = sym2poly(R(k));
    end
    C1 = double(C1);
     j(z) = sign(subs(Vi,[kp,w],[kp_f,0.01]));
```

```
        A = [-C1(1,1) 0;C1(2,1) 0];        % [ki kp]
        b = [C1(1,2);-C1(2,2)];
        min(z) = 0;
        max(z) = (-C1(2,2)/C1(2,1));
      z = z + 1;
end
va = [Range];
h = fill([min flip(max)],[va flip(va)],'y');
xlabel('k_i');
ylabel('k_p');
```

After running this MATLAB code, the resulting stabilizing set is shown in Fig. B.1.

B.1.2 PI Controller Gain and Phase Margin Design Curves

The following MATLAB sample code is used to compute the PI gain and phase margin design curves of the following plant:

$$P(s) = \frac{s-5}{s^2+1.6s+0.2}. \tag{B.2}$$

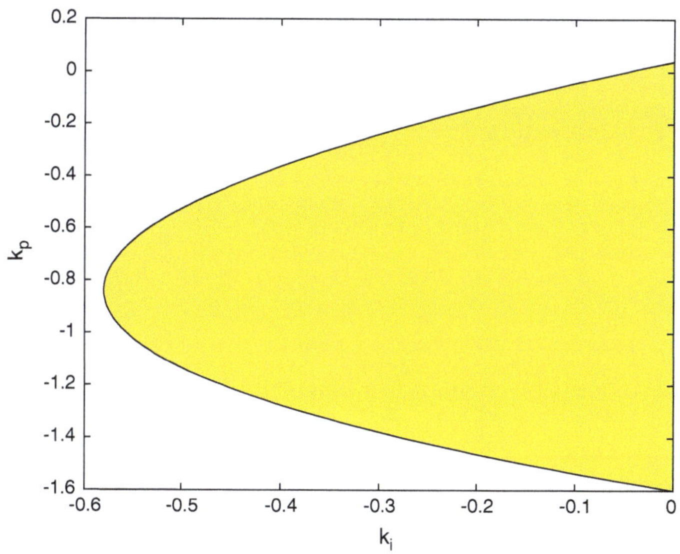

Fig. B.1 PI stabilizing set

```
%% Gain and Phase Margin Design Curves
N = [1 -5];           % Numerator of the plant
D = [1 1.6 0.2];      % Denominator of the plant
P = tf(N,D);          % Plant
PM = [1:90];          % Phase margin range
z = 1;
z2 = 1;
z4 = 1;
r = size(PM);
for wg = 0.1:0.1:3  % Gain crossover frequency range
z = 1;
  for k = 1:1:r(2)
    [MP,PP] = bode(P,wg);
    syms kp;
    syms ki;
    phi = pi + PM(k)*pi/180 - PP*pi/180;
    m2 = 1/(MP^2);
    m = sqrt(m2);
    a2 = m2;
    b2 = m2*(wg^2);
    c = wg*tan(phi);
    f1(z4) = ((kp)^2)/(a2) + (ki^2)/(b2) - 1;
    f2(z4) = kp*c + ki;
    hold on;
    X = solve([f1(z4) , f2(z4) ], [kp, ki]);
    z4 = z4 + 1;
    x1_1 = double(X.kp(1));
    x1_2 = double(X.ki(1));
    x2_1 = double(X.kp(2));
    x2_2 = double(X.ki(2));
    C1 = tf([x2_1 x2_2],[1 0]);
    S1 = allmargin(C1*P);
    E2 = S1.Stable;
    C2 = tf([x1_1 x1_2],[1 0]);
    S2 = allmargin(C2*P);
    E1 = S2.Stable;
    if E1 == 1
      R_E1 = allmargin(C2*P);
      [Gm2(z),Pm2(z),Wgm2(z),Wpm2(z)] = margin(C2*P);
      Gm_dB2(z) = 20*log10(Gm2(z));
      info_M(z,:,z2) = ...
          [Gm_dB2(z),Pm2(z),Wgm2(z),Wpm2(z),R_E1.DelayMargin];
    end
    if E2 == 1
      [Gm1(z),Pm1(z),Wgm1(z),Wpm1(z)] = margin(C1*P);
      R_E2 = allmargin(C1*P);
      Gm_dB1(z) = 20*log10(Gm1(z));
      info_M(z,:,z2) = ...
          [Gm_dB1(z),Pm1(z),Wgm1(z),Wpm1(z),R_E2.DelayMargin];
    end
    z = z + 1;
  end
z2 = z2 + 1;
end
%% Plot of the PM vs GM design curves
figure(1)
r2 = size(info_M);
con2 = 1;
for c = 1:1:r2(3)
    con = 1;
```

```
    for c2 = 1:1:r2(1)
        if info_M(c2,4,c) == 0
           break;
        end
        con = con + 1;
    end
    g(c) = plot(info_M(1:con-1,2,c),info_M(1:con-1,1,c),'-');
    hold on;
    con2 = con2 + 1;
end
wg = 0.1:0.1:3;
for c = 1:28
    legend([g(c)],['wg = ',num2str(wg(c))]);
    hold on;
end
legend('off')
legend(gca,'show')
xlabel('PM');
ylabel('GM');
```

After running this MATLAB code, the resulting stabilizing set is shown in Fig. B.2.

B.1.3 PI Controller with an H_∞ criterion

We present the MATLAB code that constructively determines all stabilizing PI controllers satisfying H_∞ norm of the error transfer function less than $\gamma = 2.0$ for the following plant:

$$P(s) = \frac{s-2}{s^2+4s+3}. \tag{B.3}$$

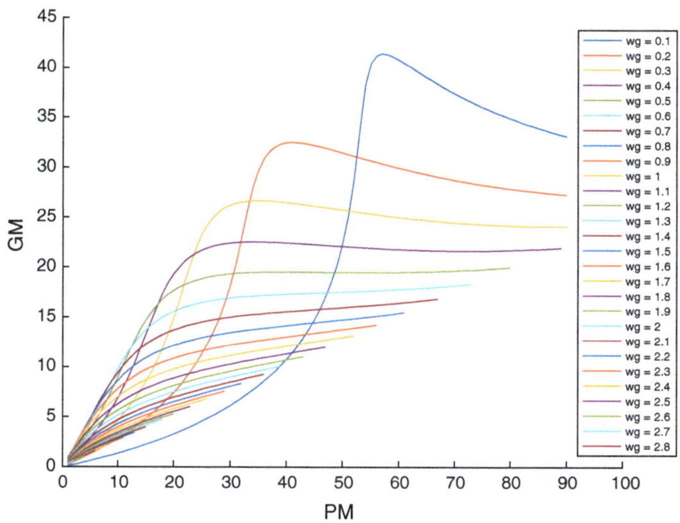

Fig. B.2 Gain and phase margin design curves

```
close all;
clear;
syms s;
% plant
N = [1 -2];              % Numerator
D = [1 4 3];             % Denominator
P_tf=tf(N,D);
N_s=poly2sym(N,s);
D_s=poly2sym(D,s);
P=N_s/D_s;
theta_rad=-pi:0.01:pi;
resol=0.01;

% 1. Load the stabilizing set
% We assume the stabilizing set is readily available
load stab_pi_plant_00.mat;
stabset = [[Ki_bounds(:,1); flipud(Ki_bounds(:,2))], ...
    [Kp_range; flipud(Kp_range);]];
poly_stabset = ...
    simplify(polyshape(stabset,'Simplify',false));
clear Ki_bounds Kp_range;

% 2. Initial Sgamma is the stabilizing set
poly_Sgamma = poly_stabset;

% 3. Fix gamma and constructively determine S gamma set
gamma=2.0;
w_freqs=linspace(0.01,3.0,300);
for idx = 1:numel(w_freqs)
    w=w_freqs(idx);
    Pr=double(real(subs(P,s,1i*w)));
    Pi=double(imag(subs(P,s,1i*w)));
    % (x - c1)^2 / a^2 + (y - c2)^2 / b^2 = 1
    c1 = -(w*Pi)/(Pr^2 + Pi^2);
    c2 = -Pr/(Pr^2 + Pi^2);
    bb = 1/((gamma^2)*(Pr^2 + Pi^2));
    aa = (w^2)*bb;
    x=c1+sqrt(aa)*cos(theta_rad);
    y=c2+sqrt(bb)*sin(theta_rad);

    poly_Sgamma = ...
        subtract(poly_Sgamma,polyshape(x,y,'Simplify',true));
end

% 4. Collect sample ellipses for display (optional)
clear idx w_freqs w Pr Pi;
gamma=2;
w_freqs=0.1:0.2:5;
coords_x = NaN(numel(w_freqs),numel(theta_rad));
coords_y = NaN(numel(w_freqs),numel(theta_rad));
parfor idx = 1:numel(w_freqs)
    w=w_freqs(idx);
    Pr=real(subs(P,s,1i*w));
    Pi=imag(subs(P,s,1i*w));
    % (x - c1)^2 / a^2 + (y - c2)^2 / b^2 = 1
    c1 = -(w*Pi)/(Pr^2 + Pi^2);
    c2 = -Pr/(Pr^2 + Pi^2);
    bb = 1/((gamma^2)*(Pr^2 + Pi^2));
    aa = (w^2)*bb;
    x=c1+sqrt(aa)*cos(theta_rad);
    y=c2+sqrt(bb)*sin(theta_rad);
    coords_x(idx,:)=x;
    coords_y(idx,:)=y;
end

figure;
```

```
fill(poly_stabset.Vertices(:,1), ...
     poly_stabset.Vertices(:,2),'y');
hold on;
plot(poly_Sgamma.Vertices(:,1), ...
     poly_Sgamma.Vertices(:,2),'-',...
     'LineWidth',2, 'Color', [0 0 0]);
for idx = 1:size(coords_x,1)
     x=coords_x(idx,:)';
     y=coords_y(idx,:)';
     plot(x,y,'k:');
end
xlabel('$$k_i$$','interpreter','latex');
ylabel('$$k_p$$','interpreter','latex');
axis([-4.4 1.2 -4.5 2.0]);
str = '$$\mathcal{S}_{\gamma = 2.0}$$';
loc = mean(poly_Sgamma.Vertices);
text(loc(1),loc(2),str, ...
     'Interpreter','latex',...
     'FontSize',14);
hold off;
```

The result of the MATLAB code generates the set of all PI controllers, the family of ellipses, and all stabilizing PI controllers $\mathcal{S}_\gamma$ satisfying the H_∞ norm less than $\gamma = 2.0$ as shown in Fig. B.3.

B.1.4 PID Controller Stabilizing Set

The following MATLAB sample code is used to compute the PID stabilizing set of the following plant:

$$P(s) = \frac{s-3}{s^3 + 4s^2 + 5s + 2}. \tag{B.4}$$

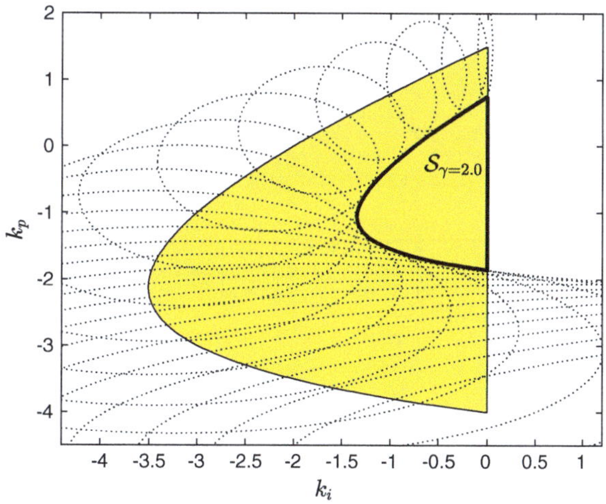

Fig. B.3 $\mathcal{S}_\gamma$ satisfying the H_∞ norm less than $\gamma = 2.0$

```
% PID stabilizing set
clear all;
syms s;
kp = sym('kp','real');
ki = sym('ki','real');
kd = sym('kd','real');
w = sym('w','real');
N = [1 -3];                     % Numerator
D = [1 4 5 2];                  % Denominator
tam_N = size(N);
tam_D = size(D);
n = tam_D(2)-1;
m = tam_N(2)-1;
ze = roots(N);                  % Zeros of N(s)
l = 1;
nz = 0;
for k = 1:m
    if real(ze(k)) > 0
        nz = 1;                 % RHP Zeros of N(s)
        l = l + 1;
    end
end
signature = n - m + 1 + 2*nz;   % Signature number
D_s = poly2sym(D,s);            % D(s)
N_s = poly2sym(N,s);            % N(s)
N_ms = subs(N_s, -s);           % N(-s)
Delta = s*D_s + (ki+kd*s^2)*N_s + kp*s*N_s;   % ...
    Characteristic equation
Delta = collect(Delta,s);       % simplify expression in ...
    terms of s
V_s = collect(Delta*N_ms,s);    % V = Delta*N(-s)
V = subs(V_s, 1i*w);            % V(jw)
Vr = real(V);                   % Real part of V
Vi = imag(V);                   % Imaginary part of V
for kp_f = -4:0.2:0.65          % Evaluate for a fixed Kp
    f_kp = subs(Vi, kp, kp_f);  % substitute value of Kp
    f_Vi = sym2poly(f_kp);      % convert to polynomial
    r = roots(f_Vi);            % find the roots
    tam3 = size(r);
    l = 1;
    r2 = 0;
    for k = 1:tam3
        if imag(r(k)) == 0 && real(r(k)) > 0  % select ...
            real, positive roots
            r2(l) = real(r(k));
            l = l + 1;
        end
    end
    wt = [0 r2(2) r2(1)] ;
    tam_w = size(wt);
    R(1) = subs(Vr, w, wt(1));
    C1(1,:) = [0,coeffs(R(1)), 0];
    Te(1,:) = [0,ki,0];
    R(tam_w(2)) = subs(Vr, w, wt(tam_w(2)));
    for k=2:tam_w(2)
        R(k) = subs(Vr, w, wt(k));
        [C1(k,:),Te(k,:)] = coeffs(R(k));
    end
    C1 = double(C1);
    A = -[C1(1,1) C1(1,2) 0;-C1(2,1) -C1(2,2) 0;C1(3,1) ...
        C1(3,2) 0];        % [kd ki kp]
    b = -[-C1(1,3);C1(2,3);-C1(3,3)];
    lb = [-300,-300,kp_f];
    ub = [300,300,kp_f];
    plotregion(A,b,lb,ub,'y');
    hold on;
```

Fig. B.4 PID stabilizing set

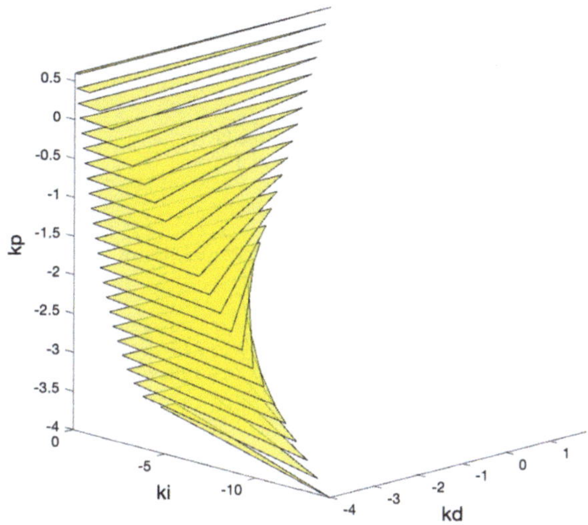


```
      axis equal
      xlabel('kd');
      ylabel('ki');
      zlabel('kp');
end
axis square;
```

After running this MATLAB code, the resulting stabilizing set is shown in Fig. B.4.

B.1.5 PID Controller Gain and Phase Margin Design Curves

The following MATLAB sample code is used to compute the PI gain and phase margin design curves of the following plant:

$$P(s) = \frac{s-3}{s^3 + 4s^2 + 5s + 2}. \tag{B.5}$$

```
% PID controller gain and phase margin design curves
clear all;
close all;
P = tf([1 -3],[1 4 5 2]);
VALC = struct([]);
wg_vector = .1:0.1:1.2;
PM_vector = 1:2:100;
k2 = 1;
k1 = 1;
for idx_wg = 1:1:length(wg_vector)
  wg = wg_vector(idx_wg)
```

```
k1 = 1;
k2 = 1;
for idx_PM = 1:1:length(PM_vector)
  k1 = 1;
  k4 = 1;
  PM = PM_vector(idx_PM)
  [MP,PP] = bode(P,wg);
  phi = pi + PM*pi/180 - PP*pi/180;
  M2 = 1/(MP^2);
  Kp1 = sqrt(M2/(1+tan(phi)^2));
  [A1,b1] = pid_set_continuous(Kp1);
  if (¬(isempty(A1)))
    for x = -6:0.1:6
      Ki = x*wg^2 - wg*tan(phi)*Kp1;
      C = tf([x,Kp1,Ki],[1 0]);
      STA = isstable(feedback(C*P,1));
      if STA == 1
        VALC{idx_PM,idx_wg}(k1,:) = [x,Ki,Kp1,wg,PM];
        k1 = k1 + 1;
      else
        VALC{idx_PM,idx_wg}(k1,:) = [0,0,0,0,0];
        k1 = k1 + 1;
      end
    end
    tam = size(VALC{idx_PM,idx_wg});
    for k3 = 1:tam(1)
      if VALC{idx_PM,idx_wg}(k3,:)≠ [0,0,0,0,0]
        C = tf([VALC{idx_PM,idx_wg}(k3,1),...
        VALC{idx_PM,idx_wg}(k3,3),...
        VALC{idx_PM,idx_wg}(k3,2)],[1,0]);
        Result = allmargin(C*P);
        if abs(Result.PhaseMargin(1)-PM) > 2
        else
          info_GMPM{idx_PM,idx_wg}(k4,:) = ...
              [Result.GainMargin(1),Result.PhaseMargin(1),...
          Result.PMFrequency(1),Result.Stable];
          k4 = k4 + 1;
        end
      end
    end
  end
  Kp2 = -sqrt(M2/(1+tan(phi)^2));
  [A2,b2] = pid_set_continuous(Kp2);
  if (¬(isempty(A2)))
    for x = -6:0.1:6
      Ki = x*wg^2 - wg*tan(phi)*Kp2;
      C = tf([x,Kp2,Ki],[1 0]);
      STA = isstable(feedback(C*P,1));
      if STA == 1
        VALC{idx_PM,idx_wg}(k1,:) = [x,Ki,Kp2,wg,PM];
        k1 = k1 + 1;
        else
        VALC{idx_PM,idx_wg}(k1,:) = [0,0,0,0,0];
        k1 = k1 + 1;
      end
    end
    tam = size(VALC{idx_PM,idx_wg});
    for k3 = 1:tam(1)
      if VALC{idx_PM,idx_wg}(k3,:)≠ [0,0,0,0,0]
        C = tf([VALC{idx_PM,idx_wg}(k3,1),...
        VALC{idx_PM,idx_wg}(k3,3),...
        VALC{idx_PM,idx_wg}(k3,2)],[1,0]);
```

```
          Result = allmargin(C*P);
          if abs(Result.PhaseMargin(1)-PM) > 2
          else
            info_GMPM{idx_PM,idx_wg}(k4,:) = ...
                [Result.GainMargin(1),Result.PhaseMargin(1),...
            Result.PMFrequency(1),Result.Stable];
            k4 = k4 + 1;
          end
        end
      end
    end
  end
end
%% Plot of PID gain and phase margin design curves
figure(1)
colors = hsv(length(wg_vector));
legendInfo=cell(length(wg_vector),1);
H = gobjects(length(wg_vector),1);
for idx_wg = 1:1:length(wg_vector)
  wg = wg_vector(idx_wg);
  for idx_PM = 1:1:length(PM_vector)-2
    h = plot3(wg*ones(1,length(info_GMPM{idx_PM,idx_wg}...
    (:,1))),info_GMPM{idx_PM,idx_wg}(:,2),...
    info_GMPM{idx_PM,idx_wg}(:,1),'-','LineWidth',1.5);
    hold on;
    set(h,'Color',colors(idx_wg,:));
    if(isempty(legendInfo{idx_wg}))
      legendInfo{idx_wg} = ['w_g = ' ...
          num2str(wg_vector(idx_wg))];
      H(idx_wg,1)=h;
    end
  end
end
legend(H,legendInfo);
set(gca,'zscale','log')
axis([0 1.2 0 100 1 40])
grid on
ylabel('Phase Margin (deg)');
xlabel('\omega_g (rad/s)');
zlabel('Gain Margin');
```

The following code corresponds to the function "pid_set_continuous()" used in the previous sample code.

```
function [A, b] = pid_set_continuous(Kp)
  syms s;
  kp = sym('kp','real');
  ki = sym('ki','real');
  kd = sym('kd','real');
  w = sym('w','real');
  if Kp < -4 || Kp > 0.65
    A = [];
    b = [];
  else
    N = [1 -3];
    D = [1 4 5 2];
    tam_N = size(N);
    tam_D = size(D);
```

```
    n = tam_D(2)-1;
    m = tam_N(2)-1;
    ze = roots(N);
    l = 1;
    nz = 0;
    for k = 1:m
      if real(ze(k)) > 0
        nz = 1;
        l = l + 1;
      end
    end
    signature = n - m + 1 + 2*nz;
    D_s = poly2sym(D,s);
    N_s = poly2sym(N,s);
    N_ms = subs(N_s, -s);
    Delta = s*D_s + (ki+kd*s^2)*N_s + kp*s*N_s;
    Delta = collect(Delta,s);
    V_s = collect(Delta*N_ms,s);
    V = subs(V_s, 1i*w);
    Vr = real(V);
    Vi = imag(V);
    for kp_f = Kp:0.2:Kp
      f_kp = subs(Vi, kp, kp_f);
      f_Vi = sym2poly(f_kp);
      r = roots(f_Vi);
      tam3 = size(r);
      l = 1;
      r2 = 0;
      for k = 1:tam3
        if imag(r(k)) == 0 && real(r(k)) > 0
          r2(l) = real(r(k));
          l = l + 1;
        end
      end
      wt = [0 r2(2) r2(1)] ;
      tam_w = size(wt);
      R(1) = subs(Vr, w, wt(1));
      C1(1,:) = [0,coeffs(R(1)), 0];
      Te(1,:) = [0,ki,0];
      R(tam_w(2)) = subs(Vr, w, wt(tam_w(2)));
      for k=2:tam_w(2)
        R(k) = subs(Vr, w, wt(k));
        [C1(k,:),Te(k,:)] = coeffs(R(k));
      end
      C1 = double(C1);
      A = -[C1(1,1) C1(1,2) 0;-C1(2,1) ...
          -C1(2,2) 0;C1(3,1) C1(3,2) 0];
      b = -[-C1(1,3);C1(2,3);-C1(3,3)];
      lb = [-300,-300,kp_f];
      ub = [300,300,kp_f];
    end
  end
end
```

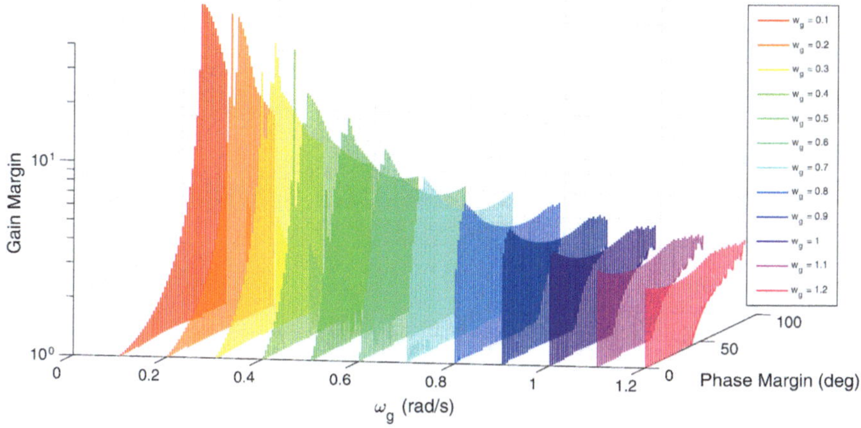

Fig. B.5 PID gain and phase margin design curves

After running this MATLAB code, the resulting PID gain and phase margin design curves are shown in Fig. B.5.

B.1.6 PID Controller with an H_∞ criterion

We present the MATLAB code to compute the set of stabilizing PID controllers satisfying H_∞ norm of the error transfer function less than $\gamma = 1.0$ for the following plant:

$$P(s) = \frac{10\,s^3 + 9\,s^2 + 362.4\,s + 36.16}{2\,s^5 + 2.7255\,s^4 + 138.4292\,s^3 + 156.471\,s^2 + 637.6472\,s + 360.1779}. \tag{B.6}$$

```
close all;
clear;
syms s;
% plant
N = [10 9 362.4 36.16]; % Numerator
D = [2 2.7255 138.4292 156.471 637.6472 360.1779]; % ...
    Denominator
P_tf=tf(N,D);
N_s=poly2sym(N,s);
D_s=poly2sym(D,s);
P=N_s/D_s;
theta_rad=-pi:0.01:pi;
resol=0.01;

% 1. Load the stabilizing set
% We assume the stabilizing set is readily available for ...
    Kd = 9
Kd = 9;
```

```
load stab_pid_hinf_plant_00.mat;
stabset = [Ki_data,Kp_data];
poly_stabset = ...
    simplify(polyshape(stabset,'Simplify',false));
clear Ki_data Kp_data;

% 2. Initial Sgamma is the stabilizing set
poly_Sgamma = poly_stabset;

% 3. Fix gamma and constructively determine S gamma set
gamma=1.0;
P_s = N_s/D_s;
syms w 'real';
P_jw= subs(P_s,s,1i*w);
P_r = simplify(real(P_jw));
P_i = simplify(imag(P_jw)/w);
P_mag2 = simplify(P_r*P_r + w*w*P_i*P_i);

Ea2 = w^2/(gamma^2*P_mag2);
Eb2 = 1/(gamma^2*P_mag2);
Ec1 = -w^2*P_i/P_mag2;
Ec2 = -P_r/P_mag2;

w_freqs=[linspace(1,20,800), linspace(1,20,800)];
for idx = 1:numel(w_freqs)
    wr=w_freqs(idx);
    % (x - c1)^2 / a^2 + (y - c2)^2 / b^2 = 1
    c1 = double(subs(Ec1,w,wr)) + wr*wr*Kd;
    c2 = double(subs(Ec2,w,wr));
    bb = double(subs(Eb2,w,wr));
    aa = double(subs(Ea2,w,wr));
    x=c1+sqrt(aa)*cos(theta_rad);
    y=c2+sqrt(bb)*sin(theta_rad);

    poly_Sgamma = ...
        subtract(poly_Sgamma,polyshape(x,y,'Simplify',true));
end

% 4. Collect sample ellipses for display (optional)
clear idx w_freqs;
gamma=1;
w_freqs=linspace(1,200,800);
coords_x = NaN(numel(w_freqs),numel(theta_rad));
coords_y = NaN(numel(w_freqs),numel(theta_rad));
parfor idx = 1:numel(w_freqs)
    wr=w_freqs(idx);
    % (x - c1)^2 / a^2 + (y - c2)^2 / b^2 = 1
    c1 = double(subs(Ec1,w,wr)) + wr*wr*Kd;
    c2 = double(subs(Ec2,w,wr));
    bb = double(subs(Eb2,w,wr));
    aa = double(subs(Ea2,w,wr));
    x=c1+sqrt(aa)*cos(theta_rad);
    y=c2+sqrt(bb)*sin(theta_rad);

    coords_x(idx,:)=x;
    coords_y(idx,:)=y;
end

figure;
fill(poly_stabset.Vertices(:,1), ...
    poly_stabset.Vertices(:,2),'y');
hold on;
plot(poly_Sgamma.Vertices(:,1), ...
    poly_Sgamma.Vertices(:,2),'-',...
    'LineWidth',2, 'Color', [0 0 0]);
for idx = 1:size(coords_x,1)
    x=coords_x(idx,:)';
```

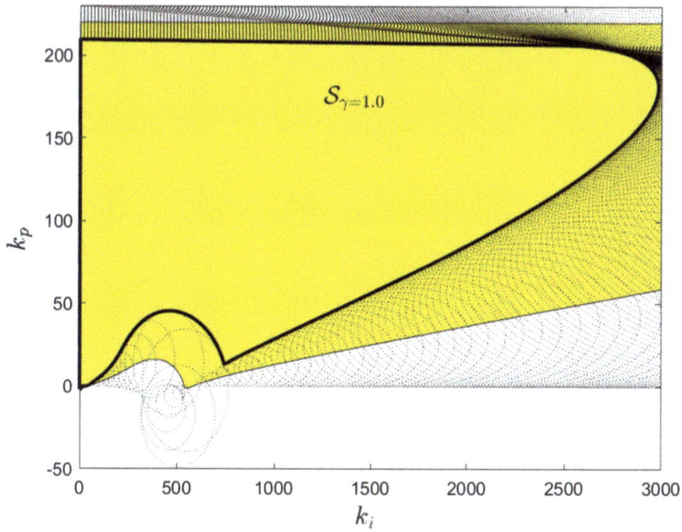

Fig. B.6 $\mathcal{S}_\gamma$ satisfying the H_∞ norm less than $\gamma = 1.0$ with $k_d = 9$

```
     y=coords_y(idx,:)';
     plot(x,y,'k:');
end
xlabel('$$k_i$$','interpreter','latex');
ylabel('$$k_p$$','interpreter','latex');
axis([0 3000 -50 230]);
str = '$$\mathcal{S}_{\gamma = 1.0}$$';
loc = mean(poly_Sgamma.Vertices);
text(loc(1),100+loc(2),str, ...
     'Interpreter','latex',...
     'FontSize',14);
hold off;
```

Figure B.6 shows the set $\mathcal{S}_\gamma$ of stabilizing PID controllers satisfying the H_∞ norm less than $\gamma = 1.0$ with $k_d = 9$.

B.2 Notes and References

The plotregion function used in the PID stabilizing set was obtained from [1].

Reference

1. Bergström, P.: Plot 2D/3D Region. File Exchange, Matlab Central, Mathworks (2009). https://la.mathworks.com/matlabcentral/fileexchange/9261-plot-2d-3d-region.

Index

I. D. Díaz-Rodríguez et al., *Analytical Design of PID Controllers*,
https://doi.org/10.1007/978-3-030-18228-1

MIX
Papier aus verantwortungsvollen Quellen
Paper from responsible sources
FSC® C105338

If you have any concerns about our products,
you can contact us on
ProductSafety@springernature.com

In case Publisher is established outside the EU,
the EU authorized representative is:
Springer Nature Customer Service Center GmbH
Europaplatz 3, 69115 Heidelberg, Germany

Printed by Libri Plureos GmbH
in Hamburg, Germany